FOOD SECURITY, SAFETY, AND SUSTAINABILITY

Validated Innovative Approaches and Technological Advancements

FOOD SECURITY, SAFETY, AND SUSTAINABILITY

Validated Innovative Approaches and Technological Advancements

José Juan Buenrostro-Figueroa, PhD
Julio César Tafolla-Arellano, PhD
Editors

AAP | APPLE ACADEMIC PRESS

First edition published 2025

Apple Academic Press Inc.
1265 Goldenrod Circle, NE,
Palm Bay, FL 32905 USA

760 Laurentian Drive, Unit 19,
Burlington, ON L7N 0A4, Canada

CRC Press
2385 NW Executive Center Drive,
Suite 320, Boca Raton FL 33431

4 Park Square, Milton Park,
Abingdon, Oxon, OX14 4RN UK

Library and Archives Canada Cataloguing in Publication

Title: Food security, safety, and sustainability : validated innovative approaches and technological advancements / José Juan Buenrostro-Figueroa, PhD, Julio César Tafolla-Arellano, PhD, editors.
Names: Buenrostro-Figueroa, José Juan, 1985- editor | Tafolla-Arellano, Julio César, editor.
Description: First edition. | Includes bibliographical references and index.
Identifiers: Canadiana (print) 20240537963 | Canadiana (ebook) 2024053803X | ISBN 9781774917084 (hardcover) | ISBN 9781774917206 (softcover) | ISBN 9781003518440 (ebook)
Subjects: LCSH: Food industry and trade—Environmental aspects. | LCSH: Food industry and trade—Technological innovations. | LCSH: Food security—Environmental aspects. | LCSH: Food supply—Environmental aspects. | LCSH: Food industry and trade—Safety measures.
Classification: LCC TD195.F57 F66 2025 | DDC 338.1/9—dc23

Library of Congress Cataloging-in-Publication Data

CIP data on file with US Library of Congress

ISBN: 978-1-77491-708-4 (hbk)
ISBN: 978-1-77491-720-6 (pbk)
ISBN: 978-1-00351-844-0 (ebk)

About the Editors

José Juan Buenrostro-Figueroa, PhD
Researcher, Research Center for Food and Development, A.C., Mexico

José Juan Buenrostro-Figueroa, PhD, is Titular Researcher at the Research Center for Food and Development, A.C., Mexico. He is a food engineer. His MSc and PhD degrees were in Food Science and Technology. Dr. Buenrostro has experience in bioprocess development, including microbial processes for enzyme production and recovery of bioactive compounds, valorization of agroindustrial byproducts, and extraction and characterization of bioactive compounds from plants and agroindustrial wastes. He has published 40 papers in indexed journals, 25 book chapters, six books as editor, and more than 70 contributions at scientific meetings, with over 1160 citations and h-index of 18 (Google Scholar database). He is a member of the Mexican Association of Food Science (AMECA), Mexican Society for Food Protection affiliate of the International Association for Food Protection, and Food Safety Thematic Network: Agroindustrial By-products Valorization and Reduction of Food Losses and Waste (PDA). Dr. Buenrostro is a member of SNII (National System of Researchers, Mexico), since 2014.

Julio César Tafolla-Arellano, PhD
Research Professor, Department of Basic Sciences, Autonomous Agrarian University Antonio Narro, Coahuila, Mexico

Julio César Tafolla-Arellano, PhD, is a full-time Research Professor in the Department of Basic Sciences of the Autonomous Agrarian University Antonio Narro in Coahuila, Mexico. He is the Head of the Biotechnology and Molecular Biology Laboratory and a member of the Mexican Association of Food Engineering and Biotechnology. He is editor in-chief of the journals *Universitas Agri* and *Revista Agraria*. Dr. Tafolla-Arellano is a member of the Mexican National System of Researchers (level I), and his research is focused on the study of the biotechnological use of plants through an interdisciplinary omics approach, biocatalysis, bioinformatics, and food biotechnology. He became an engineer in food science and technology at the

Autonomous Agrarian University Antonio Narro. He earned his PhD degree in Physiology and Plant Molecular Biology at the Center for Research and Food and Development A.C. of Sonora, Mexico. Also, he performed post-doctoral research on the postharvest shelf life of fruit and transcriptomics analysis. He has published several books chapters and articles in indexed journals and developed eight research projects. He has been advisor of several PhD, MSc, and BSc theses.

Contents

Contributors

Palacio-Márquez Alejandro
Faculty of Agricultural and Forestry Sciences, Autonomous University of Chihuahua, Delicias, Chihuahua, Mexico

Sandra Pérez Álvarez
Faculty of Agricultural and Forestry Sciences, Autonomous University of Chihuahua, Delicias, Chihuahua, Mexico

Eduardo Fidel Héctor Ardisana
Graduate Institute, Technical University of Manabí, Portoviejo, Ecuador

Ramiro Baeza-Jiménez
Biotechnology and Bioengineering Laboratory, Food and Development Research Center, Delicias, Chihuahua, Mexico

José Juan Buenrostro-Figueroa
Food and Development Research Center, Delicias Unit, Chihuahua, Mexico

Diego Eloy Carballo-Carballo
Facultad de Ciencias Químicas, Universidad Autónoma de Chihuahua, Circuito Universitario s/n, Campus II, C.P., Chihuahua, Mexico

Alejandro Alberto Camacho Dávila
Faculty of Chemical Sciences, Autonomous University of Chihuahua, Campus UACH II, Chihuahua, Chih., Mexico

Daniel Díaz-Rodríguez
Bioprocess Engineering Laboratory, School of Exact and Natural Sciences, State Distance University, San José, Costa Rica

Edwin Rojo-Gutiérrez
Food and Development Research Center, Delicias Unit, Chihuahua, Mexico

Laura María Navarro Flores
Vizcaya University of the Americas, Campus Delicias, Delicias, Chihuahua, Mexico

Félix Garcia-Ochoa
Chemical and Materials Engineering Department, Faculty of Chemical Sciences, Universidad Complutense de Madrid, Madrid, Spain

Nuria Gómez
Cellulose and Paper Group, Forest Product Department, Forest Research Center (ICIFOR-CSIC), Madrid, Spain

Mariana González-Torres
Postgraduate College Campus Córdoba, Congregation Manuel León, Municipality Amatlán de los Reyes, Veracruz, Mexico

Rodolfo Antonio Hernández-Chaverri
Bioprocess Engineering Laboratory, School of Exact and Natural Sciences, State Distance University, San José, Costa Rica

Francisco Hernández-Centeno
Antonio Narro Autonomous Agrarian University, Food Science and Technology Department, Buenavista, Saltillo, Coahuila, Mexico

María Hernández-González
Antonio Narro Autonomous Agrarian University, Food Science and Technology Department, Buenavista, Saltillo, Coahuila, Mexico

Ricardo Hernández-Martínez
Postgraduate College Campus Córdoba, Congregation Manuel León, Municipality Amatlán de los Reyes, Veracruz, Mexico

Francisco Hernández-Rosas
Postgraduate College Campus Córdoba, Congregation Manuel León, Municipality Amatlán de los Reyes, Veracruz, Mexico

José Carlos Espinoza Hicks
Faculty of Chemical Sciences, Autonomous University of Chihuahua, Campus UACH II, Chihuahua, Mexico

Francisco Javier Choix Ley
Faculty of Chemical Sciences, Autonomous University of Chihuahua, Campus UACH II, Chihuahua, Mexico

Oskar Alejandro Palacios López
Faculty of Chemical Sciences, Autonomous University of Chihuahua, Campus UACH II, Chihuahua, Mexico

Leticia Xóchitl López-Martínez
Laboratory of Antioxidants and Functional Foods, CONACYT–Center for Research in Food and Development, Col. La Victoria, CP., Hermosillo, Sonora, Mexico

Rosa María Yánez Muñoz
Faculty of Agrotechnological Sciences, Autonomous University of Chihuahua, Chihuahua, Mexico

Ezequiel Muñoz-Márquez
Center for Research in Food and Development A.C., Chihuahua, Mexico

Néstor Gutiérrez-Méndez
Faculty of Chemical Sciences, Autonomous University of Chihuahua, Campus II, Chihuahua, Mexico

Neith Aracely Pacheco-López
Center for Research and Assistance in Technology and Design of the State of Jalisco, Yucatán Scientific and Technological Park, Mérida, Yucatán, Mexico

Haydeé Yajaira López-De la Peña
Antonio Narro Autonomous Agrarian University, Food Science and Technology Department, Buenavista, Saltillo, Coahuila, Mexico

Claudio Rios-Velasco
Food and Development Research Center, A.C., Cuauhtémoc Unit, Food Physiology and Technology of the Temperate Zone, Cuauhtémoc City, Chihuahua, Mexico

Juan Daniel Rivaldi
Industrial Applications Department, Faculty of Chemical Sciences, National University of Asunción, San Lorenzo, Paraguay; Cellulose and Paper Group, Forest Product Department, Forest Research Center (ICIFOR-CSIC), Madrid, Spain; Chemical and Materials Engineering Department, Faculty of Chemical Sciences, Complutense University of Madrid, Madrid, Spain

Claudia Alejandra Rodríguez-González
Autonomous University of Juárez City, Engineering and Technology Institute, Ciudad Juárez, Chihuahua, Mexico

Dely Rubi Chávez-Garay
Technological University of Camargo, Colonia la Lagunita, Chihuahua, Mexico

Hilda Karina Saenz-Hidalgo
Biotechnology and Bioengineering Laboratory, Food and Development Research Center, Delicias, Chihuahua, Mexico

Esteban Sánchez
Center for Research in Food and Development A.C., Chihuahua, Mexico

Daniela Sánchez-Aldana
Faculty of Chemical Sciences, Autonomous University of Chihuahua, Campus UACH II, Chihuahua, Mexico

Juan Manuel Soto-Parra
Faculty of Agrotechnological Sciences, Autonomous University of Chihuahua, Chihuahua, Mexico

Juan Manuel Tirado-Gallegos
Autonomous University of Chihuahua, Zootechnics and Ecology Faculty, Chihuahua, Chihuahua, Mexico

Crescencio Urías-García
Vizcaya University of the Americas, Campus Delicias, Delicias Chihuahua, Mexico

Priscilla Vergara
Cellulose and Paper Group, Forest Product Department, Forest Research Center (ICIFOR-CSIC), Madrid, Spain

Juan Carlos Villar
Cellulose and Paper Group, Forest Product Department, Forest Research Center (ICIFOR-CSIC), Madrid, Spain

Paul Baruk Zamudio-Flores
Food and Development Research Center, A.C. – Cuauhtémoc Unit, Food Physiology and Technology of the Temperate Zone, Ciudad Cuauhtémoc, Chihuahua, Mexico

Preface

Sustainability is one of the most important issues facing food security and safe production today. Using interdisciplinary perspectives, this research-oriented volume provides a global outlook on the problems we face in the food sector. It explores integrated, context-specific approaches to food security and food safety and examines the measures that will be required to guarantee the long-term sustainability of the food supply chain by reviewing the existing data to encourage action and identify knowledge gaps. This will make it a valuable reference source for food science professionals, including food engineers, food scientists, and post-graduate researchers working in food security and safety. It also provides a detailed account of all the impacts of the food supply chain along with the basic concepts of environmental sustainability.

This research-oriented volume reports important factors that impact both food safety and nutrition security by providing insights into effective policy development and implementation. It provides a timely resource on new materials available in the food industry, along with innovative validated approaches on how to develop sustainable food for a growing world population. Using validated innovative approaches and case studies, the book reviews several types of risks associated with each step of the food chain, making it an ideal resource. This new title provides practical examples used to enhance food security, safety, and sustainability.

CHAPTER 1

Spirulina as a Novel Bioactive Material in Edible Films and Coatings as Sustainable Packaging Systems

DANIELA SÁNCHEZ-ALDANA*, OSKAR ALEJANDRO PALACIOS LÓPEZ, and FRANCISCO JAVIER CHOIX LEY

Facultad de Ciencias Químicas, Universidad Autónoma de Chihuahua, Circuito Universitario s/n, Campus UACH II, Chihuahua, Chih., México

*Corresponding author.

ABSTRACT

Packaging systems are necessary in the food industry for operations that are involved throughout the production chain and in particular when the products are presented to the final consumer. However, packing materials are usually made of non-biodegradable or renewable polymeric materials that accumulate in the environment and generate an increased carbon footprint. An alternative that has been proposed in recent years is the use of biodegradable and edible packaging materials, such as edible films and coatings, which can be used as carriers of food additives such as antioxidants, antimicrobials, color agents, flavors, etc. Thus, many natural additives have been demonstrated to act as bioactive agents in food systems which can be incorporated into a film or a coating matrix. *Spirulina maxima* or *platensis*, which is a blue-green cyanobacterium, has been consumed as a food and nutritional supplement for a long time, owing to its chemical composition, and antibacterial and antifungal properties, also, due to its antioxidant compounds (vitamins, phenolics, and pigments), minerals, polyunsaturated fatty acids, and high-quality proteins. Hence, *Spirulina* represents a potential source of preservative bioactive compounds that can constitute a novel and promising ingredient to enhance the storage

Food Security, Safety, and Sustainability: Validated Innovative Approaches and Technological Advancements.
José Juan Buenrostro-Figueroa and Julio César Tafolla-Arellano (Eds.)

quality of different food matrices. The present chapter explores the use of diverse food packaging materials that can incorporate *Spirulina* as a bioactive agent, with the consequent added health benefits. Likewise, the use of this microalgae is emphasized since it can be produced in accordance with circular economy and biorefining principles, contributing to environmental protection and sustainable food protection systems.

1.1 INTRODUCTION

Food production has a large environmental impact in many ways. It requires great amounts of water, half of the world's habitable land is used for agriculture, and it accounts for over a quarter of the world's greenhouse gas emissions and extensive land use, to name a few [1]. A study by Crippa et al. [2] estimates that food systems were responsible for greenhouse gas (GHG) emissions of around 34% (about 18 Gigatons of CO_2 equivalent per year globally (Gt $CO_2e\ yr^{-1}$)) by 2015. Regarding the food supply chain, where the transformation of raw materials into finished food products until it reaches consumers takes place, food packaging contributes to 5% of GHG emissions, representing 1.0 Gt of $CO_2e\ yr^{-1}$ to this environmental problem [1, 3].

Industrial packaging and food containers are made from different sources such as paper, cardboard, metal, glass, and plastic [4]. Also, the combination of these materials is possible to improve their functionality. Although not all food packaging materials contribute to GHG emissions equally [2], the accumulation of plastic waste in the environment and oceans due to the use of single-use packaging materials and the generation of microplastics [5, 6] are also adverse effects that have a significant impact on all the surroundings of living beings. It is estimated that by 2050, around 12,000 Mt of plastic garbage will be in landfills or the environment if manufacturing and waste management trends continue [7]. Consequently, in an environment where sustainability efforts are focused on limiting environmental effects, it is imperative to discover alternatives for various types of food packaging materials that help to reduce GHG emissions and plastic buildup.

Conventional plastics for food packaging are widely used due to their many beneficial properties, including high mechanical resistance, ability to communicate product information such as content, nutritional value, ingredients, expiration date or shelf life, storage conditions, ease of distribution, and low-cost ratio [6, 8]. However, the development of these materials must be conducted using sustainable development practices to ensure that they do not harm the environment or human health, especially since most food packaging is single-use. Nowadays, renewable, biodegradable, and active packaging are currently very important topics of research and development in the food and packaging industry. This can be observed in Figure 1.1, which shows a Scopus search by year from 2010 to 2022 with more than 3,800 publications for keywords linked to edible food packaging, active edible packaging, and sustainable food packaging, where films and coatings research has shown remarkable growth in recent decades. It can be observed from 2010 to 2019 how the number of publications related to these topics occupied around 45% of the publications for 10 years and in the last 3 years, it is observed how the number of publications doubled, occupying 55% from 2020 to 2022. In this way, it is projected that the development of edible films and coatings, active and sustainable packaging which has shown rapid progress in recent decades, will have a substantial influence on food quality in the next years.

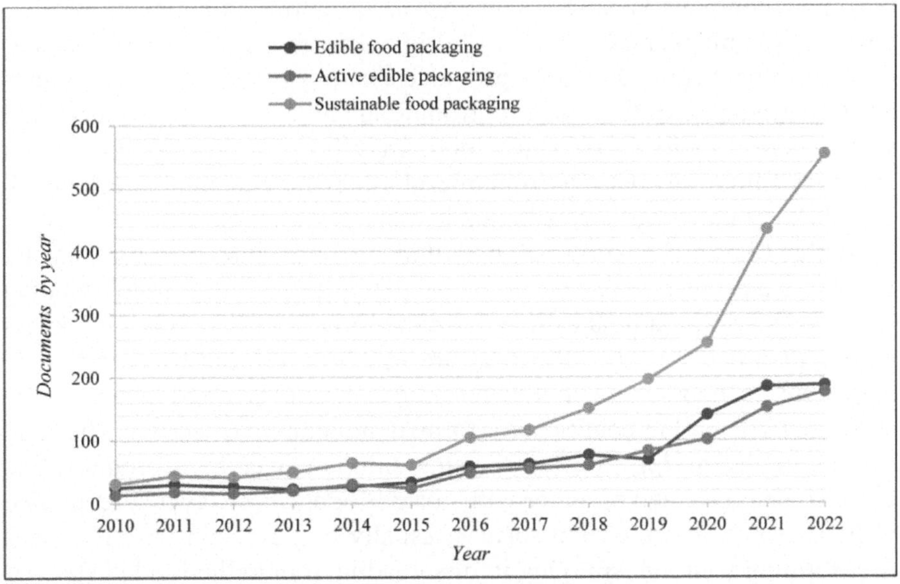

FIGURE 1.1 Annual publications reported on Scopus search for edible food packaging, active edible packaging, and sustainable food packaging keywords.

1.2 EDIBLE FILM AND COATINGS AS SUSTAINABLE FOOD PACKAGING

Food packaging is one of the most important packaging systems due to the protection against chemical, biological, and physical hazards to which foods are exposed. Nevertheless, sustainable food packages must maintain food quality and improve safety by reducing food losses and waste [9]. In this context, the requirement for fully compostable and biodegradable materials, less food waste, losses, and extended shelf life propose the use of edible materials such as films and coatings [10]. Edible materials for food packaging are used in order to reduce synthetic packaging that can also play a role as an eco-friendly biodegradable package and a protective coating on the food surface [11].

One of the major developments in the flexible packaging sector for foods is the use of renewable resources for packaging materials, such as proteins, polysaccharides, and lipids with a biological origin to avoid the use of synthetic polymers derived from petroleum [8, 12] that are edible and provide foods mechanical and barrier properties increasing their shelf life [13].

In addition to edible properties, flexible films and coatings prepared by biopolymers may also provide a structural matrix for carrying different bioactive compounds such as antioxidants, antimicrobials, probiotics, flavors, light blockers, etc. [10, 14], originating materials that are considered active packaging due to the provided capacity of bioactive compounds used to interact with food or its surrounding environment to preserve and sustain food quality [14]. Thus, foods may be protected by edible and bioactive films and coatings from microbiological, chemical, mechanical, and physical harm that are sustainable.

Edible films and coatings are both thin layers of edible material that are formed as a protective barrier in foods [12]. Edible films are self-standing structures or solid laminates that wrap the food products with a thickness of less than 0.3 mm [15, 88]; they are usually produced by a casting or extrusion method where a film-forming solution is dried over a solid surface, while edible coatings are formed and adhere directly to the food surface [4] by drying the film-forming solution over food products and can also be consumed together with coated food. These layers are applied in liquid form onto the food surface, usually by immersing the food in a film-forming solution, spraying it, or spreading it onto the food [11].

1.2.1 EDIBLE PACKAGING COMPONENTS AND PROPERTIES

Edible packaging is consumable and made from food-grade biopolymers from plants, animals, and marine life or food processing by-products. Many matrices have been used for this; finally, hydrocolloids, both polysaccharides, and proteins, are the most common group of biopolymers used in the production of edible materials [11, 12].

Table 1.1 shows the main biomaterials employed in edible films and coatings formulations and some of their functional properties. These properties explain the need to combine in some cases different biomaterials to provide better mechanical characteristics, water or gas barrier, opacity/transparency, and flexibility, among others [16]. Also, it is important to select these biomaterials due to biocompatibility and the nature of the food to which they are applied. Thus, composite films and coating blending different biomacromolecules are desirable.

TABLE 1.1 Common Biomaterials Used in Edible Films and Coating for Foods and General Properties

Biomaterials	General Properties	References
	Polysaccharides	
• Cellulose derivates • Carrageenan • Chitosan • Pectin • Starch • pullulan • Gums • Alginate	• Strong hydrogen bond with active additives. • O_2 and aroma barrier. • Strength properties. • Poor water vapor barrier.	[17]
	Proteins	
• Soybean • Wheat gluten • Corn zein • Sunflower • Gelatin • Whey • Casein • Keratin	• Hydrophilic. • Flexible and transparent. • Carrier active antimicrobial and antioxidant agents. • Poor water vapor barrier.	[16, 18]
	Lipids	
• Waxes • Triglycerides • Acetylated monoglycerides • Free fatty acids • Vegetable oils	• Opaque films. • Excellent water barrier. • Extra brittle and thicker due to their hydrophobicity.	[19]

The most common polysaccharides utilized in the formulation of edible films and coatings are cellulose derivatives, starch, alginate, pectin, chitosan, and carrageenan, whereas the most common proteins are soy, wheat, maize, sunflower, zein, whey, casein, and keratin proteins. However, those materials are hydrophilic in nature [17–20]. To improve the hydrocolloid matrix's abilities as a water vapor barrier, several types of oils and fats are added. Typically used vegetable oils, waxes, triglycerides, acetylated monoglycerides, free fatty acids, and waxes [19, 21, 22]. Also, lipids are used to prepare composite emulsion-based edible materials from hydrocolloids and lipids resulting in better functionality than films or coatings produced with only one component [23].

The above examples comprise the main structural matrix of edible films or coatings but in many cases depending on the nature of the biomacromolecules employed, the addition of low molecular weight plasticizer agents such as glycerol, sorbitol, or polyethylene glycol is needed, mention a few [24, 25] to reduce intermolecular forces to provide flexibility and permeability to the edible matrix [4]. Glycerol is the most used plasticizer in edible films and coatings because of its high availability, biocompatibility, and low cost [26]. Besides, it has been demonstrated that glycerol increases flexibility, elongation, and peel ability for edible films, and also decreases brittleness and fragility [27] among other desirable effects and the low cost.

Edible films and coatings must possess a variety of characteristics depending on the type of food to which they will be applied. However, the main properties that are sought to be addressed to generate efficient edible films and coatings are mechanical properties, such as flexibility, elongation, tensile strength, resistance, and physical stability, barrier properties to water vapor, gases such as oxygen and carbon dioxide, and odors, editability, and sensory compatibility [28, 29]. Clearly, all these properties aim to prolong the shelf life of the packaged or coated food.

The water solubility of the film or coating is also an important factor to consider as this will affect its stability and that of the packaged or coated food. This means the design of low water solubility structural matrices will benefit the resistance to water or high relative humidity during food storage [30]. Thus, higher water solubility films or coatings will be only appropriate for foods with low water content and low water activity. In addition, it is necessary to control the mass transfer which prevents the loss of solutes or volatile compounds from the film or coating matrix to food and conversely [31].

Another property of the materials used is the sensory compatibility that must exist between the structural matrix of the film or coating and the packed or coated food. In very few studies, this property is reported.

There is still a need to delve into the migration of certain components into the food that could give unpleasant tastes, odors, or colors. This property will be discussed later when the main characteristics of *Spirulina* are described.

Edible films and coatings can also provide active properties through the addition of different active compounds, as will be addressed in succeeding sections.

1.2.2 EDIBLE PACKAGING AS A CARRIER FOR FUNCTIONAL BIOACTIVE COMPOUNDS

To improve the quality and stability of edible films and coatings, a variety of active ingredients, including antioxidants, colorants, antimicrobials, flavors, and others, have been added to many edible matrices fulfilling the function of active packaging for food products [22, 24]. Research has shown that the incorporation of bioactive compounds into the structural matrices of films and coatings may be used to enhance moisture, aromas, oxygen, and carbon dioxide barriers, provide antimicrobial or antioxidant activity, target delivery of nutrients, and prolong shelf life [12].

Incorporating bioactive compounds into edible films for food packaging can either improve or worsen the properties of these materials. This depends on the physical or chemical interactions between the biopolymers and the bioactive substances, which ultimately have an impact on the structure and, in turn, the functioning of the edible films and coatings [8].

Many bioactive compounds come from vegetal extracts and natural components from plants, spices, and herbs [31]. In many cases, they come from agro-industrial by-products or residues. Other components may come from animal sources, such as polyunsaturated fatty acids from marine animals [32, 33], peptides, proteins, or enzymes from animal sources [34, 35], and bioactive peptides from marine processing byproducts [36].

Other important sources of bioactive compounds are microorganisms, such as bacteriocins (antibacterial peptides) produced by lactic acid bacteria [34], phenolic compounds from algal cells [37], and bioactive carotenoids from microalgae [38].

Some new components have been studied to improve the properties of edible film-forming materials and coatings to generate active materials that can provide novel and functional uses in the food industry. This includes the use of biomass produced by microalgae. An example is *Spirulina*, whose importance and relevance in its use and incorporation as an ingredient for the formulation of edible films and coatings will be described in succeeding sections.

1.3 ALGAE USES FOR BIOPLASTICS PRODUCTION

The algae biomass has been reported to possess the potential to be used as an ingredient for bioplastic production, not only as an additive or bioactive component but also as a main structural matrix compound as will be described below. For example, as isolated proteins complexed with lysozyme [39, 40], microalgae starch and cyanobacteria polyhydroxyalkanoates [41], agar [42], alginate [43], carrageenan [44] from different genera such as *Chlorella, Spirulina, Chlorogloea, Phaeodactylum, Scenedesmus*, among others [45]; or only as an additive such as polyphenol extracts [46] obtained by different extraction, cell disruption, or drying methods.

1.3.1 SPIRULINA PROPERTIES AND PRODUCTION

Spirulina is a multicellular and filamentous blue-green microalgae that belongs to the phylum Cyanobacteria [47]. Today, the *Spirulina* genus has important commercial value in the food industry due to its high nutritional value, mainly its high protein content. The main species such as *S. máxima* and *S. platensis* have the capacity to attain up to 70% of the protein in dry weight. In addition, *Spirulina* is also considered an important source of vitamins, minerals, essential fatty acids, antioxidants, and pigments such as β-carotene, phycocyanin, and chlorophyll, among others [48] as shown in Table 1.2. To date, *Spirulina* production has emerged around the world due to the wide applications of its compounds in diverse economic sectors such as nutraceuticals, food, cosmetics, and agriculture, as well as its ability to grow vigorously in strong sunshine under high temperatures and highly alkaline conditions; thus, this microalga has found a place in the market as "functional foods" [49]. Although the production of this microalga is simple and can be carried out from domestic to industrial scale either in open systems (ponds, lakes, or lagoons) or closed systems (various photobioreactors) [50]. The selection of a photoreactor (open or closed system) designed to provide optimal conditions for biomass productivity (pH, temperature, and light availability) is a real challenge today [51]. Nowadays, commercial *Spirulina* production is generally carried out in open raceways (open ponds divided into two or four channels of 0.2 to 0.4 m in depth) agitated by paddlewheels or propellers (which has been demonstrated to be efficient for industrial implementation) [51]; while its production on an industrial scale mainly takes place in Australia, Chile, Cuba, Israel, Vietnam, Bangladesh, the Philippines, Peru, Martinique, Brazil, Portugal, Spain, China, and other nations, resulting in a *Spirulina* production of 3,000 tons per year among these nations. Nonetheless, the need for hygienic and efficient culture systems has led to the design and use of closed growth systems photobioreactors around the world; although there are still not many companies capable of using these systems on an industrial scale, mainly due to their high cost [50]. The main closed photoreactor used is the tubular one, which consists of glass, or plastic tubes aerated by pumps [51].

TABLE 1.2 Nutrimental Composition of *Spirulina* Biomass in 100 g

Compound	Content
Protein	60–71 g
Fatty acids	4–5 g
Carbohydrate	15–18 g
Chlorophyll	1–2 g
Carotene	180–190 mg
Phycocyanin	8–12 g
Calcium	400–600 mg
Iron	50–100 mg
Potassium	200–2,000 mg
Magnesium	200–300 mg
Zinc	1–2.0 mg
Folic acid	0.1 mg
Vitamin A (beta-carotene)	100–200 mg
Vitamin B1 (thiamine)	1.5–4.0 mg
Vitamin B2 (riboflavin)	3.0–5.0 mg
Vitamin B6 (pyridoxine)	0.5–0.7 mg
Vitamin B12 (cyanocobalamin)	0.05–0.2 mg
Vitamin E (A-tocopherol)	5.0–20 mg

Source: Selected data from: Ahsan et al. [47].

Nutritional requirements for the growth of *Spirulina* have been reported to be mainly determined by the presence of high carbonate and bicarbonate concentrations which significantly improve biomass and protein production [52]. Nowadays, the use of industrial residues as the basis for the development of economic and efficient growth media for *Spirulina* biomass production is a priority, and residues such as sugar cane molasses have been demonstrated to be a successful bicarbonate supplement [53]. On the other hand, CO_2 concentration is recognized as the limiting factor for *Spirulina* biomass production. The use of gaseous residues (rich in CO_2) from several industries has been demonstrated to be a viable CO_2 source to maintain high rates of *Spirulina* production [84] with the subsequent reduction of CO_2 emissions [54]. Although the design of bioreactors and culture media for *Spirulina* growth is one of the most important issues addressed, the most viable harvesting and drying methods have also been investigated. Due to the size and form of *Spirulina* cells, the use of floating methods for biomass harvesting has proven to be economical and efficient. Nevertheless, the drying method is still one of the more expensive steps in *Spirulina* production, with spray drying being the most used [51].

Today, efforts are being focused on increasing the productivity of *Spirulina* cultured in open ponds, where its biomass productivity is about 2.846 mg/L/day; while in closed photobioreactors, the biomass productivity can reach up to about 220 mg/L/day [55]. Regardless of the above, the global *Spirulina* market is expected to grow from US $700 million in 2016 to US $2,000 million by 2026 due to the growing applications of *Spirulina* [48].

According to Soni et al. [48], humanity has not completely utilized the natural properties of *Spirulina* until now. Several studies have demonstrated the potential of this microalga as an anticancer, antiviral, antioxidant, and antiallergic agent [48, 56]. Recently, the increased production of synthetic packaging films has led to solving environmental pollution due to their non-biodegradable nature; thus, many efforts have focused on using edible materials for food packaging purposes [57]. In this context, the composition of *Spirulina* has also shown a positive impact on this topic, attracting a lot of interest due to its wide potential of high added-value and preservative bio-compounds [58]. For instance, Cardoso et al. [59] demonstrated the technological viability of biobased films formulated with cassava by-product and *Spirulina platensis*, as well as their potential to be applied as food packages. Recently, Shafiei & Mostaghim [60] showed that applications of edible coating based on chitosan/natamycin containing *Spirulina platensis* and *Chlorella vulgaris* maintained the desired quality of calf fillets. Similarly, other research has also demonstrated the increase in functionality of synthetic packaging films using byproducts of the microalgae *Spirulina*, such as the application of proteins and pigments [39, 61, 71]. The above confirms that *Spirulina* and its byproducts are an important innovative strategy to explore new packaging for food preservation.

1.3.1.1 SPIRULINA CHEMICAL AND PHYSICAL PROPERTIES

As already mentioned, *Spirulina* is a superfood that contains large amounts of protein, lipids, and carbohydrates, vitamins, minerals, as well as highly important active compounds such as phycocyanin, which can be used efficiently for the preparation of edible packaging materials. Hence, chemical, physical, and physicochemical properties are important to highlight to be useful and successful for this application. As edible films and coatings should possess desirable functional properties, for example, mechanical and barrier properties, the knowledge of the properties of *Spirulina* is desirable in order to understand how it may interact with the other components of the packaging material.

Some *Spirulina* chemical, physical, and physicochemical properties are important to highlight, such as pH, isoelectric point (pI), water or oil solubility, water and fat absorption, emulsification capacity, foaming capacity, and thermal stability, may be taken into consideration to design efficient edible films and coatings with the aim of being in accordance or compatible both with the elements of the structural matrix and with the food in which the coating is going to be applied. These properties will be a decisive factor when adding them to the formulation of packaging material.

Devi & Venkataraman [62] carried out a study due to the lack of information in this regard, where they characterized the properties of *Spirulina* flour (SF) and *Spirulina* protein concentrate (PCS). The objective of this functional characterization was to know the water and fat absorption capacity, as well as the emulsification and foam formation capacity and its stability. They found the water sorption capacity of SF was more than twice 220 and 109 g/100 g, respectively, attributing to the carbohydrates present in SF due to its hydrophilic nature. In contrast, PCS fat sorption capacity was 1.5 times higher than SF (280 and 190 g/100 g, respectively), suggesting a more lipophilic capacity for PCS. The dependence of foam and emulsification capacity on pH was also evaluated, finding that both were also dependent on the solubilized protein at pH 10. Menegotto et al. [63] carried out the determination of these same properties and, in addition to this, determined the rheological behavior and its amino acid profile. The highest protein solubility (98.07%) occurred at pH 9. In this study, all the properties evaluated were found to be dependent on pH, due to the protein-water interactions being enhanced at pH levels above or below the pI, leading to an increase in the liquid load and repulsive forces. As a result, the proteins prefer to absorb more water. Benelhadj et al. [39] solubilized the *Arthrospira platensis* biomass in water at pH 10 and then isolated the protein by precipitation at pH 3 due to its pI. Also, in this research, they found an attractive interaction between cationic lysozyme and anionic *A. platensis* protein isolate at pH 3.

Kamaruding et al. [64] also determined water absorption capacity, oil absorption capacity, emulsion aging stability, emulsion heat stability, foam capacity, and foam stability of *Spirulina* protein isolate using the pH-shift method combined with a disruption method to break down the cell wall and facilitate protein release. In this study, it was found that the solubilization at pH 7 yielded the highest protein yield and its functional properties vary depending on the solubilization pH. In contrast, the protein isolated at pH 10 presented better foaming properties than at pH 7 but the worst emulsion heat stability.

Spirulina shelf-life stability depends on the humidity content, which must be decreased to less than 15%. Thermal drying methods intended for this purpose may affect the bioactive components of *Spirulina.* In order to investigate the effect of preservative methods on important pigments of *Spirulina platensis*, Nouri & Abbasi [65] evaluated different drying methods such as shade, sun, oven, vacuum oven, microwave, freeze, and spray drying, and also freezing to determine color properties and anthocyanins. In this study, several effects of the drying methods were found that are dependent on various factors such as pH, oxygen, light, temperature, enzymes, processing time, the presence of Fe and Cu ions, and unsaturated fatty acids. As expected, high temperature and long process time in the conventional oven drying method had a significant ($P<0.05$) effect on the destruction of *Spirulina* pigments.

For all the reasons mentioned above, the thermal stability, pH dependence, water and oil absorption capacity, and other properties of *Spirulina* and its functional components must be considered to preserve its desirable effects and contribute technologically to edible films or coatings formulation.

1.3.1.2 ORGANOLEPTIC PROPERTIES FROM SPIRULINA

Organoleptic parameters of *Spirulina* biomass have been subject to limitations due to the "musty-earth flavor" [66], and only the use of small amounts of this microalga resulted in acceptability by consumers [67]. The presence of prolamins in *Spirulina* causes a "fishy" or "seaweed" smell [68]. Also, the color provided could be a sensorial aspect that consumers may disapprove of. Encapsulation methods are the main proposals that may address this concern to suppress or mask taste, flavors, and odors [67]. Another special treatment can be implemented, but the essential nutrients may be removed [68]. In addition to increasing Spirulina production costs, these processes can remove essential nutrients that make this microalga so valuable. That is why it is important to apply edible films or coatings with foods that are organoleptically compatible. At the same time, it is important to continue searching for sustainable and profitable processes for the elimination or masking of undesirable organoleptic aspects.

1.4 SPIRULINA AS EDIBLE FILMS AND COATING MATERIAL

The nutritional composition of *Spirulina*, as well as the potential biological activity of the combination of its metabolites with other potentially active compounds, are the main characteristics that make its uses attractive in food processing. In Table 1.3, a summary of the latest research that has been carried out on the potential use of *Spirulina* as a component for edible packaging materials such as films and coatings to be applied to food is presented. The structural matrix compound, plasticizer employed, main properties, and bioactive properties are described. Additionally, the *Spirulina* extraction method is described in order to compare functional properties derived from the method of obtaining.

Phycocyanin and polyphenol content in *Spirulina* biomass are both kinds of metabolites recognized with higher biological activities [79]. Moreover, the use of dry biomass is the common way to process microalgal compounds and from this, the selection of more suitable methods to dry microalgal biomass while maintaining a high proportion of phycocyanin and polyphenol content, as well as good mechanical characteristics is desirable when it is used as a component in an edible film or coating. Nevertheless, it has been reported that a specific drying method tends to favor one characteristic but affects others. For example, Larrosa et al. [40] reported *Spirulina* films plasticized with glycerol where the spouted bed dry method was compared to tray conventional dry and different *Spirulina* disrupted cells methods showed higher phycocyanin and phenolic compounds but with lower tensile strength, water vapor permeability, and thermal stability (between 90.3 and 97.9°C) depending on the dry/disruption method applied to dry *Spirulina* biomass. Contrary to the effects observed when tray drying was applied in which the biomass was less enriched in pigments and polyphenols but presented higher thermal stability and higher water vapor permeability.

TABLE 1.3 *Spirulina* as Main Structural or Bioactive Compound in Edible Films and Coatings

Main Film or Coating Structural Matrix	Plasticizer/ Emulsifier	Compounds of Spirulina Used	Dry/Disruption/ Extraction Method	Main Properties Evaluated	Bioactive Properties	References
			Edible Films			
Arthrospira platensis protein isolate complexed with hen egg white lysozyme.	Sorbitol	*Spirulina* isolated proteins.	Protein precipitation at pH 3.	Opacity, brown color, thick with poor mechanical properties (fragile and not very elastic).	Antimicrobial	[39]
Spider crab chitosan and *Spirulina* extract.	Glycerol	*Spirulina* polyphenols	Polyphenols ethanolic extraction.	Water vapor and oxygen permeability, color, opacity, and mechanical properties (mechanical (tensile strength and elongation at break).	Antioxidant and antimicrobial.	[46]
Arthrospira Spirulina dry biomass	Glycerol	*Arthrospira Spirulina* biomass.	Dry method: Spouted bed or conventional tray. Disruption method: Milling, microwave oven, or autoclave.	Tensile strength, elongation, water vapor permeability, yellow-greenness color, higher thermal stability.	Antioxidant with high phycocyanin content.	[40]
Corn starch	Glycerol	*Spirulina*, *Nannochloropsis* or *Senedesmus*	Commercial microalgae without purification or extraction process.	SEM microstructural analysis, mechanical properties (tensile strength, elastic modulus, and elongation at break), water vapor and oxygen permeability, transparency, and color.	N/A	[69]
Cassava starch, gelatin, cassava bagasse.	Glycerol	*Spirulina platensis*	Dehydrated biomass	Color, opacity, water activity, viscosity, solubility, thickness, moisture, swelling, water vapor permeability, tensile strength, elongation.	N/A	[59]
Gelatin	Glycerol and oxidized sodium alginate (as crosslinking agent).	*Spirulina platensis* protein concentrate	Extraction method: Precipitation with HCl solution to pH 3 and freeze-dried.	Atlantic mackerel (Scomber scombrus) filets film packaged. Storage conditions 7 days at 4°C. Microbial analyses, pH, lipid hydrolysis, oxidation.	Antimicrobial and antioxidant.	[70]
				Hake muscle (*Merluccius merluccius*).		[71]

TABLE 1.3 *(Continued)*

Main Film or Coating Structural Matrix	Plasticizer/ Emulsifier	Compounds of Spirulina Used	Dry/Disruption/ Extraction Method	Main Properties Evaluated	Bioactive Properties	References
Whey protein-based	Glycerol and Tween 80	*Spirulina*	Commercial microalgae without purification or extraction process.	Film thickness, water content, solubility in water, swelling index, tensile strength, Young's modulus, elongation at break, antioxidant activity, ATR-IR) analysis.	Antioxidant	[72]
Chitosan	Glycerol	*Arthrospira platensis* polysaccharide.	Commercial polysaccharide	FTIR analysis, X-ray diffraction analysis, and microscopy scanning analysis of physical properties such as film thickness, opacity, moisture content, swelling degree, water vapor permeability, antimicrobial, and antioxidant activity, and mechanical properties (tensile strength and elongation at break).	Antimicrobial and antioxidant.	[73]
Chitosan mixed with montmorillonite.	Glycerol	*Spirulina*	Commercial powder	Thickness, FT-IR analysis, color, opacity, solubility, porosity, swelling, tensile strength, oxygen and water vapor transmission rate, antimicrobial, biodegradability.	Antimicrobial	[74]
Edible Coatings						
Cassava starch or corn starch enriched.	N/A	*Spirulina platensis*	Commercial powder	Mango fruit pH, titratable acidity and total soluble, mass loss, pulp firmness, Vitamin C. Storage conditions 12 days at 10 ± 0.5°C and 63 ± 2%.	N/A	[75]
Cassava starch, cassava bagasse, gelatin.	Glycerol	*Spirulina* biomass	N/A	Cambuci pepper (*Capsicum* sp.) Color, general appearance, mass loss, texture profile, titratable acidity, pH, total soluble solids. Storage conditions 14 days at 19.8 ± 2.8°C.	N/A	[76]

TABLE 1.3 *(Continued)*

Main Film or Coating Structural Matrix	Plasticizer/ Emulsifier	Compounds of Spirulina Used	Dry/Disruption/ Extraction Method	Main Properties Evaluated	Bioactive Properties	References
Spirulina platensis	Mineral oil	*Spirulina platensis*	N/A	Pomegranate (*Punicagranatum* sp.) Color, titratable acidity, pH, total soluble solids, ascorbic acid content, total phenolic and anthocyanins content, flavonoids, total sugars, and reducing sugars. Storage conditions 15 days at 24°C and 60% RH.	Antioxidant	[77]
Guar gum enriched with Aloe vera.	Tween 80 and glycerol	*Spirulina platensis* polyphenols	Polyphenols ethanolic and aqueous extraction.	Mango fruit Firmness, weight loss, acid ascorbic content, total phenolic and flavonoid content, antioxidant activity, color, pH, titratable acidity and total soluble solids and overall visual acceptability. Storage conditions 3 weeks at 25 ± 2°C and 55–65% RH.	Antioxidant	[89]
Alginate	N/A	*Spirulina platensis* polyphenols	Polyphenols ethanolic extraction.	Mango fruit Respiration rate, weight loss, firmness, total phenolic and flavonoid content, total soluble solids, titratable acidity, peel and pulp color, antioxidant activity. Storage conditions 1 month at 12°C and 85–90% RH.	Antioxidant	[92]
Spirulina platensis	Glycerol and carboxyl methyl cellulose.	*Spirulina platensis*	Commercial powder	Strawberry Weight loss, pH, and vitamin C Storage conditions for 7 days at 4–7°C.	N/A	[78]

Nowadays, the search for effective antimicrobial coatings/films against spoilage flora and/or to decrease the risk of pathogenic microorganisms' growth is one of the consumer demands. Benelhadj et al. [39] reported that the combination of microalga proteins with antimicrobial compounds has been observed as a valuable strategy to improve the nutrient value of a food product, ensuring food safety. In this sense, different proportions of proteins from *Spirulina* complexed with a well-known food-grade antimicrobial enzyme (lysozyme) have been proposed as an alternative to Na-caseinate. It was found that films based on *Spirulina* proteins were opaque, brown, thick, and had more water affinity than films based on sodium caseinate. Additionally, the microalgae-based films were more fragile and less elastic than sodium caseinate films. Nevertheless, films based on *Spirulina* proteins showed a slower release of lysozyme kinetics, which is one of the important mechanical characteristics of an antimicrobial film.

Antioxidant activity is another important property of active films and coatings. The combination of *Spirulina* biomass with compounds well-studied as chitosan from crabs has been shown to have an improved effect on antioxidant activity. Additionally, Gram-positive antibacterial activities were found when *Spirulina* was applied as food edible films than the chitosan films without microalga, with the polyphenol content of *Spirulina* discussed as a possible responsible for this effect [46]. The study of chitosan-*Spirulina* films by X-ray diffraction suggests the production of a semi-crystalline structure with improved water resistance, vapor barrier properties, and anti-oxidant activity, but lower transparency than chitosan films without *Spirulina* [73]. Other compounds have been proposed to be combined with microalgal biomass, as is the case of gelatine-based films to package refrigerated fish and hake. The latter has demonstrated a bacteriostatic effect on psychrotrophic microorganisms, proteolytic and lipolytic bacteria growth, as well as an inhibitory effect on the production of free fatty acids, which has a direct relation with microbial enzymes activities and lower lipid oxidation [70, 71]. For fruits, the use of gelatine-based films enriched with *Spirulina* has shown an improved effect on shelf-life, as was observed in refrigerated Cambuci peppers, reducing the mass loss, and ensuring their color [76]. Alginate, which is a product from brown algae, is another compound proposed as a valuable food packaging material, improving mechanical properties, and reducing the dehydrating effect of stored fruits [80]. The combination of alginate coating with *Spirulina* biomass improves the total phenol, flavonoid, and antioxidant activity of the film used for coated fruits [87]. Nonetheless, the elaboration of edible coating based on *Spirulina* biomass has also been proposed in a simpler way in which the microalgal biomass is dissolved in mineral oil and then used to cover fruits such as pomegranate, demonstrating a positive effect of a 1% solution on organoleptic properties such as color, brightness, maintaining sugar, and reducing astringency [77].

Phycocyanin from *Spirulina* has also been proposed as a valuable source for natural blue pigments, and its interaction with coating materials is of interest in order to improve their stability. In this sense, the use of mixtures of maltodextrin with alginate or carrageenan has shown good phycocyanin dispersing and a bluer color [81].

Recently, an additional value was added to the use of *Spirulina* compounds in food processing. The possible use of the residues produced after the drying and extraction processes of *Spirulina* biomass, residues commonly higher in inorganic salts, has been proposed to be used as raw material to produce environmentally friendly bioplastics, using poly(vinyl alcohol) as a cross-linker and demonstrating higher mechanical properties and water resistance [82]. Clearly, this is not an edible material, but it is important to highlight the potential that the same residues obtained from microalgae production can generate added value to other packaging materials by providing them with properties that contribute to their subsequent biodegradability.

The nutritional value of *Spirulina* biomass, as well as the positive interaction between microalgal metabolites with other bioactive compounds (producing a synergistic effect in the desired biological activity), increases the repertoire of possible uses for these microalgae. The increment of basic studies addressing the nutritional, chemical, and mechanical properties of microalgal-based materials ensures the effectiveness of these for food processing.

1.5 CONCLUSIONS AND FUTURE PERSPECTIVE

Spirulina by itself represents a novel and attractive way to be incorporated into edible films or coatings either as the main structural matrix due to its high protein content, lipids, and carbohydrates or its active compounds as already mentioned. In addition to this, extracts of bioactive *Spirulina* compounds with biological activity can be incorporated to form blends or composites with other structural compounds to provide the edible packaging material with various functional properties.

Incorporating *Spirulina* into edible films and coatings might be a viable alternative to enhance the storage quality of food, as some research has demonstrated the benefits of *Spirulina,* or its functional components applied to various edible matrices over food.

The stability of the packaging material combined with *Spirulina* should also be assessed in comparison to the changes that the food undergoes during storage. This entails consideration of temperature, pH fluctuations, microbial development, as well as the evaluation of sensory biocompatibility and exploring alternatives to mask unpleasant odors or tastes.

Although *Spirulina* turns out to be a promising element for the formulation of highly sustainable active edible packaging materials, there is still a long way to go as it is an area of opportunity for research.

KEYWORDS

- **edible coating**
- **edible films**
- **functional bioactive compounds**
- **microbial development**
- **packaging**
- ***Spirulina***
- **sustainable food packages**

REFERENCES

1. Ritchie, H., Rosado, P., & Roser, M. (2022) *Environmental Impacts of Food Production*. OurWorldInData.org. Retrieved May 27, 2024 from, https://ourworldindata.org/environmental-impacts-of-food (accessed on 27 May 2024).
2. Crippa, M., Solazzo, E., Guizzardi, D., Monforti-Ferrario, F., Tubiello, F. N., & Leip, A. J. N. F. (2021). Food systems are responsible for a third of global anthropogenic GHG emissions. *Nature Food, 2*(3), 198–209.
3. Poore, J., & Nemecek, T. (2018). Reducing food's environmental impacts through producers and consumers. *Science, 360*(6392), 987–992.
4. Díaz-Montes, E., & Castro-Muñoz, R. (2021). Edible films and coatings as food-quality preservers: An overview. *Foods, 10*(2), 249.
5. Revel, M., Châtel, A., & Mouneyrac, C. (2018). Micro (nano)plastics: A threat to human health? *Current Opinion in Environmental Science and Health, 1*, 17–23. Available at: https://www.sciencedirect.com/science/article/pii/S2468584417300235 (accessed on 27 May 2024).
6. Kan, M., & Miller, S. A. (2022). Environmental impacts of plastic packaging of food products. *Resources, Conservation and Recycling, 180*, 106–156.

7. Geyer, R., Jambeck, J. R., & Law, K. L. (2017). Production, use, and fate of all plastics ever made. *Science Advances, 3*(7), e1700782. Available at: http://advances.sciencemag.org/content/3/7/e1700782 (accessed on 27 May 2024).
8. Nakamoto, M. M., Assis, M., de Oliveira Filho, J. G., & Braga, A. R. C. (2023). Spirulina application in food packaging: Gaps of knowledge and future trends. *Trends in Food Science & Technology*, *133*, 138–147.
9. Sharma, P., Shehin, V. P., Kaur, N., & Vyas, P. (2019). Application of edible coatings on fresh and minimally processed vegetables: a review. *International Journal of Vegetable Science, 25*(3), 295–314.
10. Summo, C., & De Angelis, D. (2022). The importance of edible films and coatings for sustainable food development. *Foods, 11*(20), 3221.
11. Galus, S., Arik Kibar, E. A., Gniewosz, M., & Kraśniewska, K. (2020). Novel materials in the preparation of edible films and coatings—A review. *Coatings, 10*(7), 674.
12. Kumar, L., Ramakanth, D., Akhila, K., & Gaikwad, K. K. (2022). Edible films and coatings for food packaging applications: A review. *Environmental Chemistry Letters, 20*, 875–900.
13. Arnon-Rips, H., & Poverenov, E. (2018). Improving food products' quality and storability by using layer by layer edible coatings. *Trends in Food Science and Technology, 75*, 81–92.
14. Versino, F., Ortega, F., Monroy, Y., Rivero, S., López, O. V., & García, M. A. (2023). Sustainable and bio-based food packaging: A review on past and current design innovations. *Foods, 12*(5), 1057.
15. Pavlath, A. E., & Orts, W. (2009). Edible Films and Coatings: Why, What, and How?. In: Huber, K., & Embuscado, M. (eds.). Edible Films and Coatings for Food Applications. Springer, New York, NY. https://doi.org/10.1007/978–0-387–92824–1_1.
16. Calva-Estrada, S. J., Jiménez-Fernández, M., & Lugo-Cervantes, E. (2019). Protein-based films: Advances in the development of biomaterials applicable to food packaging. *Food Engineering Reviews, 11*, 78–92.
17. Mohamed, S. A., El-Sakhawy, M., & El-Sakhawy, M. A. M. (2020). Polysaccharides, protein and lipid-based natural edible films in food packaging: A review. *Carbohydrate Polymers, 238*, 116178.
18. Mihalca, V., Kerezsi, A. D., Weber, A., Gruber-Traub, C., Schmucker, J., Vodnar, D. C., & Pop, O. L. (2021). Protein-based films and coatings for food industry applications. *Polymers, 13*(5), 769.
19. Hassan, B., Chatha, S. A. S., Hussain, A. I., Zia, K. M., & Akhtar, N. (2018). Recent advances on polysaccharides, lipids, and protein-based edible films and coatings: A review. *International Journal of Biological Macromolecules, 109*, 1095–1107.
20. Song, D. H., Hoa, V. B., Kim, H. W., Khang, S. M., Cho, S. H., & Ham, J. S., Seol, K. H. (2021). Edible films on meat and meat products. *Coatings, 11*(11), 1344.
21. Zubair, M., Pradhan, R. A., Arshad, M., & Ullah, A. (2021). Recent advances in lipid-derived bio-based materials for food packaging applications. *Macromolecular Materials and Engineering, 306*(7), 2000799.
22. Kaur, J., Gunjal, M., Rasane, P., Singh, J., Kaur, S., & Poonia, A., & Gupta, P. (2022). Edible packaging: An overview. In *Edible Food Packaging: Applications, Innovations and Sustainability* (pp. 3–25). Springer, Singapore. https://doi.org/10.1007/978–981–16–2383–7_1.
23. Galus, S., & Kadzińska, J. (2015). Food applications of emulsion-based edible films and coatings. *Trends in Food Science and Technology, 45*(2), 273–283.

24. Vukić, M., Grujić, S., & Odzaković, B. (2017). Application of edible films and coatings in food production. In Pellicer, E., et al. *Advances in Applications of Industrial Biomaterials* (pp. 121–138). Springer, Cham. https://doi.org/10.1007/978–3-319–62767–0_7.
25. Armghan Khalid, M., Niaz, B., Saeed, F., Afzaal, M., Islam, F., Hussain, M., & Al-Farga, A. (2022). Edible coatings for enhancing safety and quality attributes of fresh produce: A comprehensive review. *International Journal of Food Properties, 25*(1), 1817–1847.
26. Sothornvit, R., & Krochta, J. M. (2005). Plasticizers in edible films and coatings. In *Innovations in Food Packaging* (pp. 403–433). Academic Press.
27. Tarique, J., Sapuan, S. M., & Khalina, A. (2021). Effect of glycerol plasticizer loading on the physical, mechanical, thermal, and barrier properties of arrowroot (*Maranta arundinacea*) starch biopolymers. *Scientific Reports, 11*(1), 1–17.
28. Šuput, D. Z., Lazić, V. L., Popović, S. Z., & Hromiš, N. M. (2015). Edible films and coatings: Sources, properties, and application. *Food and Feed Research, 42*(1), 11–22.
29. Petkoska, A. T., Daniloski, D., D'Cunha, N. M., Naumovski, N., & Broach, A. T. (2021). Edible packaging: Sustainable solutions and novel trends in food packaging. *Food Research International, 140*, 109981.
30. Tóth, A., & Halász, K. (2019). Characterization of edible biocomposite films directly prepared from psyllium seed husk and husk flour. *Food Packaging and Shelf Life, 20*, 100299.
31. Silva-Weiss, A., Ihl, M., Sobral, P. D. A., Gómez-Guillén, M. C., & Bifani, V. J. F. E. R. (2013). Natural additives in bioactive edible films and coatings: functionality and applications in foods. *Food Engineering Reviews, 5*, 200–216.
32. Hernández-Ledesma, B., & Herrero, M. (Eds.). (2013). *Bioactive Compounds from Marine Foods: Plant and Animal Sources*. John Wiley and Sons.
33. Teixeira-Costa, B. E., & Andrade, C. T. (2022). Natural polymers used in edible food packaging—History, function and application trends as a sustainable alternative to synthetic plastic. *Polysaccharides, 3*(1), 32–58.
34. Coma, V. (2008). Bioactive packaging technologies for the extended shelf life of meat-based products. *Meat Science, 78*(1–2), 90–103.
35. Fernandes, S. S., Coelho, M. S., & de las Mercedes Salas-Mellado, M. (2019). Bioactive compounds as ingredients of functional foods: polyphenols, carotenoids, peptides from animal and plant sources new. In *Bioactive Compounds* (pp. 129–142).
36. Hayes, M., & Flower, D. (2013). Bioactive peptides from marine processing by products. In B. Hernández-Ledesma & M. Herrero (Eds.), *Bioactive Compounds from Marine Foods: Plant and Animal Sources* (pp. 57–71). Chichester, UK: John Wiley & Sons Ltd.
37. Freile-Pelegrín, Y., & Robledo, D. (2013). Bioactive phenolic compounds from algae. In *Bioactive Compounds from Marine Foods: Plant and Animal Sources* (pp. 113–129).
38. Guedes, C. A., Amaro, H. M., Sousa-Pinto, I., & Xavier Malcata, F. (2013). Bioactive carotenoids from microalgae. In *Bioactive Compounds from Marine Foods: Plant and Animal Sources* (pp. 131–151).
39. Benelhadj, S., Fejji, N., Degraeve, P., Attia, H., Ghorbel, D., & Gharsallaoui, A. (2016). Properties of lysozyme/*Arthrospira platensis* (*Spirulina*) protein complexes for antimicrobial edible food packaging. *Algal Research, 15*, 43–49.
40. Larrosa, A. P., Camara, A. S., Moura, J. M., & Pinto, L. A. (2018). *Spirulina* sp. biomass dried/disrupted by different methods and their application in biofilm production. *Food Science and Biotechnology, 27*, 1659–1665.

41. Dang, B. T., Bui, X. T., Tran, D. P., Ngo, H. H., Nghiem, L. D., Nguyen, P. T., & Varjani, S. (2022). Current application of algae derivatives for bioplastic production: A review. *Bioresource Technology, 347, 126698.*
42. Lee, W. K., Lim, Y. Y., Leow, A. T. C., Namasivayam, P., Abdullah, J. O., & Ho, C. L. (2017). Biosynthesis of agar in red seaweeds: A review. *Carbohydrate Polymers, 164*, 23–30.
43. Rhein-Knudsen, N., Ale, M. T., & Meyer, A. S. (2015). Seaweed hydrocolloid production: an update on enzyme assisted extraction and modification technologies. *Marine Drugs, 13*(6), 3340–3359.
44. Necas, J., & Bartosikova, L. (2013). Carrageenan: a review. *Veterinarni Medicina, 58*(4), 187–205.
45. Onen Cinar, S., Chong, Z. K., Kucuker, M. A., Wieczorek, N., & Cengiz, U., Kuchta, K. (2020). Bioplastic production from microalgae: A review. *International Journal of Environmental Research and Public Health, 17*(11), 3842.
46. Balti, R., Mansour, M. B., Sayari, N., Yacoubi, L., Rabaoui, L., Brodu, N., & Massé, A. (2017). Development and characterization of bioactive edible films from spider crab (*Maja crispata*) chitosan incorporated with *Spirulina* extract. *International Journal of Biological Macromolecules, 105*, 1464–1472.
47. Ahsan, M., Habib, B., Parvin, M., Huntington, T. C., & Hasan, M. R. (2008). A review on culture, production, and use of spirulina as food for humans and feeds for domestic animals. *FAO Fisheries and Aquaculture Circular*. No. 1034. Rome, FAO. 33p.
48. Soni, R. A., Sudhakar, K., Rana, R. S., Baredar, P. (2021). Food Supplements Formulated with *Spirulina*. In: Mandotra, S. K., Upadhyay, A. K., Ahluwalia, A. S. (eds.). *Algae: Multifarious Applications for a Sustainable World* (pp. 201–226). Algae. Springer, Singapore. https://doi.org/10.1007/978-981-15-7518-1_9.
49. AlFadhly, N. K., Alhelfi, N., Altemimi, A. B., Verma, D. K., & Cacciola, F. (2022). Tendencies affecting the growth and cultivation of genus *Spirulina*: An Investigative Review on Current Trends. *Plants, 11*(22), 3063.
50. Soni, R. A., Sudhakar, K., & Rana, R. S. (2017). *Spirulina*–from growth to nutritional product: A review. *Trends in Food Science and Technology, 69*, 157–171.
51. Lafarga, T., Sánchez-Zurano, A., Villaró, S., Morillas-España, A., & Acién, G. (2021). Industrial production of spirulina as a protein source for bioactive peptide generation. *Trends in Food Science and Technology, 116*, 176–185.
52. Magwell, P. F. R., Djoudjeu, K. T., Minyaka, E., Tavea, M. F., Fotsop, O. W., Tagnikeu, R. F., & Lehman, L. G. (2023). Sodium bicarbonate ($NaHCO_3$) increases growth, protein and photosynthetic pigments production and alters carbohydrate production of *Spirulina* platensis. *Current Microbiology, 80*(2), 1–13.
53. Al Mahrouqi, H., Dobretsov, S., Avilés, A., & Díaz, R. T. A. (2022). *Spirulina* optimization using cane molasses as the cost-effective alternative of sodium bicarbonate. *Biology Bulletin*, *49*(Suppl 1), S60-S68.
54. Tzachor, A., Smidt-Jensen, A., Ramel, A., & Geirsdóttir, M. (2022). Environmental impacts of large-scale *Spirulina* (*Arthrospira platensis*) production in Hellisheidi geothermal park Iceland: life cycle assessment. *Marine Biotechnology, 24*(5), 991–1001.
55. Klanchui, A., Khannapho, C., Phodee, A., Cheevadhanarak, S., & Meechai, A. (2012). A genome-scale metabolic model of *Spirulina platensis* C1. *BMC Systems Biology, 6*, 1–15.

56. Mu, N., Mehar, J. G., Mudliar, S. N., & Shekh, A. Y. (2019). Recent advances in microalgal bioactives for food, feed, and healthcare products: commercial potential, market space, and sustainability. *Comprehensive Reviews in Food Science and Food Safety, 18*(6), 1882–1897.
57. Alizadeh-Sani, M., Ehsani, A., Moghaddas Kia, E., & Khezerlou, A. (2019). Microbial gums: Introducing a novel functional component of edible coatings and packaging. *Applied Microbiology and Biotechnology, 103*, 6853–6866.
58. Hamed, I., Jakobsen, A. N., & Lerfall, J. (2022). Sustainable edible packaging systems based on active compounds from food processing byproducts: A review. *Comprehensive Reviews in Food Science and Food Safety, 21*(1), 198–226.
59. Cardoso, T., Esmerino, L. A., Bolanho, B. C., Demiate, I. M., & Danesi, E. D. G. (2019). Technological viability of biobased films formulated with cassava by-product and *Spirulina* platensis. *Journal of Food Process Engineering, 42*(5), e13136.
60. Shafiei, R., & Mostaghim, T. (2022). Improving shelf life of calf fillet in refrigerated storage using edible coating based on chitosan/natamycin containing *Spirulina platensis* and *Chlorella vulgaris* microalgae. *Journal of Food Measurement and Characterization, 16*(1), 145–161.
61. Moghaddas Kia, E., Ghasempour, Z., & Alizadeh, M. (2018). Fabrication of an eco-friendly antioxidant biocomposite: Zedo gum/sodium caseinate film by incorporating microalgae (*Spirulina platensis*). *Journal of Applied Polymer Science, 135*(13), 46024.
62. Devi, M. A., & Venkataraman, L. V. (1984). Functional properties of protein products of mass cultivated blue-green alga *Spirulina* platensia. *Journal of Food Science, 49*(1), 24–27.
63. Menegotto, A. L. L., de Souza, L. E. S., Colla, L. M., Costa, J. A. V., Sehn, E., Bittencourt, P. R. S., & Colla, E. (2019). Investigation of techno-functional and physicochemical properties of *Spirulina platensis* protein concentrate for food enrichment. *LWT, 114*, 108267.
64. Kamaruding, N. A., Muhammad Daud, N. A., Ismail, N., & Shaharuddin, S. (2022). Effect of Different solubilization pH Values on the functional properties of protein *Spirulina platensis* isolated through acidic precipitation. *Journal of Aquatic Food Product Technology, 31*(10), 1025–1037.
65. Nouri, E., & Abbasi, H. (2023). Effect of different processing methods on stability of anthocyanin and phycocyanin of *Spirulina platensis*. *Journal of Food Biosciences and Technology, 13*(2), 13–25.
66. Grahl, S., Strack, M., Mensching, A., & Mörlein, D. (2020). Alternative protein sources in Western diets: Food product development and consumer acceptance of spirulina-filled pasta. *Food Quality and Preference, 84*, 103933.
67. Nourmohammadi, N., Soleimanian-Zad, S., & Shekarchizadeh, H. (2020). Effect of *Spirulina* (*Arthrospira platensis*) microencapsulated in alginate and whey protein concentrate addition on physicochemical and organoleptic properties of functional stirred yogurt. *Journal of the Science of Food and Agriculture, 100*(14), 5260–5268.
68. Sahay, N., Chaudhari, N., Trivedi, R., Patel, T., & Kommu, S. (2022). *Spirulina* the super food; regulation of foul-fishy smell. *Academic Letters, 2*, 1–6.
69. Fabra, M. J., Martínez-Sanz, M., Gómez-Mascaraque, L. G., Gavara, R., & López-Rubio, A. (2018). Structural and physicochemical characterization of thermoplastic corn starch films containing microalgae. *Carbohydrate Polymers, 186*, 184–191.

70. Stejskal, N., Miranda, J. M., Martucci, J. F., Ruseckaite, R. A., & Aubourg, S. P., Barros-Velázquez, J. (2020a). The effect of gelatine packaging film containing a *Spirulina platensis* protein concentrate on Atlantic mackerel shelf life. *Molecules, 25*(14), 3209.
71. Stejskal, N., Miranda, J. M., Martucci, J. F., Ruseckaite, R. A., Barros-Velázquez, J., & Aubourg, S. P. (2020b). Quality enhancement of refrigerated hake muscle by active packaging with a protein concentrate from *Spirulina platensis*. *Food and Bioprocess Technology, 13*, 1110–1118.
72. Kontogianni, V. G., Chatzikonstantinou, A. V., Mataragas, M., Kondyli, E., Stamatis, H., & Bosnea, L. (2021). Evaluation of the antioxidant and physicochemical properties of Microalgae/whey protein-based edible films. In *Biology and Life Sciences Forum, 6*(1), 97.
73. Luo, A., Hu, B., Feng, J., Lv, J., & Xie, S. (2021). Preparation, and physicochemical and biological evaluation of chitosan-*Arthrospira platensis* polysaccharide active films for food packaging. *Journal of Food Science, 86*, 987–995.
74. Ramji, V., & Vishnuvarthanan, M. (2022). Chitosan ternary bio nanocomposite films incorporated with MMT K10 nano clay and spirulina. *Silicon, 14*(3), 1209–1220.
75. Onias, E. A., Rocha, R. H. C., Lima, J. D., & Onias, E. A., Furtunato, T. S. (2016). Organic 'Tommy Atkins' mango postharvest quality when treated with biofilms enriched by *Spirulina platensis*. *Rev. Científica., 44*, 286–293.
76. Cardoso, T., Demiate, I. M., & Danesi, E. D. G. (2017). Biodegradable films with *Spirulina platensis* as Coating for Cambuci Peppers (*Capsicum* sp.). *American Journal of Food Technology, 12*, 236–244.
77. de Oliveira, L. M., de Oliveira, Á. M. F., Araújo, R. H. C. R., Dias, G. A., de Medeiros Teodósio, A. E. M., de Lima, J. F., & da Silva Barbosa, L., & Guedes, W. A. (2020). *Spirulina platensis* coating for the conservation of pomegranate. *AIMS Agriculture and Food, 5*, 76–85.
78. Byantara, P., & Dianursanti. (2021). Utilization of *Spirulina platensis* microalgae as edible coating to maintain the quality of fresh strawberry (*Fragaria* sp.). In *AIP Conference Proceedings (Vol. 2344, No. 1, p. 020016).* AIP Publishing LLC.
79. Costa, B. R., Rocha, S. F., Rodrigues, M. C., Pohndorf, R. S., Larrosa, A. P., & Pinto, L. A. (2015). Physicochemical characteristics of the *Spirulina* sp. dried in heat pump and conventional tray dryers. *International Journal of Food Science and Technology, 50*, 2614–2620.
80. Senturk Parreidt, T., Müller, K., & Schmid, M. (2018). Alginate-Based Edible Films and Coatings for Food Packaging Applications. *Foods, 7*, 170.
81. Dewi, E. N., Purnamayati, L., & Kurniasih, R. A. (2017). Physical characteristics of phycocyanin from spirulina microcapsules using different coating materials with freeze-drying method. In *IOP Conference Series: Earth and Environmental Science, 55*, 012–060.
82. Zhang, C., Wang, C., Cao, G., Wang, D., & Ho, S. H. (2020). A sustainable solution to plastics pollution: An eco-friendly bioplastic film production from high-salt-contained *Spirulina* sp. residues. *Journal of Hazardous Materials, 388*, 121773.
83. Han, J. H. (2014). Edible Films and Coatings: A Review, In: Han, J.H. (eds.). *Food Science and Technology, Innovations in Food Packaging* (Second Edition), Academic Press, 213–255p.

84. Iamtham, S., & Sornchai, P. (2022). Biofixation of CO_2 from a power plant through large-scale cultivation of *Spirulina maxima. South African Journal of Botany, 147*, 840–851.
85. Kamal, I. (2019). Edible films and coatings: Classification, preparation, functionality and applications-a review. *Arc. Org. Inorg. Chem. Sci., 4*(2)–2019. AOICS. MS. ID, 184.
86. Kocira, A., Kozłowicz, K., Panasiewicz, K., Staniak, M., Szpunar-Krok, E., & Hortyńska, P. (2021). Polysaccharides as edible films and coatings: Characteristics and influence on fruit and vegetable quality—A review. *Agronomy, 11*(5), 813.
87. Rastegar, S., & Atrash, S. (2021). Effect of alginate coating incorporated with *Spirulina, Aloe vera*, and guar gum on physicochemical, respiration rate, and colour changes of mango fruits during cold storage. *Journal of Food Measurement and Characterization, 15*, 265–275.
88. Ribeiro, A. M., Estevinho, B. N., & Rocha, F. (2021). Preparation and incorporation of functional ingredients in edible films and coatings. *Food and Bioprocess Technology, 14*, 209–231.
89. Ebrahimi, F., & Rastegar, S. (2020). Preservation of mango fruit with guar-based edible coatings enriched with *Spirulina platensis* and *Aloe vera* extract during storage at ambient temperature. *Scientia Horticulturae, 265*, 109258.

CHAPTER 2

Food Grade Emulsions for Delivery of Bioactive Compounds and Functional Foods

DELY RUBI CHÁVEZ-GARAY*, NÉSTOR GUTIÉRREZ-MÉNDEZ, and DIEGO ELOY CARBALLO-CARBALLO

Facultad de Ciencias Químicas, Universidad Autónoma de Chihuahua, Circuito Universitario s/n, Campus II, C.P., Chihuahua, México

*Corresponding author.

ABSTRACT

Nowadays, it is clear the relationship between the nutritious foods we eat and our health. Currently, it is relevant to design foods that offer not only nutritional value but also health benefits as functional foods or nutraceuticals. Functional foods are any fortified, enriched, or enhanced foods that have a positive impact on health, physical performance, or state of mind when consumed on a regular basis and at certain levels. Functional foods are products that are consumed as foods and not in dosage form, while nutraceuticals, on the other hand, are healthful products that are formulated and taken in a dosage form.

Several industries such as cosmetics, pharmaceuticals, and food need to deliver functional components in an edible form. The growing demand for edible delivery systems to enhance the efficacy of bioactive components within the food has generated the application of emulsion technology for the design and fabrication of delivery systems.

The relevance of emulsions in the food area is because most of the ingredients are wholly or partly emulsions, or at some point in the production process emulsions are generated. In addition, emulsions are also used as delivery systems designed to encapsulate, protect, and control the release

Food Security, Safety, and Sustainability: Validated Innovative Approaches and Technological Advancements.
José Juan Buenrostro-Figueroa and Julio César Tafolla-Arellano (Eds.)

and absorption of ingredients such as nutraceuticals, flavors, colors, preservatives, vitamins, etc., improving their handling, efficacy, and stability.

As emulsion-based food products exhibit a wide range of physicochemical, sensory, and biological characteristics depending on the kinds of ingredients and processing conditions used to create them, it's because in this chapter, we will discuss some current studies and recent advances in the formation, properties, applications, and health benefits of a variety of emulsion systems, including conventional emulsions, nano- and micro-emulsions, Pickering emulsions, multiple emulsions, multilayered emulsions, liposomal emulsions, and niosomes.

2.1 INTRODUCTION

For some years now, the food industry has focused on obtaining and improving food products that are healthy and appealing to the consumer. This research is based on the formulation of functional food emulsions, for example, products that offer high nutritional value and provide the consumer with health benefits. In addition, this has brought with it the development and identification of novel edible ingredients that, when added, achieve these mentioned goals. Within these ingredients, the use of bioactive compounds has been chosen, which present beneficial effects for health and food quality.

However, the challenge is not only focused on obtaining these ingredients but also on their incorporation and maintenance within a complex matrix. Furthermore, investigations related to obtaining these ingredients have been broadened beyond their use in the food sector. Several industries, such as food, pharmaceuticals, and cosmetics, have opted to incorporate bioactive compounds into their formulations. Due to this, numerous delivery systems for bioactive compounds have been developed today, where emulsions are an up-and-coming option.

Specific characteristics of emulsions, complex systems formed by the dispersion of two edible and immiscible liquids (conventionally water-in-oil (W/O), oil-in-water (O/W), oil-in-water-in-oil (O/W/O), or water-in-oil-in-water (W/O/W)), allow them to encapsulate, protect, and increase the stability of compounds such as nutraceuticals, colorants, flavorings, antioxidants, preservatives, vitamins, etc. Additionally, food-grade emulsions have shown great usefulness when creating vehicles to transport and release bioactive compounds within food matrices. The following chapter explores the properties of food-grade emulsions and their application as a transport vehicle and delivery system for bioactive compounds within the formulation of functional foods.

2.2 EMULSION SYSTEMS

2.2.1 FORMATION

The use of emulsions has been widely extended within the industry, aimed at fulfilling various functions for which they were designed. Due to this, their formulation and formation will depend extensively on the ingredients that compose them, the matrix where they will be added, and the final use for which the emulsion is intended.

As previously mentioned, emulsions are a type of dispersion composed of two immiscible phases, where one of them (dispersed phase) is contained within the other (continuous phase) in the form of tiny droplets [1]. The phases usually used in the formulation of emulsions are water (W) and oil (O), which must be stabilized by adding an emulsifier or surfactant to reduce interfacial tension. The selection of the oily phase, emulsifiers, and ingredients will depend on the functional, sensory, physicochemical, and nutritional properties desired in the final product. When selecting the oil, parameters such as quality, oxidative stability, and physicochemical and sensory properties are essential. On the other hand, the Hydrophile–Lipophile difference (HLD: surfactant affinity for each phase) and the Hydrophile–Lipophile balance (HLB: surfactant solubility in each of the phases) are the two crucial factors in selecting the surfactant to use. Among the most suitable emulsifiers or surfactants are proteins, phospholipids, mono or diacylglycerols, polymers, and polyol esters of fatty acids or small particles [2, 3], whose objective is to avoid undesirable events within emulsions such as flocculation, sedimentation, creaming, phase inversion, Ostwald ripening, and coalescence.

Besides the nature of the ingredients that compose them, the formation of emulsions will be conditioned by the method by which they are produced. The first step for producing and obtaining emulsions is selecting and mixing the phases and components, followed by a process known as homogenization, which will convert the two immiscible liquids into the desired emulsion [3]. This homogenization process can be carried out in two steps: primary and secondary homogenization (*see* Figure 2.1). The primary homogenization is oriented to create the drops of the dispersed phase that will be contained within the continuous phase. In contrast, the aim of secondary homogenization is focused on reducing the size of the droplets created from the previous step [3, 4]. Equipment such as homogenizers, high-performance dispersion instruments, or ultrasonic devices helps perform the previously mentioned homogenization processes.

2.2.2 PROPERTIES

Several parameters will be directly or indirectly related to the properties of the emulsions. Factors such as the nature and concentration of the components that form them (type of oily phase and surfactants), the time or age from their preparation, the temperature, and the mechanical method by which they were obtained will determine not only the properties of the emulsion but also its stability (*see* Figure 2.2). These factors will influence the size of the droplets formed and the interfacial properties of the emulsion (adsorption, droplet charge, etc.).

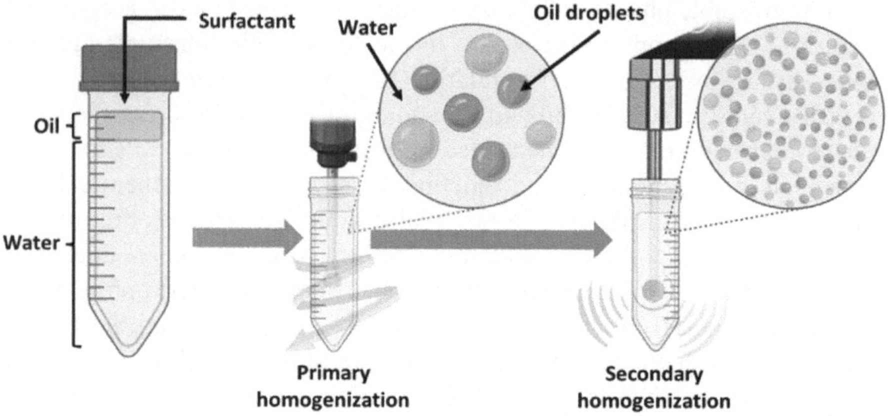

FIGURE 2.1 Homogenization steps during the emulsion preparation.

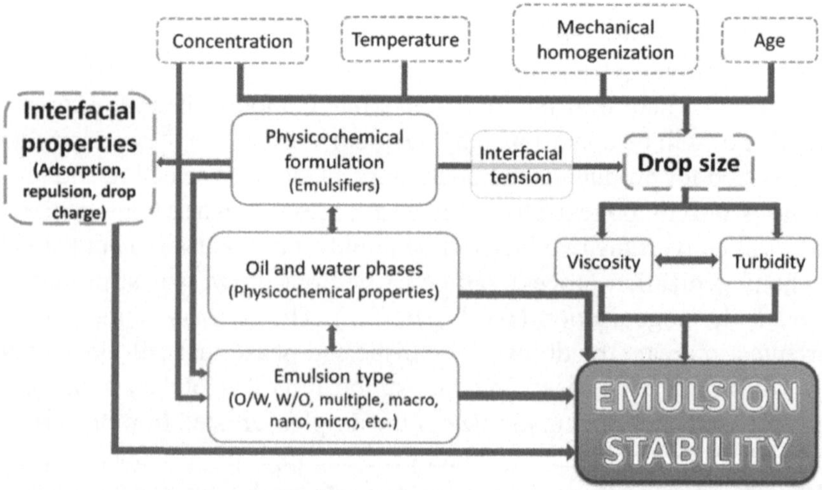

FIGURE 2.2 Factors that affect the properties of emulsions and their relationship with stability.

Source: Adapted from Ref. [5].

While the appearance and rheology are strongly influenced by the particle or drop size of the dispersed phase produced during the emulsification process, the interfacial properties will not only affect this last parameter but also the stabilization of the new drops formed, avoiding the phenomenon of coalescence [6]. The interfacial properties are those found at the interface, that is, between the continuous and the dispersed phase, and will be directly affected by the type of surfactant used for emulsification.

In addition to the inherent properties of the emulsion, it is possible to add others by adding specific ingredients within it. These ingredients must be mixed with the phase to which they have the highest affinity before emulsifying. Lipophilic components such as vitamins, dyes, and some surfactants will be integrated with the oily phase, while proteins, polysaccharides, sugar, and salts will be with the aqueous phase [3].

Nowadays, specific methods are aimed at exploring or determining emulsions' properties. Droplet size can be determined by X-ray diffraction, sedimentation techniques, or microscopy, varying in size from 5 nm to 100 μm depending on the type of emulsion [7, 8]. The viscosity and turbidity of the emulsions can be determined by rheology and colorimetry procedures, respectively. Finally, conductivity determination or particle electrophoresis can be a helpful tool to assess the charge of emulsion droplets.

2.2.3 APPLICATIONS

Due to the characteristics and properties of emulsions, their use and application have spread throughout the entire industrial sector. The benefits of emulsions in the industry have covered sectors such as food, pharmaceuticals, cleaning/beauty and cosmetics, agronomy, textiles, mining, and oil, among others.

There are food products that are emulsions *per se*. Milk is a natural O/W emulsion consumed and used to make other products. Likewise, emulsions have been used in the *food industry* to prepare sauces, dressings, mayonnaise, ice cream, drinks, etc. An essential factor in the elaboration of food emulsions is the sensory attribute, which is why various investigations have focused on this [2, 7].

Within the *pharmaceutical industry*, emulsions are oriented towards the delayed release system of drugs through combined W/O/W systems. It has been observed that this type of emulsion is capable of encapsulating and protecting water-soluble drugs, ensuring their slow release [9, 10].

Some cosmetics, such as lotions, shampoos, hand creams, etc., are formulated from O/W and W/O emulsions. It is important to note that for an emulsion used in *the cosmetic or beauty industry* to be viable, all the ingredients used for its formulation must be hypoallergenic and have no irritating or sensitive effect on the skin.

Finally, several *agrochemicals* are formulated as O/W type emulsions, where lipophilic compounds are added to the oily phase before the emulsification process [2, 11].

2.3 TYPES OF EMULSIONS

As previously described, an emulsion is a biphasic system consisting of at least two immiscible liquids, one dispersed in the other as droplets. These systems can be classified by their size (macro, mini, and nano), thermodynamic stability (micro-emulsion), the number of phases (multiple and multilayered emulsions), and the emulsifier arrangement (liposomes, niosomes, and Pickering emulsions). In this chapter section, we will discuss the key differences between these emulsions.

2.3.1 MACRO, MINI, AND NANO-EMULSIONS

Macro-emulsions are the conventional emulsions formed by mixing the oil, water, and emulsifier with mechanical shear. The droplet size in macroemulsions is big, typically between 1 and 100 μm, and usually anisotropic (direction-dependent). These big droplets scatter the light substantially, generating opaque or optically turbid emulsions with a milky-yellow color. Finally, the interfacial tension and viscosity of macro-emulsions are high [12–14].

Mini-emulsions share similar characteristics to macro-emulsions but with smaller droplet sizes, typically 0.1 to 1 μm (*see* Figure 2.3). Furthermore, miniemulsion formation requires more energy than macro-emulsions. However, mini-emulsions still tend to break down over time just as macro-emulsions and are not small enough to be considered nano-emulsions [15, 16].

Nano-emulsions fall in the nanometric scale, usually between 10 and 100 nm (0.01–0.1 μm). Such tiny emulsions can only be achieved using high-energy equipment like a homogenizer, high- shear mixer, microfluidizer, and sonicator. Additionally, nano-emulsions with droplet diameters below 80 nm (0.08 μm) are optically transparent or slightly turbid. However, when the droplet sizes are above 80 nm, the nano-emulsions become whitish due to significant multiple scattering. Finally, it is worth highlighting that nano-emulsions are metastable but not thermodynamically stable systems. In other words, nano-emulsions eventually will show gravitational separations, flocculation, coalescence, or Ostwald ripening over time. Although, these changes will occur more slowly than in macro- and mini-emulsions [13, 14, 17].

2.3.2 MICRO-EMULSIONS

Micro-emulsions are transparent and thermodynamically stable systems (*see* Figure 2.3) that do not require high-energy equipment since they self-assemble at specific surfactant/co-surfactant/co-solvent concentrations. These self-assembled structures are tiny mixed micelles or solubilized micelles that incorporate extra surfactant or co-surfactant at the interface. Under certain conditions, these mixed micelles may create random bi-continuous structures (denominated Winsor IV structures), enhancing the formation entropy (ΔSf). Furthermore, the right combination of surfactant, co-surfactant, and co- solvent drops the interfacial tension substantially (10^{-1} to 10^{-2} mN/m), minimizing the formation enthalpy (ΔHf). Consequently, the free energy of formation (ΔGf) is almost zero or negative, considering the equation $\Delta Gf = \Delta Hf - (T\,\Delta Sf)$.

2.3.3 MULTIPLE AND MULTILAYERED EMULSIONS

If the liquid dispersed (discontinuous or internal phase) is a lipid, and the external phase (continuous phase) is water, it is called an oil-in-water emulsion (O/W). Whether the discontinuous phase is water and the continuous phase is a lipid, it is referred to as a water-in-oil emulsion (W/O). However, preparing more complex mixtures such as oil-in-water-in-oil (O/W/O) or water-in-oil-in-water (W/O/W) emulsions is also possible. These types of emulsions are named double or multiple emulsions [3, 18]. Double emulsions are more challenging to produce than simple emulsions since the first emulsion (let us say W/O) must be stable enough to be incorporated in the second emulsion (O/W) and keep both emulsions stable for a relatively long time [18].

Unlike multiple emulsions, multilayered emulsions involve building multiple surfactant layers mainly by electrostatic interactions. For this purpose, emulsifier layers in the emulsion are sequentially formed (layer-by-layer) using both anionic and cationic polyelectrolytes. At first, an (O/W) emulsion with a single emulsifier layer is created using oil, water, and an ionic emulsifier positively or negatively charged. Next, a second emulsifier with an opposite charge is added in excess to the emulsion (saturated method). Then, the second emulsifier interacts with the first through electrostatic attraction, creating a second emulsifier layer. Finally, the surfactant excess that did not interact is removed by centrifugation or filtration. This process can be repeated by adding another surfactant with an opposite charge to the last one, forming thus a multilayered O/W emulsion (m- O/W). The advantage of m-O/W emulsions is their higher stability than simple or double emulsions. Some of the emulsifiers used in m-O/W emulsions are lecithin, fatty acid salts, citric acid esters, mono- and diglycerides, proteins, but mainly polysaccharides like starch, dextran, agar (neutral), alginate, pectin, Arabic gum (anionic), and chitosan (cationic) [19].

2.3.4 PICKERING EMULSIONS

There is a particular type of emulsion where the emulsifier is replaced with particles that are neither water- nor oil-soluble. Instead, depending on their chemical nature (hydrophilic or hydrophobic), these particles will immerse more into one phase than the other. Consequently, the interface will get a curvature toward the oil phase (contact angle below 90°) or the aqueous phase (contact angle above 90°), creating O/W or W/O emulsions. These emulsions were denoted as Pickering, honoring one of the first scientists who described these systems in 1907. Nowadays, Pickering emulsions have been extensively studied, and silica has been the particle most studied to form non-food Pickering emulsions. However, some authors have also reported the creation of food-grade Pickering emulsions using proteins (i.e., milk, soy, zein, and fungi proteins), starch granules, or fat crystals. Nevertheless, creating such emulsions is more challenging since these food ingredients modify their solubility depending on the pH or temperature and are prone to unfolding or melting. Therefore, some authors do not consider these systems as typical Pickering emulsions [20].

2.3.5 LIPOSOMES AND NIOSOMES

Phospholipids (PL) are amphipathic emulsifiers that self-organize into lamellar bilayer structures either as an ordered gel (Lβ) or as liquid crystalline (Lα) lamella. This organization derives from the cylinder-type shape (head and tail of similar diameter) that most PLs have. When dried PLs are rehydrated, they form multilamellar bilayer sheets, but when these are mechanically or sonically dispersed, they form unilamellar bilayer vesicles with an aqueous core named liposomes. It is worth mentioning that such liposomes can also be obtained by extrusion and dilution of a detergent-lipid mixture.

Technically, liposomes are not emulsions since they only comprise the emulsifier (PLs) and water, but they can be turned into liposomal emulsions. Liposomes have been widely used as a vehicle for hydrophilic or hydrophobic compounds. The hydrophilic compounds are encapsulated in the aqueous core, whereas the hydrophobic compounds are incorporated inside the unilamellar bilayer [16, 21, 24].

Bilayer vesicles can also be formed from surfactants other than phospholipids, particularly from non-ionic surfactants. Such non-ionic surfactant vesicles (NSVs) are referred to as niosomes. The self-assembly properties of non-ionic surfactants and the methods of noisome preparation are similar to phospholipids and liposomes. However, niosomes show advantages like higher stability in aqueous suspension and higher resistance to micellar solubilization than liposomes. Furthermore, non- ionic surfactants have lower costs than phospholipids and have more versatility in forming vesicular structures. Various non-ionic surfactants can form niosomes such as amino acids, fatty acids, amides, alkyl esters, and alkyl ethers (i.e., alkyl oxyethylenes, polyglycerol monoalkyl ethers, and polyoxylate analogs). However, only the non-ionic surfactants with a hydrophile-lipophile balance (HLB) between 4 and 8 form vesicles. Surfactants with high HLB values (14 to 17) are too hydrophilic and cannot form vesicles in an aqueous environment [25].

FIGURE 2.3 Examples of mini- (left) and micro-emulsions (right). The micro-emulsion includes Tween 80 as a surfactant (1%), acetic acid as a co-solvent (40%), limonene as a lipid or discontinuous phase (2%), and phosphate buffer pH 7 as aqueous or continuous phase (57%).

2.4 FOOD EMULSIONS

The development of emulsions is one of the most researched areas in recent years and has spread widely in the food industry. Food emulsions are familiar to almost everyone, and their relevance is because most of the ingredients are wholly or partly emulsions, or at some point in the production process emulsions are generated. In addition, emulsions are increasingly being used as delivery systems designed to encapsulate, protect, and control the release and absorption of ingredients such as nutraceuticals, flavors, colors, preservatives, vitamins, etc., improving their handling, efficacy, and stability [3].

Some examples of emulsions produced by the food industry are dairy products, beverage emulsions, dressings, etc. [26, 19] (Table 2.1). Each of these products can be characterized by the type of ingredients they contain, the methods used to fabricate them, and their physicochemical, sensory, and nutritional characteristics [3]. For the formulation and stabilization of this wide variety of products, it is necessary to use many ingredients, which should be approved and safe for human consumption.

2.4.1 EMULSION INGREDIENTS

All ingredients used to produce food emulsions have to be food-grade and microbiologically safe. Additionally, they have to maintain the stability of the product for long periods of time. The term "emulsion stability" refers to the ability of an emulsion to resist changes in its properties over time [72, 73]. The mechanisms that lead to emulsion instability include phase separation, flocculation, gravitational separation (creaming/sedimentation), Ostwald ripening, and coalescence [73] (*see* Figure 2.4). The stability and attributes of an emulsion are determined by the ingredients (such as water, oil, emulsifiers, flavors, colorants, preservatives, etc.) and the processing methods used in their manufacture [3]. Clearly, all of these requirements impose restrictions on how emulsions can be formulated and produced [19].

2.4.1.1 THE CONTINUOUS AND DISPERSED PHASE

An emulsion is considered a dispersion of two immiscible phases (typically oil and water). The fluid phase in which particles like solids and fluids are distributed is considered the continuous phase, while the phase that is present in the form of particles is called the dispersed phase. The volume ratio of the liquids in the medium will define whether it is an oil-in-water (O/W)

TABLE 2.1 Examples of Macroemulsions, Nano-Emulsions, and Microemulsions Produced in the Food Industry

System	Type	Product	References
Macro-emulsion	O/W	Ice-cream	[27–29]
	O/W	Coffee creamers	[30, 31]
	W/O/W	Cheese	[32, 33]
	O/W	Tofu and yogurts	[34–36]
	O/W	Functional foods	[37, 38]
	O/W	Beverages	[39–41]
	O/W	Mayonnaise, salad dressings and sauces	[42, 43]
	O/W	Milk	[44, 46]
	W/O	Cream	[44]
	W/O	Butter	[47–50]
	O/W	Infant formulas	[41, 51, 52]
	W/O	Margarines	[53, 54]
	O/W	Soups	[55–57]
	O/W	Cake batter	[58, 59]
	O/W	Cream liqueur	[60, 62]
Nano-emulsion	O/W	Fruit juices	[62, 63]
	O/W	Soft foods, creams, sauces, and salad dressings	[64, 65]
	O/W	Functional foods and beverages	[66, 67]
Microemulsion	W/O	Extraction of edible vegetable oils	[68, 69]
	W/O	Functional foods and soft drinks	[70, 71]

emulsion (oil droplets dispersed in continuous aqueous media) or water-in-oil (W/O) emulsion (water droplets dispersed in a lipidic media) [19].

Regardless of whether the emulsion is W/O or O/W, the water and oil will define the characteristics of the emulsion. The role of the water is important in determining the physicochemical and organoleptic characteristics of food emulsions. Its unique molecular and structural properties largely determine the solubility, conformation, and interactions of the other components present in aqueous solutions such as salt, sugar, bases, acids, alcohols, surfactants, preservatives, proteins, and polysaccharides [74].

While fats and oils present as components of continuous or discontinuous phases influence the physicochemical, nutritional, and organoleptic properties of food emulsions [64]. The ingredients that solubilize into the oil phase

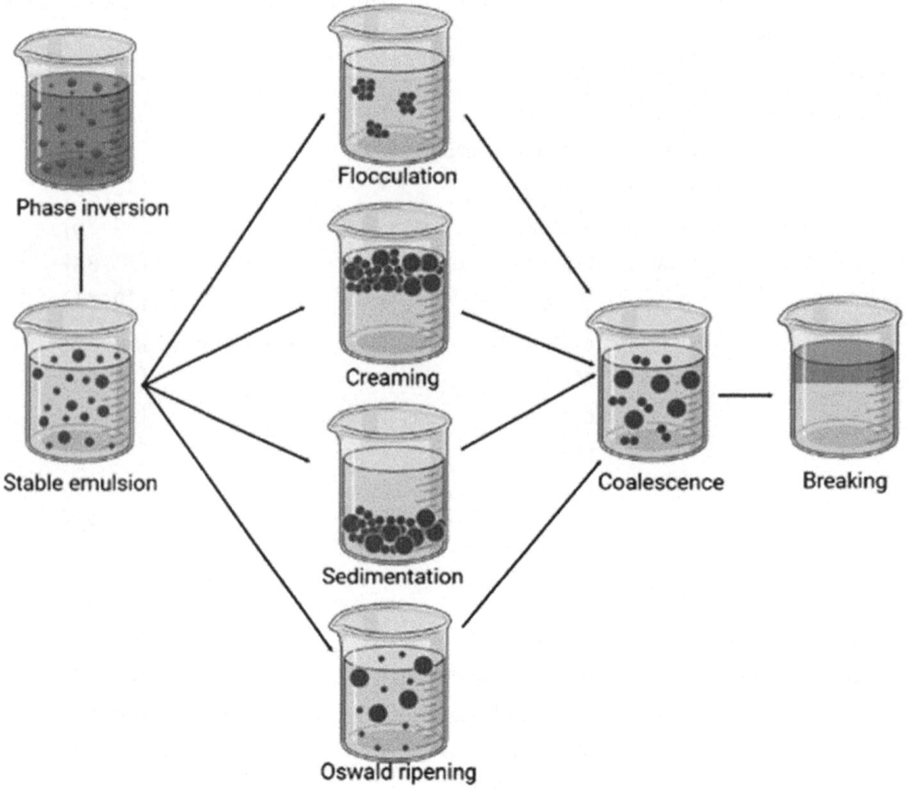

FIGURE 2.4 Mechanisms of emulsion instability.

are flavors, colors, vitamins, preservatives, fatty acids, free fatty acids, sterols, fat replacers, and weighting agents [3, 12].

2.4.1.2 EMULSIFIERS

Most food emulsions are much more complex than oil and water dispersions. They need to be added with emulsifiers that promote the formation and stabilization of an emulsion or a mixture of mutually insoluble phases by adsorbing to the surface of droplets, forming a protective interfacial layer, and preventing the droplets from merging [75]. A stabilizer will also retard the physical destabilization and promote a longer storage time by decreasing the rate of droplet encounters by increasing the viscosity of the continuous phase [75, 76].

There are emulsifiers suitable for use in food products, such as synthetic emulsifiers (e.g., Tweens and Spans), which promote excellent stability. However, they are associated with the appearance of gastrointestinal issues. In this context, the use of natural emulsifiers such as small molecules surfactants, phospholipids, proteins, polysaccharides, and saponins is preferred by food producers [3, 26, 77]. Surfactant agents are classified by their size and nature, and the interaction between the emulsion droplets and other ingredients of the food system will be promoted or hindered depending on the surfactant used.

2.4.1.2.1 *Small-Molecules Surfactants*

These emulsifiers are amphipathic molecules, which contain in their structure a hydrophilic part and a hydrophobic part. They are considered small molecules because their molecular weight ranges from approximately 250 g/mol to approximately 1,200 g/mol [77]. Surfactants that are permitted for human consumption include monoglycerides and diglycerides, sorbitan esters of fatty acids, phospholipids, polioxilen sorbitan, and others [26, 80, 79]. Due to their low molecular weight, surfactants can form nano-emulsions and interact in different ways to stabilize dispersions [80].

Lecithin is a representative of a small-molecule surfactant. It is the most widely used emulsifier in the food industry because it provides different functions, including as a wetting agent, for viscosity reduction, as a release agent, and for crystallization control [81]. Lecithin is a mixture of phospholipids and can be obtained from diverse natural sources, including milk, vegetable oils, egg yolk, meat, and fish. The structure of a phospholipid comprises a hydrophilic head holding a phosphoric acid, combined with a hydrophobic tail composed of one or two non-polar fatty acids [26].

2.4.1.2.2 *Proteins*

Proteins are the single most commonly used class of foaming and emulsifying agents in the food industry. They are natural, cheap, and non-toxic, thus making them ideal ingredients [82]. However, proteins have low surface activity in comparison with conventional emulsifiers [26].

Structurally, the presence of both hydrophilic and hydrophobic amino acids confers an amphiphilic character [26]. Because of its complex structural morphology, proteins can stabilize emulsions by forming a viscoelastic adsorbed layer which influences the stability of emulsions and foams [82].

The mechanism by which proteins protect emulsion droplets against physical destabilization is electric repulsion [77]. Proteins are adsorbed at the oil-water interface, creating a macromolecular film. Their adsorption is influenced by surface hydrophobicity and charge. Once the protein is adsorbed, the structure of the protein itself will unfold and rearrange its secondary and tertiary structure to expose hydrophobic residues to the hydrophobic phase, preventing close packing at the point of contact with the interface [79].

Concerning emulsifiers obtained from natural sources, the most common ones come from animals (e.g., milk proteins, meat, fish proteins, and egg protein) [26, 74, 77]. However, plant-based proteins are good alternatives for replacing them in products with dietary restrictions (e.g., lactose- free) and vegan and vegetarian foods. Examples include proteins derived from legumes (peas and soy) and cereals [65], which have been reported for emulsion production [26].

When comparing proteins and surfactants, the former are less surface-active than surfactants and form a viscoelastic surface.

2.4.1.2.3 Polysaccharides

Polysaccharides are composed of monosaccharide units (glucose, fructose, mannose, or galactose). These biopolymers are relatively large molecules that adsorb relatively slowly to interfaces and form thick hydrophilic interfacial layers. Their molecular weight, structural rearrangement, presence of non-polar groups (methylated groups or exogenous moieties of lipids or proteins), etc., give them functional properties such as solubility, rheology, and amphiphilic character, among others [26, 77].

Due to their molecular complexity, polysaccharides are considered amphipathic molecules when hydrophobic (glycolipids) and hydrophilic (hydroxyls) groups are presented, forming a thick stabilizing layer (e.g., gum Arabic and pectins) [26, 74]. However, when the polysaccharide is non- amphiphilic, it contributes to increasing the viscosity and stabilizing the emulsion due to its thickener role. The thickener decreases oil droplets' motion (e.g., carrageenan and alginates) [26, 77].

The most widely used polysaccharide in food emulsions is gum Arabic, which is widely used in the beverage industry to stabilize cloud and flavor emulsions. Other examples include Carrageenans, agars, alginates, galactomannans, tree gum exudates, xanthan gum, gellan gum, starch, and its derivatives [77], cellulose and its derivatives, and chitosan [74].

2.4.1.2.4 Other Food Additives

An emulsion can be formulated to contain nutritional, preservative, sensory, and processing agents. The addition of these components can improve the quality of the emulsion and therefore its stability over time. Additionally, the consumption of the product will provide the consumer with the desired nutritional and/or sensory value.

Nutritional additives are used for fortifying or enriching emulsions to correct dietary deficiencies or add nutrients to food substitutes. Examples include vitamins added to dairy products and fruit beverages [83]. Other nutritional additives include essential fatty acids and minerals.

Preservative agents include antioxidants and antimicrobials. Antioxidants delay or prevent the deterioration of food by oxidative mechanisms that cause off-flavors and off-odors characteristic of oxidative rancidity [84]. Antimicrobial agents, such as essential oils, inhibit the growth of spoilage and pathogenic microorganisms in food [85].

From a sensory perspective, flavors, fat replacers, and texture modifiers are also included. Flavors are added in biphasic systems, where the compounds partition into different phases, and the release can be modulated in various ways [1].

2.4.2 FORMATION OF FOOD EMULSIONS

Homogenization is the process of producing emulsions. Sufficient energy needs to be applied to the oil/water interface to break up the bulk oil into smaller droplets. In general, the smaller the size to be achieved, the larger the energy input required [73].

In the elaboration of macroemulsions, simple high-speed mixers or colloid mills are used to obtain mayonnaise and salad dressings where the stability of the final product depends less on the presence of very small particles than the overall composition and high viscosity of the preparation. On the other hand, high-pressure valve homogenizers, microfluidization, membrane, and microchannel homogenizers are used to produce liquid emulsions with smaller droplets (400–300 nm). The reduction in the droplet size will prevent creaming and coalescence [73, 89]. To further reduce the droplet size (300–50 nm), a larger energy input is required and could be supplied by an ultrasonic homogenizer [3, 12].

2.4.3 APPEARANCE, FLAVOR, AND RHEOLOGY OF FOOD EMULSIONS

The first impression of the consumer toward any food emulsion and other products is defined by appearance. Visual quality such as color, flow consistency, phase separation, and sedimentation can be easily observed by the eye. Emulsification gives food products the sensory and appearance properties that significantly influence the consumer's perception [73].

However, the production of food that is appetizing to the eye is not enough. Recently, there has been a growing interest in functional foods with low concentrations of fat, salt, and sugar, but also enriched with bioactive ingredients, desirable flavor, and texture profiles, to develop healthier diets for human well-being. The above is a big challenge for industrial and academic researchers to design healthy food products without sacrificing food flavor [1]. It's because researchers have been working on designing delivery systems for flavor compounds, controlling their delivery and release when they are subject to different environments during product processing, storage, and mastication in the oral cavity [87].

2.5 EMULSIONS USED AS A VEHICLE TO DELIVER BIOCOMPONENTS AND DRUGS

With the growing popularity of the functional food market, bioactive ingredients from natural sources are increasingly used because of their ability to promote better health and prevent chronic diseases. Also, to improve the organoleptic characteristics of the products.

2.5.1 BIOCOMPONENTS IN FOODS

Functional foods or nutraceuticals have been explored for their nutritional and biological value, such as antioxidants, antimicrobial, anticancer, cardiovascular implications, fat adsorption, anti- inflammatory, neuroprotection, and others [88].

The emulsion has become popular as a vehicle to promote the incorporation of bioactive components into a food system [87]. Emulsions are used as delivery systems designed to encapsulate, protect, and control the release and absorption of ingredients such as nutraceuticals, flavors/colors, preservatives, etc., improving their handling, efficacy, and stability [3].

The trends in the food sector and an increasing interest in health promotion have influenced the design of food emulsions that contain not only food ingredients but also health-promoting molecules (e.g., PUFAs, vitamins, or phytochemicals that are often lipophilic and can be combined directly with the dispersed oil phase) [72, 77]. Bioactive ingredients may be isolated from natural sources, or they may be chemically synthesized [3]. Once they are added to food, they need to be protected from natural digestive processes and be able to be released at the right moment. Vitamins, minerals, nutraceuticals, flavors, colors, and preservatives are labile enough to preserve their molecular, physicochemical, and nutritional properties through the gastrointestinal tract because environmental conditions such as pH, ionic strength, solvent type, and temperature will affect their performance. Each active ingredient has its own unique properties and functional requirements, and therefore specific delivery systems must be developed for each ingredient and application, taking these factors into account [3].

Flavors such as citrus oils and natural extracts need to be encapsulated to allow a controlled release profile, retard chemical degradation, and facilitate storage and utilization. For antimicrobials (e.g., essential oils) and antioxidants (e.g., tocopherols, carotenoids, flavonoids, phenolic compounds), the process of encapsulation will also improve matrix compatibility, mask undesirable off-flavors, and increase efficacy. Bioactive peptides obtained from plants, milk, and meat retard degradation in the stomach, reduce astringency and bitterness and also control bioactivity and the release profile. Oligosaccharides and fibers (e.g., prebiotics and chitosan) improve product texture and control delivery in the gastrointestinal tract [37].

The way in which bioactive agents are encapsulated in multiphase food systems will determine their digestive fate. The bioavailability of health-promoting compounds could be increased or can delay or prevent the digestion of certain macronutrients if it is desirable. Such as lipids, to prolong satiety signals, which could be a basis for weight-management strategies [77].

2.5.2 DELIVERY OF DRUGS

Nanoemulsion and microemulsion drug delivery systems are advanced modes for delivering and improving the bioavailability of drugs that have high first-pass metabolism [89].

A microemulsion is a dispersion made of oil, water, surfactant, and a co-surfactant. It is an isotropic and thermodynamically stable system that forms spontaneously (droplet size ranging from 1 to 100 nm). It is very easy to prepare due to its spontaneous formation ability. However, its composition and regular use of excess amounts of surfactant and co-surfactant increase the cost of operation. Besides, an excess concentration of surfactants can lead to mucosal toxicity, which is the principal reason for not using it for oral administration [12].

On the other hand, nano-emulsions are very similar to microemulsions with nano-scale particles but are obtained by mechanical forces. A nano-emulsion is more water-soluble than a conventional emulsion and has a higher bioavailability of lipophilic drugs. It is the preferred dosage form to incorporate gastrointestinal irritation-causing active drugs and first-pass metabolism-mediated degradation-prone drugs [12]. The bioavailability is due to their small droplet size, which allows the active ingredient to diffuse easily and quickly out of them, enabling efficient absorption through cells. Additionally, since nanoemulsions are transparent or slightly turbid, they can be used in food or beverages without being unpleasantly noted [88]. Stability issues like creaming, flocculation, coalescence, and sedimentation are rarely observed in nanoemulsions [12].

Nanoemulsions are included as colloidal carriers for targeted delivery of neutron capture therapy agents, photosensitizers, diagnostic agents, and anticancer drugs because their reduced droplet size can be easily targeted to the tumor area [90].

KEYWORDS

- **anticancer drugs**
- **bioactive components**
- **delivery systems**
- **emulsions**
- **lipophilic drugs**
- **nanoemulsions**

REFERENCES

1. Mao, L., Roos, Y. H., Biliaderis, C. G., & Miao, S. (2017). Food emulsions as delivery systems for flavor compounds: A review. *Critical Reviews in Food Science and Nutrition, 57*(15), 3173–3187.
2. Langevin, D. (2020). *Emulsions, microemulsions and foams* (p.195). Cham, Switzerland: Springer.
3. McClements, D. J. (2015). *Food emulsions: Principles, practices, and techniques.* Third Edition: CRC Press.

4. Trujillo-Cayado, L. A., Alfaro, M. C., Santos, J., Calero, N., & Muñoz, J. (2018). Influence of primary homogenization step on microfluidized emulsions formulated with thyme oil and Appyclean 6548. *Journal of Industrial and Engineering Chemistry, 66*, 203–208.
5. Salager, J. L. (2020). Guidelines for the formulation, composition, and stirring to attain desired emulsion properties (type, droplet size, viscosity, and stability). In A. Chattopadhyay & K. Mittal (Eds.), *Surfactants Solution* (pp. 261–295). CRC.
6. Ravera, F., Dziza, K., Santini, E., Cristofolini, L., & Liggieri, L. (2021). Emulsification and emulsion stability: The role of the interfacial properties. *Advances in Colloid and Interface Science, 288*, 102344.
7. Saifullah, M., Ahsan, A., & Shishir, M. R. I. (2016). Production, stability, and application of micro- and nanoemulsion in food production and the food processing industry. In M. G. Alexandru (Ed.), *Emulsions* (Vol. 3, pp. 405–442). Academic Press.
8. McClements, D. J., & Rao, J. (2001). Emulsion droplet size determination. *Current Protocols in Food Analytical Chemistry, D3*(3.3.1–D3.3.9).
9. Hino, T., Yamamoto, A., Shimabayashi, S., Tanaka, M., & Tsujii, D. (2000). Drug release from w/o/w emulsions prepared with different chitosan salts and concomitant creaming up. *Journal of Controlled Release, 69*(3), 413–419.
10. Raynal, S., Grossiord, J. L., Seiller, M., & Clausse, D. (1993). A topical W/O/W multiple emulsion containing several active substances: Formulation, characterization, and study of release. *Journal of Controlled Release, 26*(2), 129–140.
11. Tadros, T. F. (2016). *Emulsions: Formation, Stability, Industrial Applications*. Walter de Gruyter GmbH & Co KG.
12. Kale, S. N., & Deore, S. L. (2017). Emulsion micro emulsion and nano emulsion: A review. *Systematic Reviews in Pharmacy, 8*(1), 39–47.
13. McClements, D. J., & Rao, J. (2011). Food-grade nanoemulsions: Formulation, fabrication, properties, performance, biological fate, and potential toxicity. *Critical Reviews in Food Science and Nutrition, 51*(4), 285–330.
14. Khan, M. F., et al. (2014). Emulsion separation, classification, and stability assessment. *Journal of Pharmaceutical Sciences, 2*(2), 56–62.
15. Santana, R. C., Perrechil, F. A., & Cunha, R. L. (2013). High- and low-energy emulsifications for food applications: a focus on process parameters. *Food Engineering Reviews, 5*(2), 107– 122.
16. Gutiérrez-Méndez, N., Chavez-Garay, D. R., & Leal-Ramos, M. Y. (2022). Lecithins: A comprehensive review of their properties and their use in formulating microemulsions. *Journal of Food Biochemistry, 46*(7), e14157.
17. McClements, D. J. (2012). Nanoemulsions versus microemulsions: terminology, differences, and similarities. *Soft Matter, 8*(6), 1719–1729.
18. Friberg, S., Larsson, K., & Sjoblom, J. (2003). *Food Emulsions*. CRC Press.
19. Koubaa, M., et al. (2018). Multilayered Emulsions. In *Emulsion-Based Systems for Delivery of Food Active Compounds* (pp. 105-119)
20. Linke, C., & Drusch, S. (2018). Pickering emulsions in foods – opportunities and limitations. *Critical Reviews in Food Science and Nutrition, 58*(12), 1971–1985.
21. Dowhan, W., Bogdanov, M., & Mileykovskaya, E. (2008). Chapter 1 – Functional roles of lipids in membranes. In D. E. Vance & J. E. Vance (Eds.), *Biochemistry of Lipids, Lipoproteins and Membranes* (Fifth Edition) (pp. 1–37). Elsevier.

22. Muderhwa, J. M., et al. (1999). Oil-in-water liposomal emulsions: Characterization and potential use in vaccine delivery. *Journal of Pharmaceutical Sciences, 88*(12), 1332–1339.
23. Li, J., et al. (2015). A review on phospholipids and their main applications in drug delivery systems. *Asian Journal of Pharmaceutical Sciences, 10*(2), 81–98.
24. Stuart, M. C. A., & Boekema, E. J. (2007). Two distinct mechanisms of vesicle-to-micelle and micelle-to-vesicle transition are mediated by the packing parameter of phospholipid–detergent systems. *Biochimica et Biophysica Acta (BBA) – Biomembranes, 1768*(11), 2681– 2689.
25. Marianecci, C., et al. (2014). Niosomes from 80s to present: The state of the art. *Advances in Colloid and Interface Science, 205*, 187–206.
26. Santamaria-Echart A., Fernandes, I. P., Silvia, S. C., Rezende, S., Colucci, G., Días, M. M., &Barreiro, M. F., (2021). New trends in natural emulsifiers and emulsion technology for the food industry. Natural Food additives, 1-31.
27. Goff, H. D. (1997). Colloidal aspects of ice cream—a review. *International Dairy Journal, 7*(6–7), 363–373.
28. Goff, H. D. (2016). Milk proteins in ice cream. In *Advanced Dairy Chemistry* (pp. 329-345).
29. Wang, W., et al. (2022). Effects of soybean oil body as a milk fat substitute on ice cream: Physicochemical, sensory and digestive properties. *Foods, 11*(10), 1504.
30. Chung, C., et al. (2017). Use of natural emulsifiers in model coffee creamers: Physical properties of quillaja saponin-stabilized emulsions. *Food Hydrocolloids, 67*, 111–119.
31. Hedayatnia, S., et al. (2016). Effect of different fat replacers and drying methods on thermal behavior, morphology and sensory attributes of reduced-fat coffee creamer. *LWT-Food Science and Technology, 72*, 330–342.
32. Urgu, M., et al. (2019). Milk fat substitution by microparticulate protein in reduced-fat cheese emulsion: The effects on stability, microstructure, rheological and sensory properties. *Food Science of Animal Resources, 39*(1), 23.
33. Urgu, M., Unluturk, S., & Koca, N. (2018). Effects of fat reduction on the stability, microstructure, rheological, and colour characteristics of white-brined cheese emulsion with different emulsifying salt amounts. *Korean Journal for Food Science of Animal Resources, 38*(5), 866.
34. Chen, J., et al. (2022). Physicochemical and nutritional properties of yogurt emulsion with lycopene during chilled storage. *Journal of Food Science and Technology, 59*(10), 4037– 4044.
35. Li, H., et al. (2021). Utilization of thermal-denatured whey protein isolate-milk fat emulsion gel microparticles as stabilizers and fat replacers in low-fat yogurt. *LWT, 150*, 112045.
36. Lalou, S., El Kadri, H., & Gkatzionis, K. (2017). Incorporation of water-in-oil-in-water (W1/O/W2) double emulsion in a set-type yogurt model. *Food Research International, 100*, 122–131.
37. Jiménez-Colmenero, F. (2013). Emulsiones múltiples; compuestos bioactivos y alimentos funcionales. *Nutr. Hosp., 28*(5), 1413–1421.
38. Jiménez-Colmenero, F. (2013). Potential applications of multiple emulsions in the development of healthy and functional foods. *Food Research International, 51*(1), 64–74.

39. Piorkowski, D. T., & McClements, D. J. (2014). Beverage emulsions: Recent developments in formulation, production, and applications. *Food Hydrocolloids, 42*, 5–41.
40. Molet-Rodríguez, A., Salvia-Trujillo, L., & Martín-Belloso, O. (2018). Beverage emulsions: key aspects of their formulation and physicochemical stability. *Beverages, 4*(3), 70.
41. Drapala, K. P., Mulvihill, D. M., & O'Mahony, J. A. (2018). A review of the analytical approaches used for studying the structure, interactions, and stability of emulsions in nutritional beverage systems. *Food Structure, 16*, 27–42.
42. Yildirim, M., Sumnu, G., & Sahin, S. (2016). Rheology, particle-size distribution, and stability of low-fat mayonnaise produced via double emulsions. *Food Science and Biotechnology, 25*, 1613–1618.
43. Mirzanajafi-Zanjani, M., Yousefi, M., & Ehsani, A. (2019). Challenges and approaches for the production of a healthy and functional mayonnaise sauce. *Food Science & Nutrition, 7*(8), 2471–2484.
44. Singh, H., & Gallier, S. (2017). Nature's complex emulsion: The fat globules of milk. *Food Hydrocolloids, 68*, 81–89.
45. Ferreiro, T., Gayoso, L., & Rodriguez-Otero, J. L. (2015). Milk phospholipids: Organic milk and milk rich in conjugated linoleic acid compared with conventional milk. *J. Dairy Sci., 98*(1), 9–14.
46. Liang, Y., et al. (2017). The heat stability of milk protein-stabilized oil-in-water emulsions: A review. *Current Opinion in Colloid & Interface Science, 28*, 63–73.
47. Sullo, A., Arellano, M., & Norton, I. T. (2014). Formulation engineering of water in cocoa– butter emulsion. *Journal of Food Engineering, 142*, 100–110.
48. Prosapio, V., & Norton, I. T. (2019). Development of fat-reduced chocolate by using water- in-cocoa butter emulsions. *Journal of Food Engineering, 261*, 165–170.
49. Rønholt, S., et al. (2014). The effect of butter grains on physical properties of butter-like emulsions. *Journal of Dairy Science, 97*(4), 1929–1938.
50. Buldo, P., Kirkensgaard, J. J., & Wiking, L. (2013). Crystallization mechanisms in cream during ripening and initial butter churning. *J. Dairy Sci., 96*(11), 6782–6791.
51. Drapala, K. P., et al. (2017). Influence of emulsifier type on the spray-drying properties of model infant formula emulsions. *Food Hydrocolloids, 69*, 56–66.
52. Shanmugam, A., Chandrapala, J., & Ashokkumar, M. (2012). The effect of ultrasound on the physical and functional properties of skim milk. *Innovative Food Science and Emerging Technologies, 16*, 251–258.
53. Silva, T. J., Barrera-Arellano, D., & Ribeiro, A. P. B. (2021). Margarines: Historical approach, technological aspects, nutritional profile, and global trends. *Food Research International, 147*, 110486.
54. Dadalı, C., & Elmacı, Y. (2019). Characterization of volatile release and sensory properties of model margarines by changing fat and emulsifier content. *European Journal of Lipid Science and Technology, 124*(6), 1900003.
55. Chen, B., et al. (2019). Intact starch granules for Pickering emulsion: Exploring mechanism of cleaning with washing rice water and floury soup. *Colloids and Surfaces A: Physicochemical and Engineering Aspects, 561*, 155–164.
56. Lad, M., Hewson, L., & Wolf, B. (2012). Enhancing saltiness in emulsion-based foods. *Flavour, 1*(1), 1–8.

57. Kambabazi, M. R., et al. (2022). Physicochemical properties and sensory evaluation of a bean-based composite soup flour. *Legume Science, 4*, 157–164.
58. Shyu, Y. S., & Sung, W. C. (2010). Improving the emulsion stability of sponge cake by the addition of polyglutamic acid. *Journal of Marine Science and Technology, 18*(6), 895–900.
59. Pan, H., et al. (2021). Xanthan gum-assisted fabrication of stable emulsion-based oleogel structured with gelatin and proanthocyanidins. *Food Hydrocolloids, 115*, 106596.
60. Heffernan, S. P., et al. (2011). Efficiency of a range of homogenisation technologies in the emulsification and stabilization of cream liqueurs. *Innovative Food Science & Emerging Technologies, 12*(4), 628–634.
61. Ekene, U. S., & Okey, O. E. (2022). Studies on the production of cream liqueur using whiskey and milk cream. *SAR Journal of Pathology and Microbiology, 3*(5), 73–80.
62. Ghosh, V., Mukherjee, A., & Chandrasekaran, N. (2014). Eugenol-loaded antimicrobial nanoemulsion preserves fruit juice against microbial spoilage. *Colloids and Surfaces B: Biointerfaces, 114*, 392–397.
63. Dasgupta, N., et al. (2016). Fabrication of food grade vitamin E nanoemulsion by low energy approach, characterization and its application. *International Journal of Food Properties, 19*(3), 700–708.
64. Song, H. Y., & McClements, D. J. (2021). Nano-enabled-fortification of salad dressings with curcumin: Impact of nanoemulsion-based delivery systems on physicochemical properties. *LWT, 145*, 111299.
65. Ribes, S., Fuentes, A., & Barat, J. M. (2019). Effect of oregano (*Origanum vulgare* L. ssp. Hirtum) and clove (*Eugenia* spp.) nanoemulsions on *Zygosaccharomyces bailii* survival in salad dressings. *Food Chemistry, 295*, 630–636.
66. Bush, L., Stevenson, L., & Lane, K. E. (2019). The oxidative stability of omega-3 oil-in-water nanoemulsion systems suitable for functional food enrichment: A systematic review of the literature. *Critical Reviews in Food Science and Nutrition, 59*(7), 1154–1168.
67. Sanguansri, L., Oliver, C. M., & Leal-Calderon, F. (2013). Nanoemulsion technology for delivery of nutraceuticals and functional-food ingredients. In *Bio-Nanotechnology: A Revolution in Food, Biomedical and Health Sciences* (pp. 1–22).
68. Abbasi, S., & Radi, M. (2016). Food grade microemulsion systems: Canola oil/lecithin: n- propanol/water. *Food Chemistry, 194*, 972–979.
69. Abbasi, S., & Scanlon, M. G. (2023). Microemulsion: A novel alternative technique for edible oil extraction a mechanistic viewpoint. *Critical Reviews in Food Science and Nutrition*, 63(30), 10461-10482.
70. Elshareif, O. (2012). Preparation, stability and application of lycopene-containing microemulsion. In *XII International Symposium on the Processing Tomato 971* (pp. 167–174).
71. Dutta, D., et al. (2023). Encapsulation: Microemulsion. In *Principles of Biomaterials Encapsulation* (Vol. 1, pp. 157–195). Woodhead Publishing.
72. Berton-Carabin, C., & Schroën, K. (2019). Towards new food emulsions: Designing the interface and beyond. *Current Opinion in Food Science, 27*, 74–81.
73. Hu, Y.-T., et al. (2017). Techniques and methods to study functional characteristics of emulsion systems. *Journal of Food and Drug Analysis, 25*, 16–26.
74. McClements, D. J. (2004). *Food Emulsions: Principles, Practices and Techniques. CRC Press*

75. Zembyla, M., Murray, B. S., & Sarkar, A. (2020). Water-in-oil emulsions stabilized by surfactants, biopolymers and/or particles: A review. *Trends in Food Science & Technology, 104*, 49–59.
76. Guzey, D., & McClements, D. J. (2006). Formation, stability and properties of multilayer emulsions for application in the food industry. *Advances in Colloid and Interface Science, 128–130*, 227–248.
77. Berton-Carabin, C. C., Sagis, L., & Schroen, K. (2018). Formation, structure, and functionality of interfacial layers in food emulsions. *Annual Review of Food Science and Technology, 9*, 551–587.
78. Hasenhuettl, G. L., & Hartel, R. W. (2008). *Food Emulsifiers and Their Applications* (2nd ed.). Springer.
79. Ozturk, B., & McClements, D. J. (2016). Progress in natural emulsifiers for utilization in food emulsions. *Current Opinion in Food Science, 7*, 1–6.
80. Walstra, P. (2001). *Physical Chemistry of Foods* (1st ed.). CRC Press.
81. List, G. R. (2015). Soybean lecithin: Food, industrial uses, and other applications. In *Polar Lipids* (pp. 1–33).
82. Wilde, P., Mackie, A., Husband, F., Gunning, P., & Morris, V. (2004). Proteins and emulsifiers at liquid interfaces. *Advances in Colloid and Interface Science, 108*, 63–71.
83. Farahmand, S., Tajerzadeh, H., & Farboud, E. S. (2006). Formulation and evaluation of a vitamin C multiple emulsion. *Pharmaceutical Development and Technology, 11*(2), 255–261.
84. Decker, E. A., Warner, K., Richards, M. P., & Shahidi, F. (2005). Measuring antioxidant effectiveness in food. *Journal of Agricultural and Food Chemistry, 53*(10), 4303–4310.
85. Hao, R., et al. (2022). Development of essential oil-emulsion based coating and its preservative effects on common carp. *LWT, 154*, 112582.
86. Dalgleish, D. G. (1997). Adsorption of protein and the stability of emulsions. *Trends In food Science and Technology, 8*(1), 1–6.
87. Zhu, Y., et al. (2020). A review of the rheological properties of dilute and concentrated food emulsions. *Journal of Texture Studies, 51*, 45–55.
88. Nevárez-Moorillón, G. V., Prado-Barragán, A., Martínez-Hernández, J. L. & Aguilar, C. N. (Eds.). (2020). *Food Microbiology and Biotechnology: Safe and Sustainable Food Production.* CRC Press.
89. Kumar, M. B. R., Shukla, A. K., & Jain, C. P. (2019). Techniques for formulation of nanoemulsion drug delivery system: A review. *Preventive Nutrition and Food Science, 24*(3), 225–234.
90. Bhatt, P., & Madhav, S. (2011). A detailed review on nanoemulsion drug delivery system. *International Journal of Pharmaceutical Sciences and Research, 10*, 2482–2489.

CHAPTER 3

Nanotechnology Applied to Agriculture for the Efficient Use of Nitrogen

EZEQUIEL MUÑOZ-MÁRQUEZ,[1] JUAN MANUEL SOTO-PARRA,[2] ROSA MARÍA YÁNEZ MUÑOZ,[2] and ESTEBAN SÁNCHEZ[1*]

[1]*Centro de Investigación en Alimentación y Desarrollo A. C., Chihuahua, Mexico*

[2]*Facultad de Ciencias Agrotecnológicas, Universidad Autónoma de Chihuahua, Chihuahua, México*

[*]Corresponding author.

ABSTRACT

As the climate is changing globally and the population is increasing, agriculture faces great challenges, such as the demand for food and the excessive use of fertilizers that seriously affect the environment. Faced with this problem, nanotechnology offers effective solutions that have been shown to increase crop yields through the efficient use of fertilizers. Nanofertilizers are an important tool to improve the growth and productivity of crops; they are focused on the efficiency of nutrient use, especially nitrogen. Their use has managed to minimize losses and reduce production costs. Foliar nanofertilizers, in particular, have the advantage of facilitating the entry of nutrients through the leaves, leading to higher efficiency and rapid response in the crop. Nanofertilizers are more reactive and penetrate through the cuticle, ensuring controlled release and targeted distribution within the different plant tissues. However, there are few reviews that expose nanotechnological aspects focused on the efficient use of fertilizers, and especially on the efficient use of nitrogen. In this review, current research on this particular topic is summarized, which is a very important issue for the sustainable development of the agro-industry.

Food Security, Safety, and Sustainability: Validated Innovative Approaches and Technological Advancements.
José Juan Buenrostro-Figueroa and Julio César Tafolla-Arellano (Eds.)

3.1 INTRODUCTION

The nanotechnology of the new century has promised a true industrial change. The countries with the vision of adopting this new technology will be able to maintain a level of competition that will allow them to remain within the action table where the great technological-industrial powers of the world interact [1]. This branch of science has the capacity to improve the different productive sectors that have serious economic effects on the countries. In essence, the agro-industry is one of the most important and promising sectors that play a crucial role in the global economy. This activity supplies food for people and animals, in addition to producing raw materials for various industries [2]. Currently, it is well known that with the exponential increase in population, the agricultural sector will need to provide twice as much food to meet the demand of the world population in the coming decades [3]. Taking into account the increase in population, along with the massive rate of deforestation, the accelerated decline of arable land, and climate change, it is important to highlight and emphasize the need to develop new techniques that can improve the yield and productivity of crops grown in both favorable environmental conditions and challenging conditions [4].

Nanotechnology has the capacity to improve and potentiate the agricultural sector through the use of innovative technological tools that involve the absorption and assimilation of nutrients, the detection of diseases, and their subsequent treatments [5]. In the coming years, the trend in this sector is the use of lower doses of pesticides, herbicides, and other compounds that are harmful to humans. This can be achieved thanks to the development of nanostructured catalysts that will increase their efficiency. Additionally, to improve productivity, it is intended to use intelligent nanosensors that provide accurate information on soil conditions, climate, and other important parameters that allow the best initial and corrective decisions to be made, guaranteeing high yields [6].

A significant improvement in nanotechnology is the production of chemical and biological nanofertilizers (nanobiofertilizers). These nanotechnological discoveries have the potential to slowly and sustainably release the active nutrient that will be absorbed by plants, and thus prevent it from being lost by interaction with other nutrients and microorganisms present in the soil, or the contact surface [7]. In addition, these innovative agricultural inputs improve crop growth and productivity without causing damage to the environment [8]. These nanoactive compounds have special characteristics that allow them to be effectively absorbed and easily assimilated into plant metabolism. Some relevant characteristics are a diameter between 1 and

100 nanometers (nm) and a very large specific surface area that has great reactivity [9] (Figure 3.1). This type of nanofertilizer is applied to plants in the form of emulsions, or nanoparticles that contain encapsulated nutrients or with thin protective film coatings [10]. Their controlled release can be "programmed" based on the different fluctuations of humidity, environmental temperature, and soil acidity in the case of nanofertilizers applied to the rhizosphere; enhancing their effect compared to traditional fertilizers [11]. It is well known that most of the traditional fertilizers used by farmers for crop production have high nutrient losses. The reason is their very nature is strongly affected by hydrolysis, decomposition, leaching, and degradation by photolysis [12].

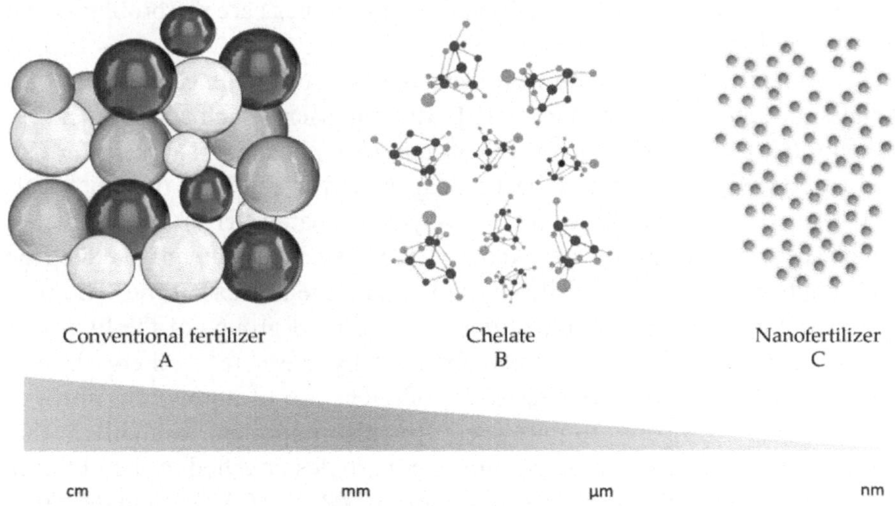

FIGURE 3.1 Particle size diagram of fertilizers used in the agricultural sector. (A) Conventional fertilizer in powder, pearls, crystals, granules, or large particles between 0.84 and 4.7 mm [13]. (B) Chelated particles are made up of crystals of different shapes and sizes (from 0.2 to 1 mm), highly soluble in water [14]. And (C) nanofertilizer, nanoparticles, or nanoencapsulates between 1 and 100 nm [15].

In this context, the scientific-technological advance represented by nanotechnology has demonstrated the ability to transform agriculture by providing innovative tools that focus on the detection and correction of biotic and abiotic stress, improving the ability of plants to absorb and assimilate nutrients, and enabling the early detection of diseases [16]. All with the aim of enhancing yields and nutritional values [17]. Scientific research focused on nanotechnology applied to the agricultural sector has increased exponentially in recent years. However, there are only a few reviews that discuss nanotechnological aspects related to the effective use of fertilizers, particularly focusing on the efficient utilization of nitrogen. This review summarizes current research on this specific topic, which is crucial for the development of sustainable agriculture.

3.1.1 NANO-FERTILIZERS

Since the discovery of chemical fertilizers, it has been certain that these are essential for agricultural production, and today this determination has been further strengthened. It is well known that the efficiency of synthetic chemical fertilizers has decreased as soil and water pollution and greenhouse gas emissions have increased [18], and precisely these are the problems that nanotechnology solves.

The manufacture of nano-fertilizers is based on the encapsulation of a mineral nutrient in nanoparticles [19]. Macronutrients such as nitrogen (N), calcium (Ca), magnesium (Mg), potassium (K), phosphorus (P), sulfur (S), and carbon (C) have already been encapsulated in different nanomaterials [20]. An example of these are urea-hydroxyapatite (HA) nanoparticles, which according to studies carried out by [21] have been shown to reduce the consumption of nitrogen fertilizers and extend their release time. The urea nanoparticles interact with the amine and carbonyl groups of the hydroxy-apatite nanoparticles. The results obtained by these researchers showed that urea-hydroxyapatite increases the efficient use of agronomic nitrogen by about 30% compared to pure urea. It is also important to highlight that nitrogen, phosphorus, and potassium nanoparticles attached to the chitosan biopolymer increase their absorption by 17.04%, 16.31%, and 67.50%, respectively, in coffee plants [22]. Cottonseed production was also increased by 42.2% when fertilizing plants with foliar magnesium oxide nanoparticles [23]. These mentioned nanohybrids have two very important characteristics: a smaller volume and abundant pores, which makes them resistant to rain and wind, and at the same time easily transferable to plant cells [18]. Another benefit of nanofertilizers that is important to highlight is that due to their chemical reactions and the physical and chemical interaction with traditional fertilizers, they manage to reduce and homogenize to a certain extent the size of the fertilizer particles with which they come into contact, which significantly increases the efficiency of these fertilizers [19].

In addition to macronutrients, micronutrients have also been included in the elaboration of nanofertilizers, due to their obvious essentiality in plant physiology. Nutrients such as zinc (Zn), iron (Fe), manganese (Mn), molybdenum (Mo), boron (B), copper (Cu), and nickel (Ni) are applied foliar or edaphically to plants. These fertilizer nanotypes have shown their effectiveness in significantly increasing yields and promoting growth [26]. As has been mentioned in different investigations around the world, micronutrients participate in important physiological processes. Applying these nutrients in nanofertilizers improves their assimilation so that they are integrated into the metabolic pathways involved in photosynthesis, the activity of antioxidant enzymes, and other metabolic processes involved in plant development [27].

To achieve the above, the effective parameter that is directly related to fertilization efficiency is the reduced size of the nanoparticles [28]. All nanofertilizers have this characteristic that allows a greater number of particles per application unit, and a coverage of larger specific plant surface areas [19].

We have then, that the greater the coverage and interaction of the nanofertilizer with the leaves and roots, the better absorption of the fertilizer is promoted. Additionally, the unique properties of these fertilizers reduce their loss by weathering and lead to long-lasting effects [18]. Their porous structure helps the active nutrient to enter the cells through molecular transporters or ionic channels and manage to activate signaling pathways related to phytohormones or other growth factors [29] (Figure 3.2).

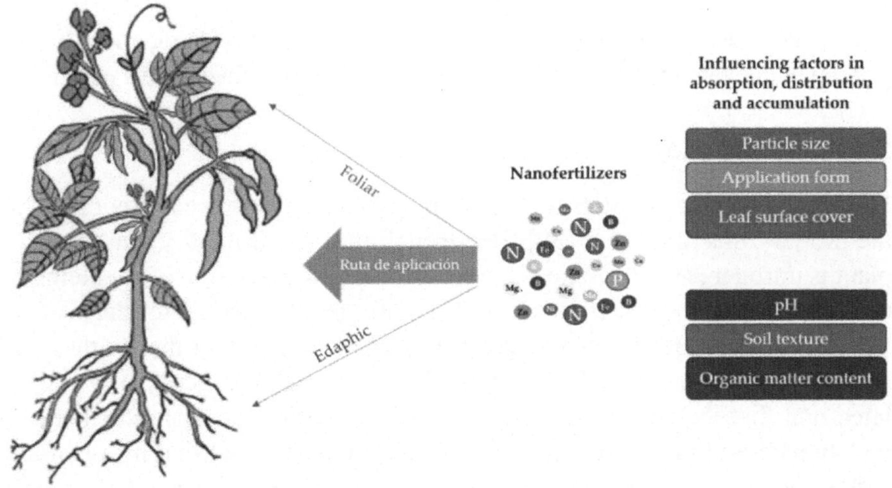

FIGURE 3.2 Simplified scheme of forms of application of nano-fertilizers. Its application by foliar spraying and edaphic irrigation promotes its absorption and reduces losses, promoting agricultural and environmental sustainability.

As previously mentioned, the unique characteristics of nano-fertilizers influence the metabolic activities of plants to different degrees compared to traditional fertilizers and also have the potential to mobilize nutrients in the rhizosphere [30]. Below is a comparative scheme of conventional fertilization and fertilization based on the application of nano-fertilizers (Figure 3.3).

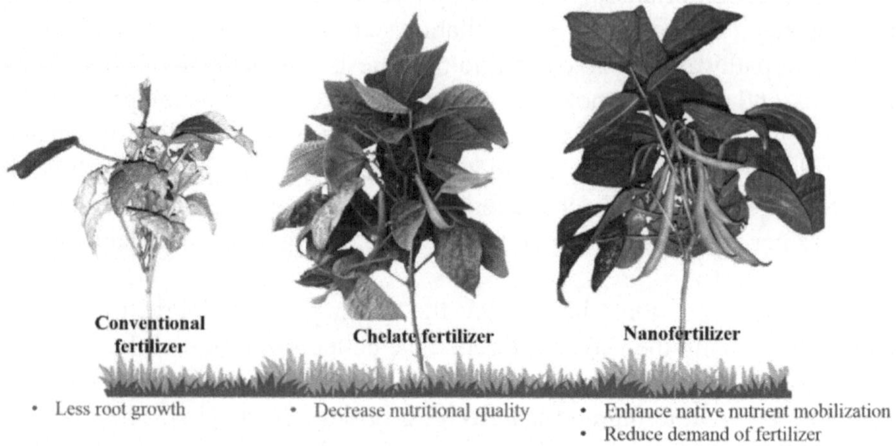

FIGURE 3.3 Comparative scheme of some pros and cons in the use of conventional fertilizers and nanofertilizers. The red text represents potential negative impacts. It should be considered that the influence of the nanofertilizer depends on the composition and the concentration of application.

To place the efficient use of nitrogen in the context, it is important to have a general understanding of some key factors that directly influence this parameter.

3.1.2 NITROGEN

Nitrogen is a determining factor for plant growth and has been categorized as the primary macronutrient [30]. Between 1 and 5% of the dry matter of the plant is nitrogen, and it occupies the second place after carbon of the elements that are required in greater quantity [31]. Nitrogen is essential for the production of different metabolites of great importance, including the synthesis of amino acids and proteins. This is why the amount of nitrogen that is assimilated and distributed throughout the plant has significant consequences on metabolism and plant growth. It is also important to note that nitrogen and carbon metabolism are closely interrelated [32]. It is convenient to specify that photosynthesis and respiration provide the carbon skeletons and the energy necessary for the synthesis of amino acids [33], while most of the nitrogen in the leaf is present in the proteins essential for carbon assimilation and metabolism. The main enzyme in carbon fixation, Rubisco, alone can represent up to 30% of the total nitrogen and 60% of the soluble proteins in mature leaves [32]. For these and many other important characteristics, nitrogen is irreplaceable in crop production systems; and to considerably improve agricultural productivity, the use of nitrogen fertilizers is necessary. However, the efficiency of use of this nutrient is very low, as a consequence of the factors and processes of nitrogen loss such as denitrification, volatilization, leaching, fixation in soil particles, immobilization by the soil microbiota and the rhizosphere, and the runoffs. It is said then that from 45 to 70% of the nitrogen that is applied is not used by the crop and is lost, which causes great economic losses and serious environmental pollution [34, 35].

In this context, it must also be taken into account how plants eliminate part of the nitrogen when it is applied in excess, and are subjected to stress due to nitrogen toxicity, without yet reaching the death threshold. This elimination occurs through absorption and assimilation; the absorption by the substrate where there are high concentrations of the nutrient, the volatilization of ammonia, and the emissions of N_2 and N_2O by simultaneous nitrification and denitrification (Figure 3.4).

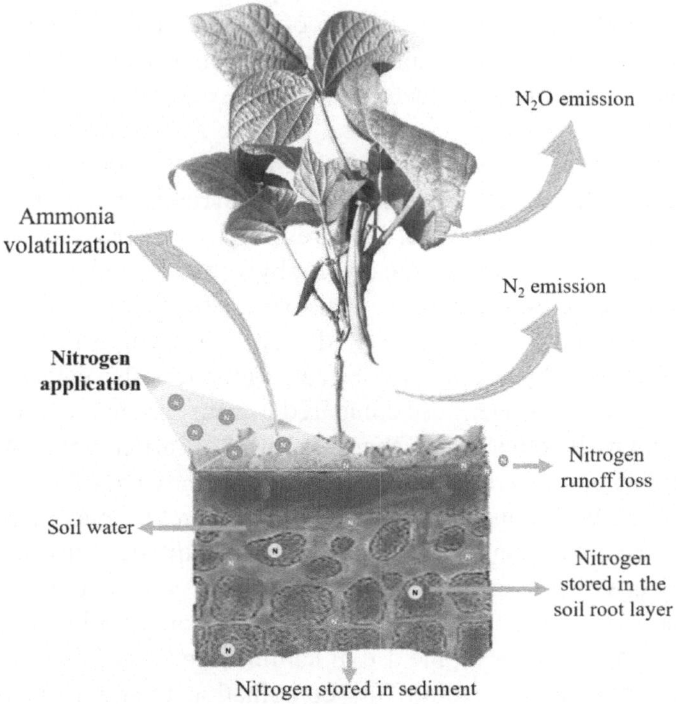

FIGURE 3.4 Type of nitrogen losses applied to crops.

3.1.3 NITROGEN NANO-FERTILIZERS

For the vast majority of crops, the macronutrients nitrogen, phosphorus, and potassium are the most important nutrients. However, the use efficiency of applied chemical macronutrients is very low; approximately 50–70% of nitrogen and potassium, and 80–90% of phosphorus are lost and cannot be absorbed by plants. Consequently, all these unused fertilizers are released into the environment [22].

Recent studies have focused on evaluating macronutrient-based nanofertilizers. An example is the urea-based nanofertilizer with sulfur coating, which effectively increased potato crop yield and reduced nitrate leaching [36]. Studies have also been carried out in which NPK nutrients are loaded in chitosan nanoparticles, and these in turn have an external coating of acrylic acid-co-acrylamide. The effect of controlled release of NPK nanoparticles suspended in polyacrylic hydrogel with positive effects on performance was also investigated [22]. In the same way, the stability of a colloidal suspension of nanoparticles of chitosan-methacrylic acid combined with the macroelements nitrogen, phosphorus, and potassium has been evaluated; the results showed a higher addition of nitrogen with promising effects for crops [34]. Sarkar et al. [37] compared polymeric compounds such as kaolinite, mica, smectite, and amorphous aluminosilicates, as well as nanoclay. In this case, the positive effect was accentuated with the use of polymeric compounds plus nanoclay, since the release of the nutrient was slower, benefiting the crop. The presence of the aluminosilicates had no effect on this parameter.

Wang et al. [38] carried out an investigation in rice cultivation by applying conventional urea and nano-urea. Their results showed that with the application of nano-urea, the biomass, the chlorophyll content, and mainly the grain production increased significantly. In this parameter, the application of nano-urea increased grain yield by 44.5% and only 10.2% with conventional urea. The loss of nitrogen by leaching in rice cultivation was also evaluated; where nano-urea had less nitrogen loss compared to conventional urea that had 70 to 74% loss. On the other hand, Zhang & Lin [39] carried out slow nitrogen release studies using carbon graphene nanomaterials with KNO_3 granules. The nanofertilizer prolonged the release of nitrogen for up to 8 hours. Similarly, other researchers have evaluated montmorillonite-urea nanocomposites as slow-release fertilizers and for their ecological and economic efficiency, which obtained good results with a ratio of 1:20 montmorillonite-urea [40].

There are also studies on the use of kaolinite and granulated urea coated with water-based epoxy resin. This combination retarded the release of nitrogen up to three times compared to the sample that did not contain these compounds. Furthermore, this same combination of agglomerates with chitosan reduced the volatilization of ammonia by up to 12% compared to the use of conventional urea [34].

Urea-hydroxyapatite nanohybrids have been tested. This nanocomposite allowed to reduce the concentration of nitrogen by up to 50% compared to the application of conventional urea according to the results published by the researchers. This was attributed to the slow release of nitrogen by urea-hydroxyapatite, unlike conventional urea, which is consumed in a few minutes [41].

A urea combined with bentonite was also tested, which showed greater efficiency than the use of conventional urea. In addition, a nanocomposite of *Gliricidia sepium* encapsulated with urea-modified hydroxyapatite nanoparticles can maximize the efficient use of nitrogen and minimize adverse effects on the environment. Similarly, mesoporous silica nanoparticles have been synthesized, which have the particularity of retaining urea molecules and releasing them in a controlled manner, making them potential nanocarriers for nitrogen fertilizers.

In investigations by Zareabyaneh & Bayatvarkeshi [36], they tested nanonitrogen-chelated, sulfur-coated nanonitrogen-chelated, and sulfur-coated urea fertilizers. Their results showed a reduction in nitrate leaching by 35.72%, 41.56%, and 9.94%, respectively. Similarly, the increase in yield was 56.10%, 59.61%, and 49.76% more compared to the application of conventional urea.

3.1.4 MOLYBDENUM

Molybdenum is a vital trace element for the growth of plants. This element is used by specific enzymes as a cofactor in redox reactions within plant metabolism. It has an effect on chlorophyll production, plant growth, and yield. It is also of great importance for atmospheric nitrogen-fixing bacteria and the plants that live in symbiosis with them [42]. Molybdenum deficiency causes thinning of the leaves of the plant and mottling between the veins as a consequence of chlorosis due to the reduction of the content of chlorophyll and ascorbic acid [43]. In addition, it has a key function in mononuclear molybdoenzymes that regulate metabolic progression and ascorbate-glutathione activity. It should be noted that molybdenum is essential in the formation of the enzyme Nitrate Reductase, which plays a fundamental role in the assimilation of nitrogen [44]. It also occurs as a vital part of a complex organic pterin called molybdenum cofactor (MoCo) (Figure 3.5). It should also be noted that along with xanthine dehydrogenase/oxidase, another molybdoenzyme has also been recognized as having a very important role in ureide biosynthesis and purine catabolism. In legumes, during abscisic acid (ABA) biosynthesis, the conversion of sulfite to sulfate by the action of sulfite oxidase and aldehyde oxidase is of vital importance for the metabolism of sulfur-containing amino acids [45]. Similarly, the molybdoenzyme aldehyde oxidase (AO) plays a key role in the relationship between plant development and environmental stress [46]. Molybdenum is also involved in xanthine dehydrogenase and aldehyde oxidase molybdenum hydroxylases. These enzymes within plant metabolism have distinctive signatures of reactive oxygen species that are induced by abscisic acid and drought. Phytohormone biosynthesis is catalyzed by AO molybdoenzymes, such as ABA and IAA, by converting abscisic aldehyde and indoleacetoaldehyde to their respective phytohormones. The vital IAA in auxins has a crucial role in different plant processes such as abscission, phototropism, root initiation, fruit development, apical dominance, and gravitropism. Similarly, ABA exerts a crucial response to environmental stress [45, 47].

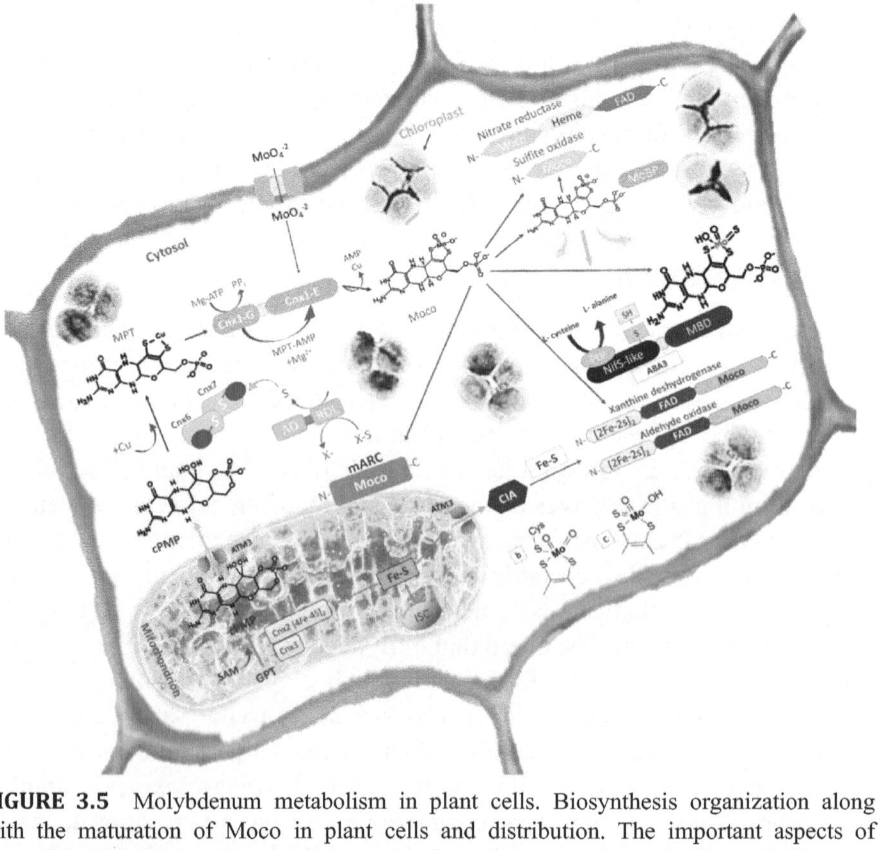

FIGURE 3.5 Molybdenum metabolism in plant cells. Biosynthesis organization along with the maturation of Moco in plant cells and distribution. The important aspects of Moco biosynthesis are demonstrated starting from the conversion of GTP to cPMP in the mitochondria, to the cytosolic steps via MPT and Moco.

Moco biosynthesis enzymes and Moco-binding proteins, which are not typically classified as Mo-enzymes, are shown in blue. The Cnx2 dependence on [Fe–S] and S-adenosyl methionine (SAM) is shown. MPT-synthase, consisting of Cnx6 and Cnx7, is sulfurated by Cnx5, along with the principal sulfur donor (X–S) organized by the rhodanese-like domain of Cnx5 (RLD) being unclear. The adenylation domain of Cnx5 (AD) is required for adenylation and activation of the small MPT synthase subunit Cnx7. It is supposed that copper (Cu) is placed immediately after dithiolene formation. The separate reactions of Cnx1 and its outcomes (Moco, pyrophosphate PPi, AMP, copper) are observed. Mature Moco can be either bound to a Moco-binding protein (MoBP), to the Mo-enzymes mARC, AO, XDH, NR, and SO, or to the Moco-binding C-terminal domain of the ABA3 protein (MBD). ABA3 generates a protein-bound persulfide, which is the cause of the terminal sulfur ligand of Moco in AO and XDH. Like Cnx2, AO, and XDH depend on [Fe–S]-clusters, which are synthesized in mitochondria by the iron–sulfur cluster biosynthesis machinery (ISC) prior to export by ATM3 and assembly by the cytosolic iron–sulfur cluster assembly machinery (CIA). (b) Structure of the Mo-center in enzymes of the SO family (SO and NR). (c) Structure of the Mo-center in enzymes of the XO family (AO and XDH) [46, 47].

It should be mentioned that molybdenum is considered a key element in the execution of the nitrogen cycle. The particularity of this metal lies in the oxidation-reduction reactions in which it is involved. Its absence and that of the enzymes that contain it trigger various adverse reactions that can lead to the death of living beings in all the kingdoms of nature in which it acts [48].

3.1.5 MOLYBDENUM NANOFERTILIZERS

There is little literature in relation to the evaluation of molybdenum nanofertilizers applied to crops. Molybdenum applications are conventional products. The studies that have currently been published are focused on the use of molybdenum nanoparticles. The effect of molybdenum disulfide (MoS_2) in plants has been investigated, where a significant increase in the length and biomass of roots and stems was obtained. In addition, the chlorophyll content was increased [49]. Nano-molybdenum trioxide (MoO_3) has been shown to disrupt the cell walls of harmful bacteria and has the potential to treat phytopathogenic bacteria [50]. Commercial molybdenum-based nanocomposites such as molybdenum dioxide, nano-molybdenum,

molybdenum trioxide (MoO_3), or molybdenum disulfide (MoO_2) were also tested in soybean culture, which showed direct effects on the symbiotic rhizobia of soybean, in addition to influencing the growth and development of plants. Similarly, the effect on the enzymatic activity of superoxide dismutase (SOD) and peroxidase (POD) has been reported, which are activated to eliminate reactive oxygen species [51]. In recent studies where foliar applications of Molybdenum Nanofertilizer were made at low doses and combined with edaphic nitrogen fertilization at low concentrations, it was possible to increase the yield in green beans by 43.17% compared to the highest doses of these same elements [52].

3.1.6 FORMS OF APPLICATION OF NANOFERTILIZERS

The new formulations of nanofertilizers are made by means of encapsulation. This process is carried out taking into account the interaction of the nanofertilizer and its active nutrient with the environment, the plant, and the microorganisms present. This is why nanofertilizers must be applied in such a way that they do not lose their vital properties of efficiency, stability, controlled release, and solubility, as well as simple and safe handling and application. The delivery of the nutrient to the plants can be done in a foliar and edaphic way; nanofertilizers have the versatility of being able to formulate up to 77.9% of the nutrient of interest and with lower percentages of other compatible nutrients [53]. Also, the elaboration and application of nanoencapsulates of combined macronutrients such as nitrogen, phosphorus, calcium, potassium, magnesium, and sulfur in proportions ranging from 0.2 to 0.8% [54]. Similarly, nanofertilizers based on micronutrients such as Zn, B, and Cu in concentrations of 1 to 6% [55] and Mo [56]. In addition, there are already commercial products called PPC nano that contain proteins, ammonium sulfate, potassium oxide, and sodium in nutrient concentrations of 2.1% and 19.6% of proteins [57]. On the other hand, nanofertilizers are marketed that have macro- and micronutrients such as nitrogen, phosphorus, potassium, sulfur, and zinc in their formulation, in addition to containing organic acids, organic carbon, amino acids, probiotics, and vitamins [58]. Another type of nanoformulation is called Nano-Green. This type of nanofertilizer is made from extracts of soybean, corn, potato, and coconut plants [57]. Finally, there are Biozar Nanofertilizers, which are a combination of organic materials with macro- and micronutrients, as well as other macromolecules [59].

3.1.7 FOLIAR APPLICATION OF NANOFERTILIZERS TO INCREASE THEIR EFFICIENCY

With the foliar application of nutrients, it is intended to use all the pathways offered by the foliar area of the plant to adequately nourish it. Assimilation can occur by adsorption on the leaf surface, stomatal and cuticular penetration, through other epidermal structures, cellular absorption, and the penetration of nutrients into metabolically active compartments in the leaf. For foliar application to be efficient, the active nutrient must penetrate the plant surface so that it can be metabolized in the cells where it is required. This is why a fertilizer must have the ability to "deposit" the nutrient in the plant tissues so that it can be quickly assimilated [60]. The formulation of the nutrient solution to be applied is the main process involved in nutrient absorption; in addition to the atomization of the solution, spreading, coverage, and retention on the leaf surface [61].

All fertilizers that are foliar applied to crops face the challenge of penetrating various structural barriers. This is because the fertilizers are made from salts (cations/anions) that may have difficulties penetrating the cells of the internal plant tissue. It is important to highlight that the cell wall has a pore size that oscillates between 5 and 20 nm, which is why this difficulty of penetration occurs for most fertilizers [62]. It is precisely because of this pore size that nanoparticles, having a diameter smaller than the pore size, can more efficiently penetrate the cell wall and reach the plasma membrane [63]. In itself, the application of nanofertilizers is a promising and efficient option to bring nutrients to the desired part of the plants. Pre-engineered nanoparticles can penetrate stomatal pores with a size smaller than 50 nm [64, 65]. The nanoparticles are transported from the site of application to the cells, through the phloem vessels probably through the plasmodesmata [66]. Nanoparticles are also transported to the plant by forming membrane transporter complexes that move through the vascular system and redistribute to plant parts [64].

In studies carried out by Abdel-Aziz, Hasaneen, & Omar [67] on wheat plants, nanoparticles were detected in the phloem tissues. This indicates that the nanoparticles, once absorbed, were transported through the phloem route from the leaves to the stem and roots in plants. This same behavior was reported by Wang et al. [64] in watermelon plants, where they found that nanoparticles smaller than 100 nm can easily penetrate through the stomata of the leaves and could be distributed to the stems through the phloem sieve elements (Figure 3.6). Particles already within the plant can be transported between cells via plasmodesmata and transported via ion channels, aquaporins, endocytosis, or by binding to organic chemicals [68].

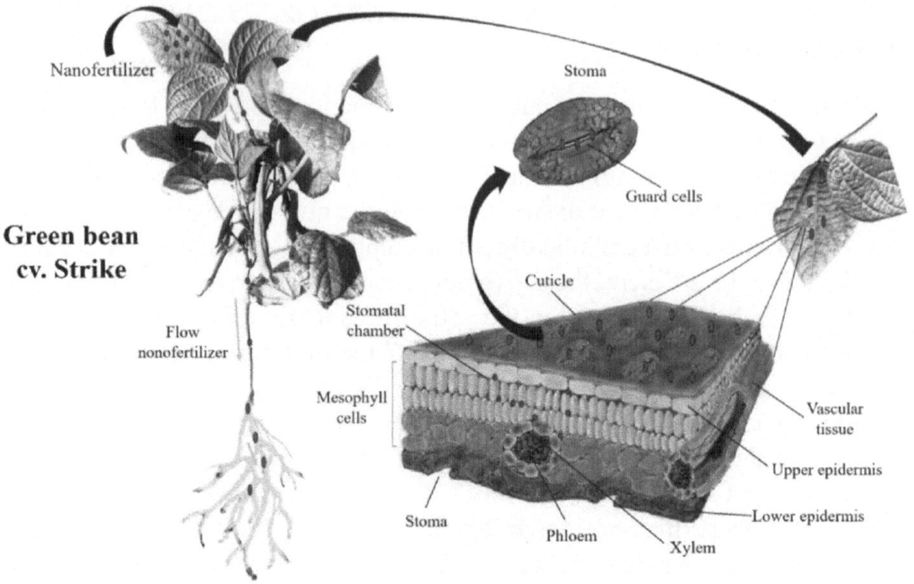

FIGURE 3.6 Schematic diagram of the transport of nanofertilizers within green bean plants cv strike. Stoma openings are approximately ~6.7 μm in diameter.

Regarding growth parameters, nanofertilizers play an important role in the biochemical and physiological processes of plants by increasing the availability of nutrients. These readily available nutrients improve metabolic processes and meristematic activity, leading to greater apical development. Documented studies have shown that foliar applications of nitrogen nanofertilizers increased the dry weight of mint leaves by 165%. Similarly, foliar fertilization of NPK nanoformulates in wheat significantly increased leaf growth [69]. Additionally, foliar applications of zinc nanofertilizer in millet crops increased shoot length and plant dry biomass. Similarly, in cotton cultivation, the biomass of the plants increased, and the chlorophyll content and antioxidant capacity were improved [68].

Derived from the different investigations into the use of nanofertilizers, there are already different commercial products on the market that combine two or three elements. Some examples are Agriklik, Richfield, and Nualgi. Also, single-nutrient nanofertilizers such as Nanomag, Nanozinc, Potacristal, Fosfoona, and Nanobor are slowly beginning to be used by producers and evaluated for crops such as cotton, wheat, and corn with favorable results in growth and yield [67, 70].

3.1.8 ESTIMATION OF NITROGEN USE EFFICIENCY IN PLANTS

As previously mentioned, nitrogen use efficiency (NUE) has been defined as the maximum yield produced per unit of fertilizer applied, accumulated, absorbed, and used by plants for total biomass production and fruit production. All this with the objective of determining the capacity of the plant to convert an applied chemical input into products (biomass, fruits, and seeds), which consequently will translate into the economic performance expected by the producers, within the agricultural production system [71].

All these definitions of efficiency have been classified in Figure 3.7.

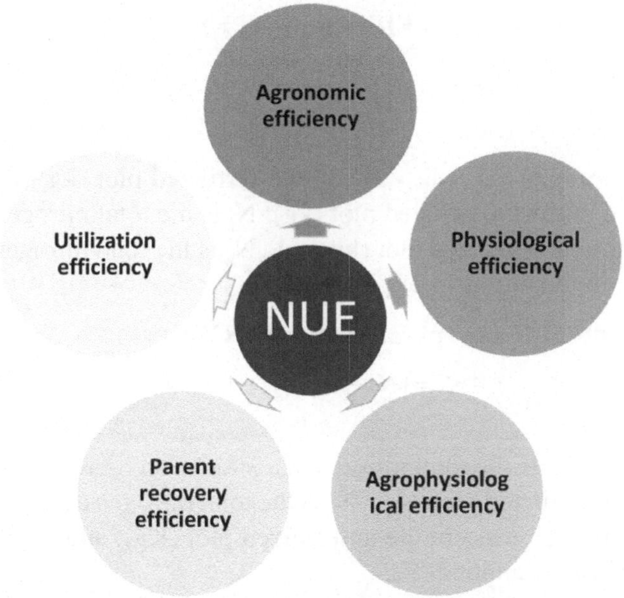

FIGURE 3.7 Classification of the different types of nitrogen use efficiency in plants. Focused on evaluating the final destination of applied nitrogen fertilizers.

In order to calculate the NUE, it is necessary to use different formulas [72]:

1. **Agronomic Efficiency (AE):**

$$AE = (P_f - P_u / N_a) = Kg\,Kg^{-1} \tag{1}$$

where; P_f is the fruit yield of the fertilized plot (kg); P_u is the fruit yield of the unfertilized plot (kg); and N_a is the amount of nitrogen applied (kg).

2. **Physiological Efficiency (PE):**

$$PE = (Y_f - \frac{Y_u}{N_f} - N_u) = Kg\,Kg^{-1} \tag{2}$$

where; Y_f is the total biomass of the fertilized plot (kg); Y_u is the total biomass of the unfertilized plot (kg); N_f is the total nitrogen accumulation in the fertilized plot (kg); and N_u is the total nitrogen accumulation nitrogen in the unfertilized plot (kg).

3. **Agro-Physiological Efficiency (APE):**

$$APE = (P_f - \frac{p_u}{N_f} - N_u) = Kg\,Kg^{-1} \tag{3}$$

where; P_f is the fruit yield in the fertilized plot (kg); P_u is the fruit yield in the unfertilized plot (kg); N_f is the total nitrogen accumulation in the fertilized plot (kg); and N_u is the total nitrogen accumulation in the unfertilized plot (kg).

4. **Apparent Recovery Efficiency (ARE):**

$$APE = (N_f - N_u / N_a)100 = \% \tag{4}$$

where; N_f is the total nitrogen accumulation times the total biomass in the fertilized plot (Kg); N_u is the total nitrogen accumulation times the total biomass in the unfertilized plot (Kg); and N_a is the amount of nitrogen applied (Kg).

5. **Utilization Efficiency (UE):**

$$UE = PE\,x\,ARE = Kg\,Kg^{-1} \tag{5}$$

Physiological efficiency is multiplied by the apparent recovery efficiency.

It is necessary to highlight that the nitrogen use efficiency parameters must also be taken into account. These parameters specify the ability of plant roots to absorb nitrogen from the soil [nitrogen uptake efficiency parameter

(NUpE)] and define the fraction of nitrogen acquired by the plant that will be converted into total biomass or plant yield fruits [parameter of nitrogen utilization efficiency (NUtE)] [73].

The nitrogen use efficiency (NUE) parameters are obtained as follows:

- Total nitrogen accumulation (TNA) is calculated using the total nitrogen concentration multiplied by the total plant biomass [74].
- Nitrogen uptake efficiency (NUpE) is calculated as TNA divided by root dry weight (RDW) and the result is expressed in mg N g^{-1} RDW [74].
- The nitrogen utilization efficiency (NUtE) is calculated as the dry weight (DW) of the leaf tissue divided by the nitrogen concentration, and the result is expressed in g^2 LDW mg^{-1} N [12].

The nitrogen use efficiency parameters are of vital importance to reduce the excessive use of nitrogen fertilizers, as well as when the availability of nitrogen limits the growth of a crop; and always with the specific objective of benefiting the environment and farmers [56].

It is necessary to emphasize that crops with higher NUE have higher yields with lower amounts of nitrogen applied and require less nitrogen to produce the same yield as those with lower NUE capacity and high N applications. Therefore, when the nitrogen use efficiency, production costs, and NO_3^- pollution to the environment are reduced [75].

3.1.9 FUTURE PROSPECTS FOR THE USE OF NANOFERTILIZERS

Thanks to the benefits of nanofertilizers and their potential applications for the entire agricultural sector, great interest was aroused in increasing crop productivity through the use of this new technology. The great economic benefits for farmers that are guaranteed with the use of nanofertilizers is the significant reduction in the amount of fertilizer applied; by avoiding huge losses from volatilization and leaching [29]. Simultaneously, they have an impact on the increase in yield and the quality of the fruits with enormous potential to increase the profits of producers compared to conventional fertilizers [76]. However, although the results obtained by applying nanofertilizers show their enormous relevance and potential, they have not yet been fully released on the market. Firstly, the uncertainty generated by the use of nanomaterials on humans and the environment must be dissipated. This will be achieved with scientific studies that are carried out in the coming years, and that focus on little-explored topics so that sufficient knowledge is generated to allow the true development of sustainable agriculture [25].

A key aspect that should be investigated in detail about the use of nanofertilizers is to improve the efficient use of nitrogen (NUE) as it is the most widely applied fertilizer and the one that represents the highest cost, and consequently the productivity of the crops that this entails. However, most of the research focuses on the analysis of the properties of nanofertilizers, leaving aside additional evaluations such as the efficient use of nitrogen, which is a critical point in determining the efficacy of nanofertilizers [24].

3.2 CONCLUSION

The manufacture of nanomaterials is a scientific advance in the development of new consumer products. Its application in agriculture is still in the initial stage; however, it should be noted that it has the potential to revolutionize the agricultural sector, especially everything related to the application of fertilizers. The application of nanofertilizers undoubtedly improves crop production. Their nanometric size and great reactivity improve penetration into plant tissues and their assimilation, significantly impacting productivity and increasing the efficient use of fertilizers, mainly nitrogen. The increase in the efficient use of nitrogen is essential for the sustainable development of agricultural systems, and nanotechnology represents a viable alternative to achieve it.

ACKNOWLEDGMENT

We acknowledge CONACyT-México (the Consejo Nacional de Ciencia y Tecnología) for the valuable help provided through the Convocatoria Atención a Problemas Nacionales: Project #1529 "Biofortification of basic agricultural crops to combat malnutrition and guarantee food security in Mexico." To the Laboratorio de Fisiología y Nutrición Vegetal del Centro de Investigación en Alimentación y Desarrollo A.C. Coordinación Delicias, Chihuahua.

KEYWORDS

- **agriculture precision**
- **fertilizers**
- **molybdenum**
- **nanofertilizers**
- **nanomaterials**
- **nanotechnology**
- **nitrogen**
- **sustainability**

REFERENCES

1. Lira-Saldivar, R. H. (2016). Agronano Tecnología: Nueva frontera de la revolución verde. In *Desarrollo de la Investigación Sobre Nanotecnología en el Centro de Investigación en Química Aplicada (CIQA-CONACYT) Agronanotecnología* (pp. 15–18). Saltillo, Coahuila, México. ISBN: 978-607-7692-37-9.
2. United Nations, Department of Economic and Social Affairs, Population Division. (2019). *World Population Prospects 2019: Highlights* (ST/ESA/SER.A/423). Department of Economic and Social Affairs, Population Division: New York, NY, USA.
3. Bellard, C., Bertelsmeier, C., Leadley, P., Thuiller, W., & Courchamp, F. (2012). Impacts of climate change on the future of biodiversity. *Ecology Letters, 15*, 365–377.
4. Ahmad, Z., Tahseen, S., Wasi, A., Ganie, I. B., Shahzad, A., Emamverdian, A., Ramakrishnan, M., & Ding, Y. (2022). Nanotechnological interventions in agriculture. *Nanomaterials, 12*(2667), 1–31. https://doi.org/10.3390/nano12152667.
5. Prasad, R., Bhattacharyya, A., & Nguyen, Q. D. (2017). Nanotechnology in sustainable agriculture: Recent developments, challenges, and perspectives. *Frontiers in Microbiology, 8*, 1014.
6. Fraceto, L. F., Grillo, R., de Medeiros, G. A., Scognamiglio, V., Rea, G., & Bartolucci, C. (2016). Nanotechnology in agriculture: Which innovation potential does it have? *Frontiers in Environmental Science, 4*, 20.
7. Janmohammadi, M., Navid, A., Segherloo, A. E., & Sabaghnia, N. (2016). Impact of nano-chelated micronutrients and biological fertilizers on growth performance and grain yield of maize under deficit irrigation condition. *Biologia, 62*, 134–147.
8. Sekhon, B. S. (2014). Nanotechnology in agri-food production: An overview. *Nanotechnology Science and Applications, 7*, 31–53.
9. Morteza, E., Moaveni, P., Farahani, H. A., & Kiyani, M. (2013). Study of photosynthetic pigments changes of maize (*Zea mays* L.) under nano TiO_2 spraying at various growth stages. *SpringerPlus, 2*, 1–5.

10. DeRosa, M. C., Monreal, C., Schnitzer, M., Walsh, R., & Sultan, Y. (2010). Nanotechnology in fertilizers. *Nature Nanotechnology, 5*(2), 91–91.
11. Naderi, M. R., & Danesh-Shahraki, A. (2013). Nanofertilizers and their roles in sustainable agriculture. *International Journal of Agriculture and Crop Sciences (IJACS), 5*(19), 2229–2232.
12. Siddiqui, M. H., Al-Whaibi, M. H., Firoz, M., & Al-Khaishany, M. Y. (2015). Role of nanoparticles in plants. In M. H. Siddiqui, M. H. Al-Whaibi, & F. Mohamed (Eds.), *Nanotechnology and Plant Sciences* (pp. 19–35). Springer: New York, NY, USA.
13. Torres, D. M. (2007). Calidad de fertilizantes: ¿Cuáles son las principales propiedades y atributos que determinan su valor agronómico? In *Simposio de "Fertilidad 2007."* IPNI Cono Sur, 1, 5–12.
14. Martínez-Sánchez, C. (2020). Fertilizantes para fertirriego: Conceptos y propiedades. In *Serie de Seminarios Virtuales 2020.* Colegio Mexicano de Ingenieros en Irrigación (COMEII), México, pp. 33.
15. Alfaro, V. M., Mejía, B. J., & Salazar, S. F. (2020). Nanofertilizante de aplicación foliar para uso en praderas. *Campo Tecnológico INIA, 17*, 14–15.
16. Nuruzzaman, M., Rahman, M., Liu, Y., & Naidu, R. (2016). Nanoencapsulation, nano-guard for pesticides: A new window for safe application. *Journal of Agricultural and Food Chemistry, 64*(7), 1447–1483.
17. Fraceto, L., Grillo, R., de Medeiros, G., Scognamiglio, V., Rea, G., & Bartolucci, C. (2016). Nanotechnology in agriculture: Which innovation potential does it have? *Frontiers in Environmental Science, 4*, 20, 1–5.
18. Shang, Y., Hasan, M. K., Ahammed, G. J., Li, M., Yin, H., & Zhou, J. (2019). Applications of nanotechnology in plant growth and crop protection: A review. *Molecules, 24*, 2558.
19. Guo, H., White, J. C., Wang, Z., & Xing, B. (2018). Nano-enabled fertilizers to control the release and use efficiency of nutrients. *Current Opinion in Environmental Science & Health, 6*, 77–83.
20. Singh, H., Sharma, A., Bhardwaj, S. K., Arya, S. K., Bhardwaj, N., & Khatri, M. (2021). Recent advances in the applications of nanoagrochemicals for sustainable agricultural development. *Environmental Science: Processes & Impacts, 23*, 213–239.
21. Kottegoda, N., Sandaruwan, C., Priyadarshana, G., Siriwardhana, A., Rathnayake, U. A., Berugoda Arachchige, D. M., & Amaratunga, G. A. J. (2017). Urea-hydroxyapatite nanohybrids for slow release of nitrogen. *ACS Nano, 11*, 1214–1221.
22. Ha, N. M. C., Nguyen, T. H., Wang, S. L., & Nguyen, A. D. (2019). Preparation of NPK nanofertilizer based on chitosan nanoparticles and its effect on biophysical characteristics and growth of coffee in greenhouse. *Research on Chemical Intermediates, 45*, 51–63.
23. Kanjana, D. (2020). Foliar application of magnesium oxide nanoparticles on nutrient element concentrations, growth, physiological, and yield parameters of cotton. *Journal of Plant Nutrition, 43*, 3035–3049.
24. Benício, L. P. F., Constantino, V. R. L., Pinto, F. G., Vergütz, L., Tronto, J., & da Costa, L. M. (2017). Layered double hydroxides: New technology in phosphate fertilizers based on nanostructured materials. *ACS Sustainable Chemistry & Engineering, 5*, 399–409.
25. Carmona, F. J., Dal Sasso, G., Ramírez-Rodríguez, G. B., Pii, Y., Delgado-López, J. M., Guagliardi, A., & Masciocchi, N. (2021). Urea-functionalized amorphous calcium phosphate nanofertilizers: Optimizing the synthetic strategy towards environmental sustainability and manufacturing costs. *Scientific Reports, 11*, 3419.

26. Saini, S., Kumar, P., Sharma, N. C., Sharma, N., & Balachandar, D. (2021). Nano-enabled Zn fertilization against conventional Zn analogues in strawberry (Fragaria × ananassa Duch.). *Scientia Horticulturae, 282*, 110016.
27. Meier, S., Moore, F., Morales, A., González, M. E., Seguel, A., Meriño-Gergichevich, C., & Alarcón, D. (2020). Synthesis of calcium borate nanoparticles and its use as a potential foliar fertilizer in lettuce (*Lactuca sativa*) and zucchini (*Cucurbita pepo*). *Plant Physiology and Biochemistry, 151*, 673–680.
28. Seleiman, M. F., Almutairi, K. F., Alotaibi, M., Shami, A., Alhammad, B. A., & Battaglia, M. L. (2020). Nano-fertilization as an emerging fertilization technique: Why can modern agriculture benefit from its use? *Plants, 10*(1), 2–27.
29. Zulfiqar, F., Navarro, M., Ashraf, M., Akram, N. A., & Munné-Bosch, S. (2019). Nanofertilizer use for sustainable agriculture: Advantages and limitations. *Plant Science, 289*, 110270.
30. Raliya, R., Saharan, V., Dimkpa, C., & Biswas, P. (2018). Nanofertilizer for precision and sustainable agriculture: Current state and future perspectives. *Journal of Agricultural and Food Chemistry, 66*(26), 6487–6503.
31. Marschner, H. (2011). Functions of macronutrients. In *Marschner's Mineral Nutrition of Higher Plants* (pp. 135–190). Academic Press.
32. Perchlik, M., & Tegeder, M. (2018). Leaf amino acid supply affects photosynthetic and plant nitrogen use efficiency under nitrogen stress. *Plant Physiology, 178*, 174–188. https://doi.org/10.1104/pp.18.00597.
33. Nunes-Nesi, A., Fernie, A. R., & Stitt, M. (2010). Metabolic and signaling aspects underpinning the regulation of plant carbon nitrogen interactions. *Molecular Plant, 3*, 973–996.
34. Cruz Hernández, M. Á. (2018). *Desarrollo de un-Nano-Fertilizante Nitrogenado y su Evaluación en Lechuga (Lactuca sativa* L.*)*. Colegio de Postgraduados, Montecillo, Texcoco, Estado de México. *ColposDigital, 1*(1), 1–118
35. Wang, X., Fan, J., Xing, Y., Xu, G., Wang, H., Deng, J., Zhang, F., Li, P., & Li, Z. (2019). The effects of mulch and nitrogen fertilizer on the soil environment of crop plants. *Advances in Agronomy, 153*, 121–173.
36. Zareabyaneh, H., & Bayatvarkeshi, M. (2015). Effects of slow-release fertilizers on nitrate leaching, its distribution in soil profile, N-use efficiency, and yield in potato crop. *Environmental Earth Sciences, 74*, 3385–3393.
37. Sarkar, D., Liu, W., Xie, X., Anselmo, A. C., Mitragotri, S., & Banerjee, K. (2014). MoS2 field-effect transistor for next-generation label-free biosensors. *ACS Nano, 8*(4), 3992–4003.
38. Wang, X., Wang, T., Tian, X., & Ma, G. (2011). Effects of NMUrea on nitrogen runoff losses of surface water and nitrogen fertilizer efficiency in paddy field. *Transactions of the Chinese Society of Agricultural Engineering, 27*(1), 106–111.
39. Zhang, T., & Lin, W. (2014). Metal–organic frameworks for artificial photosynthesis and photocatalysis. *Chemical Society Reviews, 43*(16), 5982–5993.
40. Golbashy, M., Sabahi, H., Allahdadi, I., Nazokdast, H., & Hosseini, M. (2016). Synthesis of the montmorillonite-pomegranate (*Punica granatum* L.) peel polyphenols nanostructure as a drug delivery vehicle. *Biomedical and Pharmacology Journal, 9*(1), 385–392.

41. Kottegoda, N., Munaweera, I., Madusanka, N., & Karunaratne, V. (2017). A green slow-release fertilizer composition based on urea-modified hydroxyapatite nanoparticles encapsulated wood. *Current Science, 101*(1), 73–78
42. Bambara, S., & Ndakidemi, P. (2010). The potential roles of lime and molybdenum on the growth, nitrogen fixation, and assimilation of metabolites in nodulated legume: A special reference to *Phaseolus vulgaris* L. *African Journal of Biotechnology, 8*(17), 2482–2489.
43. Rana, M. S., Sun, X., Imran, M., Ali, S., Shaaban, M., Moussa, M. G., & Hu, C. (2020). Molybdenum-induced effects on leaf ultra-structure and rhizosphere phosphorus transformation in *Triticum aestivum* L. *Plant Physiology and Biochemistry, 153*, 20–29. https://doi.org/10.1016/j.plaphy.2020.05.010.
44. Nasar, J., & Shah, Z. (2017). Effect of iron and molybdenum on yield and nodulation of lentil. *ARPN Journal of Agricultural and Biological Science, 12*(11), 332–339.
45. Rana, M. S., Bhantana, P., Imran, M., Saleem, M. H., & Chengxiao, H. (2020). Molybdenum potential vital role in plants metabolism for optimizing the growth and development. *Annals of Environmental Science and Toxicology, 4*(1), 32–44. https://doi.org/10.17352/aest.000024.
46. Bittner, F., & Mendel, R. R. (2010). Cell biology of molybdenum. In *Cell Biology of Metals and Nutrients* (pp. 119–143). Springer.
47. Hell, R., & Mendel, R. R. (2010). Cell biology of molybdenum. In *Plant Cell Monographs* (Vol. 17, pp. 119–143). Springer. https://doi.org/10.1007/978-3-642-10613-2_6.
48. Pérez-González, A., Gómez-Peralta, J. I., Garza-Ortiz, A., & Barba-Behrens, N. (2011). Importancia del molibdeno en los sistemas biológicos y su papel en enzimas mononucleares como parte del cofactor Moco. *Química Orgánica, 1*, 24–33.
49. Li, Y., Jin, Q., Yang, D., & Cui, J. (2018). Molybdenum sulfide induces growth enhancement effect of rice (*Oryza sativa* L.) through regulating the synthesis of chlorophyll and the expression of aquaporin gene. *Journal of Agricultural and Food Chemistry, 66*(16), 4013–4021.
50. Avila-Arias, H., Nies, L. F., Gray, M. B., & Turco, R. F. (2019). Impacts of molybdenum-, nickel-, and lithium-oxide nanomaterials on soil activity and microbial community structure. *Science of the Total Environment, 652*, 202–211.
51. Yang, J., Song, Z., Ma, J., & Han, H. (2020). Toxicity of molybdenum-based nanomaterials on the soybean–rhizobia symbiotic system: Implications for nutrition. *ACS Applied Nano Materials*, 1–24. https://doi.org/10.1021/acsanm.0c00943.
52. Muñoz-Márquez, E., Soto-Parra, J. M., Pérez-Leal, R., Yánez-Muñoz, R. M., Noperi-Mosqueda, L., & Sánchez-Chávez, E. (2022). Aplicación de nanomolibdeno en frijol y su impacto sobre la eficiencia del nitrógeno. *Revista Mexicana de Ciencias Agrícolas, 28*(1), 319–329. https://doi.org/10.15835/nbha50212672.
53. Ekinci, M., Dursun, A., Yoldorom, E., & Parlakova, F. (2014). Effects of nanotechnology liquid fertilizers on the plant growth and yield of cucumber (*Cucumis sativus* L.). *Acta Scientiarum Polonorum-Hortorum Cultus, 13*, 135–141.
54. Petosa, A. R., Rajput, F., Selvam, O., Öhl, C., & Tufenkji, N. (2017). Assessing the transport potential of polymeric nanocapsules developed for crop protection. *Water Research, 111*, 10–17.
55. War, J. M., Fazili, M. A., Mushtaq, W., Wani, A. H., & Bhat, M. Y. (2020). Role of nanotechnology in crop improvement. In K. Hakeem & T. Pirzadah (Eds.),

Nanobiotechnology in Agriculture. Nanotechnology in the Life Sciences (pp. 1–24). Springer, Cham.
56. Muñoz-Márquez, E., Soto-Parra, J. M., Noperi-Mosqueda, L. C., & Sánchez, E. (2022). Application of molybdenum nanofertilizer on the nitrogen use efficiency, growth, and yield in green beans. *Agronomy, 12*(3163), 1–20. https://doi.org/10.3390/agronomy12123163.
57. Sangeetha, J., Mundaragi, A., Thangadurai, D., Maxim, S. S., Pandhari, R. M., & Alabhai, J. M. (2019). Nanobiotechnology for agricultural productivity, food security, and environmental sustainability. In D. Panpatte & Y. Jhala (Eds.), *Nanotechnology for Agriculture: Crop Production & Protection* (pp. 1–20). Springer, Singapore.
58. Panda, J., Nandi, A., Mishra, S. P., Pal, A. K., Pattnaik, A. K., & Jena, N. K. (2020). Effects of nano fertilizer on yield, yield attributes, and economics in tomato (*Solanum lycopersicum* L.). *International Journal of Current Microbiology and Applied Sciences, 9*, 2583–2591.
59. Agrimonti, C., Lauro, M., & Visioli, G. (2020). Smart agriculture for food quality: Facing climate change in the 21st century. *Critical Reviews in Food Science and Nutrition, 10*, 1–11.
60. Fernández, V., Sotiropoulos, T., & Brown, P. (2015). *Fertilización Foliar: Principios Científicos y Práctica de Campo* (Primera ed.). IFA, Paris, Francia. 1–160. ISBN 979-10-92366-03-7.
61. Borja, M., & Andrés, S. (2022). Efecto de aplicaciones de nutrientes foliares en el rendimiento y calidad físico-química de mandarina (*Citrus reticulata*) var. Willow Leaf. BS thesis. Facultad de Ciencias Agrícolas, Universidad Central del Ecuador, Quito: UCE, 1–121.
62. Schwab, F., Zhai, G., Kern, M., Turner, A., Schnoor, J. L., & Wiesner, M. R. (2015). Barriers, pathways, and processes for uptake, translocation, and accumulation of nanomaterials in plants: Critical review. *Nanotoxicology, 10*, 257–278.
63. Pérez-de-Luque, A. (2017). Interaction of nanomaterials with plants: What do we need for real applications in agriculture? *Frontiers in Environmental Science, 5*, 12.
64. Wang, W. N., Tarafdar, J. C., & Biswas, P. (2013). Nanoparticle synthesis and delivery by an aerosol route for watermelon plant foliar uptake. *Journal of Nanoparticle Research, 15*(1), 1–13. https://doi.org/10.1007/s11051-013-1417-8.
65. Deepa, M., Sudhakar, P., Nagamadhuri, K. V., Reddy, K. B., Krishna, T. G., & Prasad, T. N. V. K. V. (2015). First evidence on phloem transport of nanoscale calcium oxide in groundnut using solution culture technique. *Applied Nanoscience, 5*, 545–551.
66. Etxeberria, E., Gonzalez, P., Bhattacharya, P., Sharma, P., & Ke, P. C. (2016). Determining the size exclusion for nanoparticles in citrus leaves. *HortScience, 51*(6), 732–737.
67. Abdel-Aziz, H., Hasaneen, M. N., & Omar, A. (2018). Effect of foliar application of nano chitosan NPK fertilizer on the chemical composition of wheat grains. *Egyptian Journal of Botany, 58*(1), 87–95.
68. Mahil, E. I. T., & Kumar, B. A. (2019). Foliar application of nanofertilizers in agricultural crops–A review. *Journal of Farm Sciences, 32*(3), 239–249.
69. Rostami, M., Movahedi, Z., & Davari, M. R., & Siahpoosh, S. (2017). Effect of foliar application of biofertilizer and nano-fertilizers on morpho-physiological characteristics of peppermint (*Mentha piperita* L.). In *Tropentag 2017-Future Agriculture: Social-ecological Transitions and Bio-Cultural Shifts*. Malayer University, Iran.

70. Gomaa, M. A., Radwan, F. I., Kandil, E. E., & Al-Challabi, D. H. H. (2017). Comparison of some new maize hybrids response to mineral fertilization and some nanofertilizers. *Alexandria Science Exchange Journal, 38*(3), 506–513.
71. Fageria, N. K., & Baligar, V. C. (2005). Enhancing nitrogen use efficiency in crop plants. *Advances in Agronomy, 88*, 97–185.
72. Fageria, N. K. (2003). Plant tissue test for determination of optimum concentration and uptake of nitrogen at different growth stages in lowland rice. *Communications in Soil Science and Plant Analysis, 34*, 259–270.
73. Subramanian, K. S., Manikandan, A., Thirunavukkarasu, M., & Rahale, C. S. (2015). Nano-fertilizers for balanced crop nutrition. In *Nanotechnologies in Food and Agriculture* (pp. 69–80). Springer, Cham, Switzerland.
74. Chhipa, H. (2017). Nanofertilizers and nanopesticides for agriculture. *Environmental Chemistry Letters, 15*, 15–22.
75. Hasaan, A. E., Said, M. G., Alaa, H. R., & Radi, S. K. (2020). Effects of different nitrogen fertilization rates and foliar application of humic acid, fulvic acid, and tryptophan on growth, productivity, and chemical composition of common bean plants (Phaseolus vulgaris L.). *Alexandria Science Exchange Journal, 41*, 191–204.
76. Gaiotti, F., Lucchetta, M., Rodegher, G., Lorenzoni, D., Longo, E., Boselli, E., & Delgado-López, J. M. (2021). Urea-doped calcium phosphate nanoparticles as sustainable nitrogen nanofertilizers for viticulture: Implications on yield and quality of Pinot Gris grapevines. *Agronomy, 11*(6), 1026. https://doi.org/10.3390/agronomy11061026.

CHAPTER 4

Perspectives of Bioplastics in Food Packaging: Sustainable Alternatives of Production

MARIANA GONZÁLEZ-TORRES,[1] NEITH ARACELY PACHECO-LÓPEZ,[2] FRANCISCO HERNÁNDEZ-ROSAS,[1] and RICARDO HERNÁNDEZ-MARTÍNEZ[3*]

[1]*Colegio de Postgraduados-Campus Córdoba, Carretera Federal Córdoba-Veracruz Km 348, Congregación Manuel León, Municipio Amatlán de los Reyes, Veracruz, México*

[2]*Centro de Investigación y Asistencia en Tecnología y Diseño del Estado de Jalisco, A.C, Tablaje Catastral 31264 Km 5.5 Carr. Sierra Papacal-Chuburná Puerto, Parque Científico Tecnológico de Yucatán, Mérida, Yucatán, México*

[3]*CONACYT–Colegio de Postgraduados-Campus Córdoba, Carretera Federal Córdoba-Veracruz Km 348, Congregación Manuel León, Municipio Amatlán de los Reyes, Veracruz, México*

*Corresponding author.

ABSTRACT

The use of synthetic materials or conventional plastics manufactured from fossil-based raw materials has generated severe environmental problems. Although they are widely used in the food industry, there are sustainable alternatives such as polyhydroxyalkanoates (PHA) and polylactic acid (PLA), also known as bioplastics. Bio-based plastics are a promising alternative to reduce the negative environmental impact of plastic food packaging. Currently, environmental damage has encouraged a market for bio-based plastics, with a lower carbon footprint, improved resource efficiency, and

Food Security, Safety, and Sustainability: Validated Innovative Approaches and Technological Advancements. José Juan Buenrostro-Figueroa and Julio César Tafolla-Arellano (Eds.)

better material properties. Therefore, there is a need to develop the next generation of bio-based materials such as PHA and PLA to replace the current fossil-based plastics. Regarding PHAs, there are several reports that indicate the potential of some microorganisms to accumulate in their cells, but sustainable production systems are yet to be developed. The PLA-derived plastics have shown advantages over fossil-based ones. This review presents the most relevant parameters that need to be considered for the effective and sustainable production of PLA and PHA plastics used in food packaging.

4.1 INTRODUCTION

At present, there are serious environmental problems worldwide that have been generated by various factors such as the indiscriminate use of fossil-based disposable plastic packaging. However, at the same time, there are important benefits of plastic packaging for a cost-effective global supply of food; nevertheless, diverse consumption leads to a diverse waste stream, generating immense volumes of plastic waste. The consequent environmental and management problems can be increased due to the short lifespan of many plastic products, resulting in large amounts of waste [1, 2]. Additionally, this situation generates problems related to yield, processing, and overall cost. Above all, as these materials are non-biodegradable, inadequate disposal causes the death of millions of animals per year and the reduction of soil fertility, along with ecological risks because of the direct alteration of physicochemical processes, such as heavy metal adsorption and desorption, resulting in changes in the bioaccumulation and toxic effects on organisms, among other problems [3–5].

The global production of fossil-based plastics has increased during the last seven decades, which resulted in their circulation in the environment and further breakdown into smaller particles such as mesoplastics (10 mm), microplastics (≤5 mm), and nanoplastics (<1 μm) [6]. The issues caused by plastics are due to the accumulation of micro- and nanoplastics in the food chain via agricultural products and water [7]. The amount of plastic waste entering the oceans from land-based sources exceeds 4.8 million tons per year; by 2025, it is projected to reach 250 million tons per year [8]. As shown in Figure 4.1, toxic debris released from traditional fossil-based plastics enters the food chain in the form of microplastics, with serious consequences for human health, such as congenital diseases and malignant cancers. Dioxins released from plastic polymers are persistent organic pollutants that cause tumors and neurological damage in humans [9].

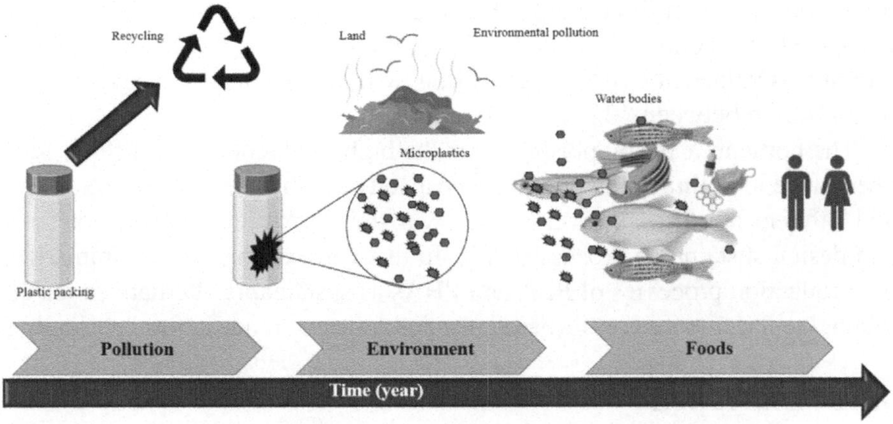

FIGURE 4.1 Schematic representation of pollution caused by fossil-based plastics.

In recent years, the production and application of biodegradable and bio-based plastics (bioplastics) have been increasingly investigated as potential alternatives to fossil-based plastics; they have the same functions as synthetic plastics and are largely environmentally friendly. Bioplastics can reduce the carbon footprint and environmental impact because they are biodegradable and may therefore contribute to global warming and environmental pollution mitigation, along with effects on land use change [5, 10]. Currently, the application of and demand for bioplastics are increasing, mainly because of the availability of raw materials, their renewability, their technical properties and advanced functionalities, their recycling potential, and their compatibility with food production processes [9]. In 2018, the European Union (EU) presented *Vision 2030* to accomplish a more sustainable plastics industry, with the goal of promoting the use of bioplastics, which resulted in increased demand for bioplastics [11]. Bioplastics are good candidates to replace fossil-based plastics in different applications, such as the use of polylactide acid (PLA) and polyhydroxyalkanoates (PHAs) [12].

4.2 BIOPLASTICS IN FOOD PACKAGING

The use of bioplastics in food packaging is growing slowly, making it necessary to increase the search for the sources and processes for producing biomaterials that can be used in the food industry [8]. Bioplastics such as PLA and PHAs are biopolymers produced by microorganisms, potentially reducing the environmental damage caused by conventional plastics. However, in 2019, bioplastics only accounted for 1% of the total plastics on the market (2.11 million tons), reaching an estimated 2.44 million tons in 2024 [13]. In economic terms, for 2019, the packaging market's economic output was valued at USD 303.26 billion, with the potential to grow by USD 60.33 billion between 2020 and 2025 [14].

The bottleneck for bioplastic use is the high production cost. Efforts have been made to optimize the production capacity via the use of microorganisms and different substrates/carbon sources, with the aim to reduce production costs and design sustainable processes [15]. In this context, this study summarizes the production processes of PLA and PHAs as sustainable alternatives for the gradual replacement of fossil-based plastics and new trends in food packaging.

4.2.1 LACTIC ACID

The production of bioplastics is limited by suitable low-cost substrates that are not in demand in the food industry. Currently, the high production costs limit the extensive application of various biodegradable materials, such as in the case of PLA synthesis using lactic acid (LA) (*see* Figure 4.2) [16]. Lactic acid can be produced via chemical and biological processes. However, there is a tendency to produce LA as a sustainable alternative through the integral use of different renewable carbon sources such as food, forestry, municipal, and agricultural wastes, as well as crop feedstocks (*see* Table 4.1). It is possible to obtain optically pure LA enantiomers with the use of adequate strains, potentially decreasing the environmental impact and increasing productivity and product purity [17].

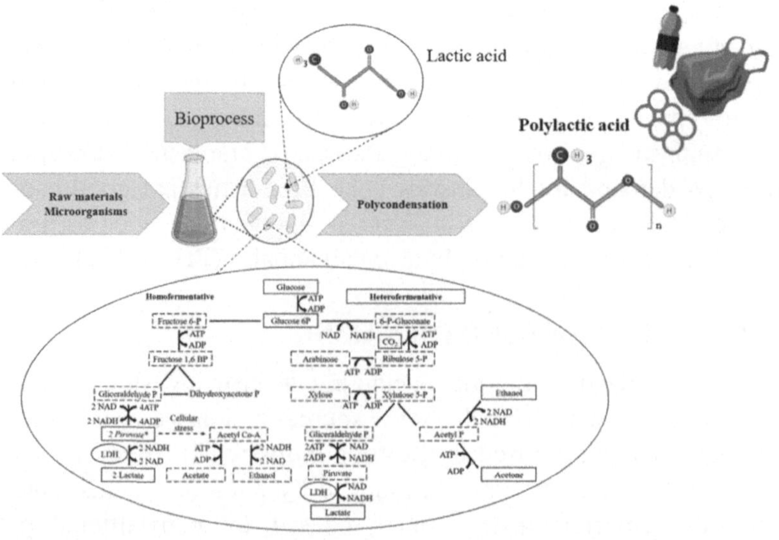

FIGURE 4.2 Schematic representation of lactic acid by bacterial strains production and polylactic acid synthesis.

TABLE 4.1 Conditions of Lactic Acid Production by Different Bacterial Strains Using Renewable Materials Sources

Microorganism	Substrate	Initial pH	Temperature (° C)	Time of maximum production (h)	Lactic acid (g/L)	Reference
Kosakonia cowanii	Cotton waste	7	30	72	28.14	[18]
Pediococcus pentosaceus, HLV1	Sugar field residues	7	40	72	62	[19]
Escherichia coli JU15	*Cistus ladanifer* residues	7	37	48	40	[20]
Bacillus amyloliquefaciens	Sugarcane molasses	6.5	30	24	178	[21]
Lactobacillus pentosus	Sugarcane bagasse	6.5	35	120	55.44	[22]
Lactobacillus amylovorus	Cassava bagasse and corn steep liquor	6	37	12	31.6	[23]
Lactobacillus delbrueckii	Molasses and corn steep liquor	5.5	39	48	112.84	[24]
Lactobacillus casei DSM 20011	Ricotta chess wey	7	37	58	65.4	[25]
Lactobacillus bulgaricus	Cheese whey powder	6.3-6.5	42	48	113.18	[26]
Lactobacillus casei 2125	Tamarind kernel powder	6	38	72	79	[27]
Weissella soli	Milk whey	6.8	37	20	7.21	[28]
Enterococcus hirae ds10	Beet molasses	7	40	24	40.6	[29]
Bacillus coagulans	Sugarcane bagasse	6	52	49	55.9	[30]

As seen in Table 4.1, LA can be produced *via* fermentation using different microorganisms. However, the genus *Lactobacillus* is highlighted, with the highest production reported for the species *Lactobacillus bulgaricus* (113 g/L), surpassed by *Bacillus amyloliquefaciens* with a maximum production of 178 g/L. The production conditions depend on the microorganisms used and range from pH 5 to 7 and from 30 to 42°C. However, the proposals shown in Table 4.1 are sustainable alternatives in which different agro-industrial wastes are valorized as renewable carbon sources, for example, molasses, cheese whey, cotton, and tamarind waste.

Most importantly, in the sustainable generation of LA, processes with adequate yields and low production costs are needed, which mostly depend on the adequate selection of the microorganisms to be used. One of the main requirements is the production of sufficient quantities of LA, along with the metabolism of low-cost carbon sources from raw materials. Homo- or hetero-fermentative bacteria are most suitable as they produce only LA through the Embden-Meyerhof-Parnas route, converting 1 mol of glucose into 2 mol of LA (*see* Figure 4.2) [23]. However, the raw material used for LA production is also important as the monosaccharides or carbon sources determine the metabolic flux; possible carbon sources are glucose, fructose, xylose, and arabinose, and the final fermentation products are shown in Figure 4.2. Additionally, it is important to know the nutritional requirements, such as nitrogen sources, vitamins, and other nutrients, as well as the adequate C/N (carbon to nitrogen) ratio for LA production, depending on the bacterial strain used [23, 31]. The LA production processes must meet certain requirements, such as allowing the use of agro-industrial waste or high productivity rates at low production costs. Nowadays, there are alternatives such as continuous bioprocesses with immobilized cells and simultaneous saccharification and fermentation (SSF).

4.2.1.1 LACTIC ACID PRODUCTION BY SIMULTANEOUS SACCHARIFICATION FERMENTATION

Another important aspect of reducing costs is the optimization of LA production to improve production rates. As the raw materials differ in their composition, the processes are diverse. In this context, SSF is a good alternative as it saves energy and prevents the adverse effects of high sugar concentrations (catabolic repression) [32]. Generally, SSF is used when the raw materials are rich in fermentable sugars and when it is necessary to remove recalcitrant lignin (lignocellulosic), which is performed in a single bioreactor containing lignocellulosic substrates, hydrolytic enzymes, and microorganisms; the fermentable sugars obtained via enzymatic reactions are rapidly consumed by the microorganisms and converted into LA [33]. Another advantage of this strategy is the lower water demand and the possibility of the physiological pre-adaptation of the production strains; in addition, the lag phase is reduced, resulting in higher growth rate and yield levels [34].

The main challenge when using SSF is to find an adequate combination of hydrolytic enzymes and microorganisms as enzymatic reaction and fermentation need to be performed in similar pH and temperature ranges; the microorganisms used have to produce a minimum of 100 g/L, with a purity of 99%, for cost-efficient downstream operations [35].

In previous studies, LA production was performed using different lignocellulosic materials such as sweet potato [36], sweet sorghum bagasse [37], bagasse sulfite pulp [38], restaurant food waste [39], sugar beet pulp [33], rice straw [40], and corn crop residues [35].

4.2.1.2 LACTIC ACID PRODUCTION BY IMMOBILIZED CELLS

Another important alternative for increased LA production and reduced production costs is an immobilized cell system in bioreactors to improve stability and facilitate reuse and continuous operation; in addition, the cells do not have to be separated from the substrate to obtain pure LA [41]. The main advantage of this approach is that the lag phase is minimized, catabolic repression is diminished (tolerance to high sugar concentrations), pH control is improved, and productivity is high [42].

In this system, LA production depends on the type of immobilization technique and support; the strategies used for LA production are mainly adsorption, entrapment, encapsulation, covalent bonding, and cross-linking [43], using Ca-alginate gels, polyethylene amine, and plastic composite materials for cell immobilization [42]. For LA production using immobilized cells, similar to SSF, when the raw material is lignocellulosic biomass, pretreatment is necessary to liberate the fermented sugar for conversion to LA. For lignocellulosic biomass saccharification for LA production, mainly enzymatic hydrolysis, alkali and diluted acids, steam explosion, wet oxidation, supercritical fluid extraction, and microwave-assisted extraction are mostly used [44, 45].

4.2.1.3 POLYLACTIC ACID (PLA) SYNTHESIS AND PROPERTIES

For PLA production, it is important to use strains that produce adequate quantities of LA. As seen in Figure 4.2, the manufacturing of PLA depends on a second step of LA or its lactide form polymerization by polycondensation, ring-opening polymerization (ROP), and azeotropic dehydration condensation. The method used directly impacts the production costs and the properties of the bioplastics, such as stiffness and transparency, defining the end uses, such as for plastic bottles, stiff packages, or degradable films for packaging [46, 47]. For their characteristics, bioplastics can be used extensively as packaging. Also, due to the short service lifetime, they would contribute to the reduction of environmental impact [48], as they easily find their way into the environment as a result of the human throwaway culture. In addition to being biodegradable, PLA can also be recycled or incinerated using steam or boiling water and can be hydrolyzed to lactic acid, leading to molecular recycling, and allowing recycling of packaging materials [49].

In recent years, the interest in using PLA films in the food packaging industry has increased because of their excellent transparency, higher permeability to carbon dioxide than to oxygen (perm selectivity) when compared to most conventional fossil-based plastics (20 times better oxygen barrier properties than polystyrene), better mechanical performance than polystyrene, good water resistance, good chemical resistance, good film sealing at temperatures below the melting temperature, and better ultraviolet light barrier properties than low-density polyethylene [49].

Polylactic acid is commercialized under different brand names and by different companies, such as Solanyl, NatureWorks by Rodenburg in the Netherlands, PLA by Cargill in the USA, Ingeo® by NatureWorks LLC (Minnetonka, MN, USA), PLA by Galactic in Belgium, L-PLA by Purac in the Netherlands, PLA-based resins: Bioflex and Biograde PLA in FKuR in Germany, and Lacty by Schimadzu in Japan [9, 50]. Generally, PLA is used in products such as coffee capsules, diaper tops or back sheets, cups, or yogurt packaging [50].

4.2.2 POLYHYDROXYALKANOATES

Polyhydroxyalkanoates are considered bioplastics and can be produced by numerous microorganisms intracellularly as granules (*see* Figure 4.3) as an energy reserve, usually as a result of nutrient limitations (nitrogen and phosphorus) and excess carbon [51]. They have characteristics similar to those of fossil-based plastics and are biocompatible, biodegradable, UV resistant, non-hydrolysable in liquid solutions, and thermoplastic, have greater biological compatibility, and are processed in solution, melt, powder, and emulsion phases through different technologies [52, 53].

The PHAs are mainly linear polyesters containing 3-hydroxy fatty acid monomers; an ester bond is formed by the reaction between the carboxyl group of one monomer and the hydroxyl group of another monomer unit (*see* Figure 4.3). Most PHAs produced by various microorganisms are of rectus (R) configuration [54]. The PHAs are classified according to their chain length as short-chain-length scl-PHAs, containing from 4 to 5 carbon atoms, medium-chain-length mcl-PHAs, containing from 6 to 14 carbon atoms, and long-chain-length lcl-PHAs, containing more than 14 carbon atoms [9, 55, 56]. According to the microorganism used is the metabolic pathway used for their synthesis, in Figure 4.3 it is possible to appreciate the way of synthesis for Poly-β-hydroxybutyrate and Poly-β-hydroxyvalerate, which are the metabolic pathways followed by the bacterial strains when the metabolism of Entner-Doudoroff or Pentose Phosphate pathways.

Several microorganisms and raw materials as carbon sources have been used for PHA production, such as *Burkholderia cepacia, Bacillus cereus, Cupriavidus necator* on brewer's spent grain [57], *Pseudomonas* sp. phDV1 on phenol [58], *Halomonas* sp. on glucose and fructose [59], *Pseudomonas* sp. strain-P(16) on rice bran and soy molasses [60], *Ancylobacter* sp. on crude paper industry effluent [61], *Pseudomonas* sp. AK-3 and AK-4 on sucrose [62], *Pseudomonas putida* NBUS12 on eco-pollutant styrene [63], *Bacillus cereus* ARY73 on glucose [64], *Burkholderia glumae* MA13 on crude glycerol [65], *Cupriavidus necator* on beet molasses [66], *Suhomyces kilbornensis* and *Bacillus thuringiensis* on sucrose [67], and mixed cultures on lignocellulosic hydrolysates, acetate co-substrate, and organic urban waste [68, 69]. They are generally produced under conditions of excess carbon source and limitations of oxygen, phosphorus, and nitrogen or even fluctuating pH of the growth media [70].

However, the production costs are generally high as the substrates account for 30% to 40% of the total production costs. Some biopolymers can be produced from cheap substrates such as sucrose, glucose, lignin, and waste material [71]. Another limitation is the low productivity, and in economically feasible and simple downstream processes, PHA recovery is crucial [72]. Therefore, polyhydroxyalkanoate coproduction with other metabolites has emerged as a good alternative to decrease production costs.

4.2.3 POLYHYDROXYALKANOATE COPRODUCTION

Similar to PLA production, the production of PHAs is limited by the high production costs. However, as a sustainable alternative, PHAs are being produced through the integral use of different renewable carbon sources such as food waste, forestry, municipal, and agricultural wastes, and energy crops (*see* Figure 4.3). A novel trend is PHA production with other metabolites, which decreases the production costs (*see* Table 4.2); this also lowers the environmental impact and increases productivity and product purity [17].

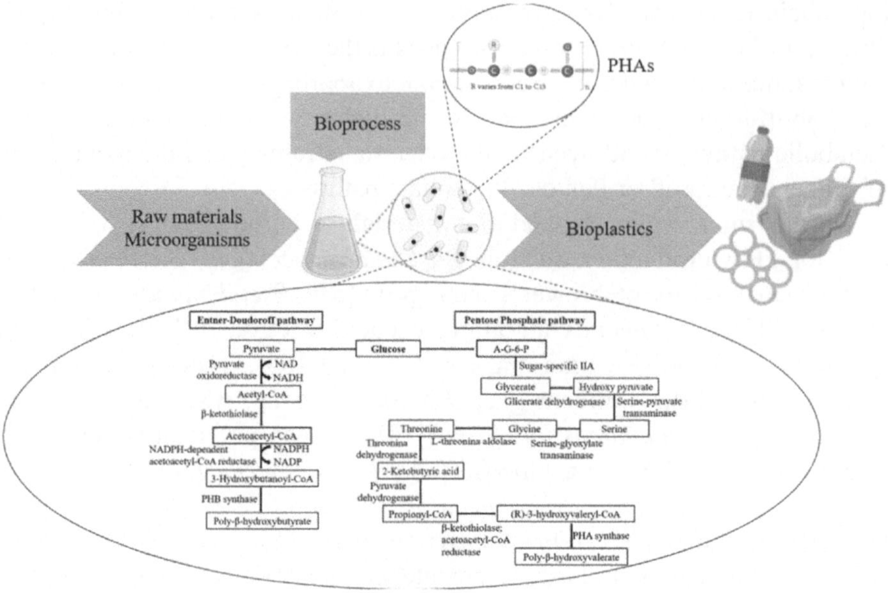

FIGURE 4.3 Schematic representation poly-β-hydroxybutyrate and poly-β-hydroxyvalerate production by bacterial strains

TABLE 4.2 Coproduction of PHA with Other Microbial Metabolites by Some Bacterial Strains

Microorganism	Substrate	Coproducts	Yields (g/L)	Reference
Bacillus megaterium ALA2	Corn cob	PHA and astaxanthin	11.5 0.91 mg/L	[73]
Halomonas smyrnensis AAD6T	Sucrose	PHA and levan	0.48 15.3	[74]
Bacillus thuringiensis HA1, and *Suhomyces kilbournensis* HD1	Sucrose	PHA and exopolysaccharides	0.058, 0.148 and 16.5, 0.086	[67]
Cupriavidus necator DSM 545	Broken rice	PHA and methane	0.95 465.28 gVS/L[a]	[75]
Pseudomonas chlororaphis	Crude glycerol	PHA and phenazines	19.05 0.079	[76]
Pseudomonas aeruginosa	Soybean oil	PHA and rhamnolipids	3.81 1.63	[77]
Yangia sp. ND199	Sucrose	PHA and exopolysaccharides	1.3 0.28	[78]

TABLE 4.2 *(Continued)*

Microorganism	Substrate	Coproducts	Yields (g/L)	Reference
Paracoccus sp. LL1	Glucose	PHA and astaxanthin	6.97 0.019	[79]
Iodobacter sp. PCH194	Glucose	PHB and violacein pigment	11.0 1.5	[80]
Enterobacter aerogenes	Oilcake	PHA and rhamnolipids	4.2 5.81	[81]
Paracoccus sp. LL1		PHA and carotenoids	0.0009 0.0023	[82]
Halomonas bluephagenesis	Glucose	PHB and ectoine	8.0 32	[83]
Paracoccus sp.	Waste cooking oil	PHA and carotenoids	1.0 0.89	[84]
Paracoccus sp. LL1	Glycerol	PHA and carotenoids	9.52 0.0071	[85]
Halomonas smyrnensis	Sucrose	PHBV and levan	0.48 1.3	[86]

[a] gVS/L: g of volatile solids/L.

As seen in Table 4.2, the production of PHAs via fermentation can be carried out with different microorganisms. However, the use of *Iodobacter* sp. PCH194 resulted in the highest PHA production; when glucose was used as a carbon source, the colorant violacein was coproduced. As seen in Table 4.2, various agro-industrial wastes and sugars, such as glucose and sucrose, can be valorized. The production of various compounds together with PHAs depends on the microorganisms used, is possible to appreciate that there is no tendency of microorganisms reported to generate a big spectrum to find new microorganisms with the capacity to coproduce PHAs, and the possibility to optimize the production of the metabolites in accordance with the best metabolism of the microorganism used. Also, it is possible to appreciate in Table 4.2 that PHAs were coproduced more with an exopolysaccharide, generating expectations for designing coproduction processes with these biomolecules.

As seen in Table 4.2, the highest production of PHAs (19.05 g/L) by fermentation was reported for *Pseudomonas chlororaphis* in coproduction with phenazines (0.079 g/L), using crude glycerol as a carbon source. On the other hand, *Enterobacter aerogenes* was reported to have an equilibrated coproduction of PHAs with rhamnolipids, with 4.2 and 5.81 g/L being two examples of coproduction processes that determine the reduction of PHAs production costs. As mentioned by Raza et al. [87], the factors that impact the cost of PHA production include the carbon source, productivity, and the recovery method, the latter of which depends on the complexity and raw materials used for purification.

4.2.3.1 PHA RECUPERATION

Some strategies for PHA recuperation have been reported. Non-halogenates have been used because of their low toxicity; however, chlorinated solvents, such as chloroform and dichloromethane, are also efficient [88]. In this sense, it is necessary to design an extraction process that is sustainable and results in a high-quality product. As indicated by Koller [89], PHA recuperation needs to be sustainable, without the loss of solvents; also, the extraction method may determine the molecular weight, purity, and recovery yield of the polymer, directly impacting the production costs. Importantly, the method of PHA recovery is determined by the microorganisms used. The yield and quality of the recovered PHA will depend on the method used, and more methods used are chemical through the use of halogenated solvents such as chloroform, dichloroethane, chloropropane, methyl chloride or solvents such as acetone, methyl isobutyl ketone, propylene carbonate, ethyl acetate and isoamyl alcohol which act in two stages: the first stage permeabilizes the membrane and the second stage solubilizes it; can also be recovered by chemical digestion using sodium dodecyl sulfate (SDS), sodium hypochlorite (NaClO), sodium hydroxide (NaOH) and sulfuric acid (H_2SO_4); however, the use of chemicals can cause negative characteristics of the end-product have been reported [90, 91]. On the other hand, Aqueous two-phase systems (ATPS) are new methods of extraction and purification of various materials, which present some advantages, e.g., the facility of separation/purification of biomolecules [91].

A new trend in PHA recovery is the use of green solvents with low or no toxicity and possibly derived from biochemical conversion. This can overcome ecological issues and limitations involving worker safety or stringent regulations on solvent traces in goods for particular applications, such as ethers, esters, carbonates, and ketones. Green solvents such as dimethyl carbonate (DMC), propylene carbonate, ethyl acetate, methyl isobutyl ketone, and ionic liquids are alternatives to chlorinated hydrocarbons as they can be recycled, increase polymer recovery yield, quality, and purity, decrease molecular weight, and have positive impacts on ecological feasibility and environmental performance [92–95].

Another alternative is enzymatic extraction/digestion using proteases such as alcalase and cell wall-degrading enzymes such as lysozyme, which liberate PHA granules [96]. Also, enzyme mixtures containing proteases, phospholipases, lysozyme, and nucleases may be used for PHA recuperation, although the costs may be higher [89].

In the food packaging industry, PHAs are popular because of their high waterproofing and gas permeability; they are also biodegradable upon disposal and can be processed via injection molding [9, 53]. The best results were obtained when bioplastics were mixed with other natural biopolymers in food packaging, such as PHA/PLA mixtures, improving the oxygen transfer rate, crystallinity, and molecular distribution of the biomaterial [50, 97]. The oxygen transfer and waterproofing properties of bioplastics are advantageous in the packaging of foods such as meat and cheese products [98].

The principal potential of PLAs and PHAs is based on their degradability in different natural environments. Under controlled conditions, several tests were performed to investigate their degradability, using physical and respirometric methods as well as reactive pyrolysis gas chromatography [50]. Compared to PLAs, PHAs are both compostable and biodegradable in marine environments [9].

Polyhydroxyalkanoates are commercialized by different companies, such as Novamont and Bio-on (PHA) in Italy, Avantium (FDCA and PEF) and Rodenburg Biopolymer (PHA) in the Netherlands, Technaro (PHA) in Germany, Kanejka (PHBH) in Tokyo, Japan, NAFIGATE corporation (P3HB) in the Czech Republic, Metabolics (PHA, PHB) in Woburn, Massachusetts, Dinamer Scientific (mcl-PHA) in Bainbridge, Georgia, Global biopolymers (Bio-PE, Bio-PP, Bio-PET, PBS, PHA, PBAT) in Thailand, TerraVerdae BioWorks Inc (PHA) in Canada, Shenzhen Ecomann Biotechnology Co., Ltd (PHA) and Newlight Technologies, LLC (PHBV) in China, Newlight Technologies, LLC (PHA) in the US [99].

4.3 CONCLUSION

Bioplastics such as PLAs and PHAs are novel alternatives to conventional plastics in food packaging. They are considered green materials and can be produced via valorizing raw materials by lignocellulolytic enzymes. Modern technologies can be applied to reduce production costs and design sustainable production processes.

KEYWORDS

- **bioplastics**
- **conventional plastics**
- **environmental problems**
- **fossil-based plastics**
- **lignocellulolytic enzymes**
- **polyhydroxyalkanoates**
- **polylactic acid**
- **raw materials**

REFERENCES

1. Molina-Besch, K., & Keszleri, H. (2023). Exploring the industrial perspective on biobased plastics in food packaging applications–Insights from Sweden. *Sustainable Production and Consumption*, *35*, 72–84.
2. Hahladakis, J. N., Velis, C. A., Weber, R., Iacovidou, E., & Purnell, P. (2018). An overview of chemical additives present in plastics: Migration, release, fate and environmental impact during their use, disposal and recycling. *Journal of Hazardous Materials, 344*, 179–199.
3. Ojha, N., & Das, N. (2021). Microbial production of bioplastics: Current trends and future perspectives. *Bioplastics for Sustainable Development, 1,* 1–60.
4. Liu, G., Dave, P. H., Kwong, R. W., Wu, M., & Zhong, H. (2021). Influence of microplastics on the mobility, bioavailability, and toxicity of heavy metals: a review. *Bulletin of Environmental Contamination and Toxicology*, *107*, 710–721.
5. Zhao, X., Wang, Y., Chen, X., Yu, X., Li, W., Zhang, S., & Zhu, H. (2023). Sustainable bioplastics derived from renewable natural resources for food packaging. *Matter, 6*(1), 97–127.
6. Sridharan, S., Kumar, M., Saha, M., Kirkham, M. B., Singh, L., & Bolan, N. S. (2022). The polymers and their additives in particulate plastics: What makes them hazardous to the fauna? *Science of the Total Environment*, *824*, 153828.
7. Acquavia, M. A., Pascale, R., Martelli, G., Bondoni, M., & Bianco, G. (2021). Natural polymeric materials: A solution to plastic pollution from the agro-food sector. *Polymers, 13*(1), 158.
8. Adeyeye, O. A., Sadiku, E. R., Babu Reddy, A., Ndamase, A. S., Makgatho, G., Sellamuthu, P. S., & Jamiru, T. (2019). The use of biopolymers in food packaging. In *Green Biopolymers and Their Nanocomposites* (pp. 137–158). Springer.
9. Naser, A. Z., Deiab, I., & Darras, B. M. (2021). Poly (lactic acid) (PLA) and polyhydroxyalkanoates (PHAs), green alternatives to petroleum-based plastics: A review. *RSC Advances*, *11*(28), 17151–17196.

10. Atiwesh, G., Mikhael, A., Parrish, C. C., Banoub, J., & Le, T. A. T. (2021). Environmental impact of bioplastic use: A review. *Heliyon, 7*(9), e07918.
11. Taib, N. A. A. B., Rahman, M. R., Huda, D., Kuok, K. K., Hamdan, S., Bakri, M. K. B., & Khan, A. (2022). A review on poly lactic acid (PLA) as a biodegradable polymer. *Polymer Bulletin, 80*(2), 1179–1213.
12. Lomwongsopon, P., & Varrone, C. (2022). Contribution of Fermentation Technology to Building Blocks for Renewable Plastics. *Fermentation, 8*(2), 47.
13. Dedieu, I., Peyron, S., Gontard, N., & Aouf, C. (2022). The thermo-mechanical recyclability potential of biodegradable biopolyesters: Perspectives and limits for food packaging application. *Polymer Testing, 111*, 107620.
14. Kumari, S. V. G., Pakshirajan, K., & Pugazhenthi, G. (2022). Recent advances and future prospects of cellulose, starch, chitosan, polylactic acid, and polyhydroxyalkanoates for sustainable food packaging applications. *International Journal of Biological Macromolecules, 221*, 163–182.
15. Dwivedi, R., Pandey, R., Kumar, S., & Mehrotra, D. (2020). Polyhydroxyalkanoates (PHA): Role in bone scaffolds. *Journal of Oral Biology and Craniofacial Research, 10*(1), 389–392.
16. Ahmad, A., Banat, F., & Taher, H. (2020). A review on the lactic acid fermentation from low-cost renewable materials: Recent developments and challenges. *Environmental Technology & Innovation, 20*, 101138.
17. Djukić-Vuković, A., Mladenović, D., Ivanović, J., Pejin, J., & Mojović, L. (2019). Towards sustainability of lactic acid and poly-lactic acid polymers production. *Renewable and Sustainable Energy Reviews, 108*, 238–252.
18. El-Sheshtawy, H. S., Fahim, I., Hosny, M., & El-Badry, M. A. (2022). Optimization of lactic acid production from agro-industrial wastes produced by *Kosakonia cowanii*. *Current Research in Green and Sustainable Chemistry, 5*, 100228.
19. Katepogu, H., Wee, Y. J., Chinni, S. V., Gopinath, S. C., Syed, A., Bahkali, A. H., & Lebaka, V. R. (2022). Lactic acid production from sugarcane field residue as renewable and economical bioresource by newly isolated *Pediococcus pentosaceus* HLV1. *Biomass Conversion and Biorefinery, 13*(16), 14927–14937.
20. Alves-Ferreira, J., Carvalheiro, F., Duarte, L. C., Ferreira, A. R., Martinez, A., & Pereira, H. (2020). D-Lactic acid production from Cistus ladanifer residues: Co-fermentation of pentoses and hexoses by *Escherichia coli* JU15. *Industrial Crops and Products, 177*, 114519.
21. Vignesh Kumar, B., Muthumari, B., Kavitha, M., John Praveen Kumar, J. K., Thavamurugan, S., Arun, A., & Jothi Basu, M. (2022). Studies on optimization of sustainable lactic acid production by *Bacillus amyloliquefaciens* from sugarcane molasses through microbial fermentation. *Sustainability, 14*(12), 7400.
22. González-Leos, A., Bustos, M. G., Rodríguez-Castillejos, G. C., Durán, L. R., & Del Ángel, J. D. Á. (2020). Kinetics of lactic acid fermentation from sugarcane bagasse by Lactobacillus pentosus. *Revista Mexicana de Ingeniería Química, 19*(1), 377–386.
23. Macedo, J. V. C., de Barros Ranke, F. F., Escaramboni, B., Campioni, T. S., Núñez, E. G. F., & de Oliva Neto, P. (2020). Cost-effective lactic acid production by fermentation of agro-industrial residues. *Biocatalysis and Agricultural Biotechnology, 27*, 101706.
24. Beitel, S. M., Coelho, L. F., & Contiero, J. (2020). Efficient conversion of agroindustrial waste into D (-) lactic acid by *Lactobacillus delbrueckii* using fed-batch fermentation. *BioMed Research International, 2020*, 1–13.

25. Costa, S., Summa, D., Semeraro, B., Zappaterra, F., Rugiero, I., & Tamburini, E. (2020). Fermentation as a strategy for bio-transforming waste into resources: Lactic acid production from agri-food residues. *Fermentation, 7*(1), 3.
26. Liu, P., Zheng, Z., Xu, Q., Qian, Z., Liu, J., & Ouyang, J. (2018). Valorization of dairy waste for enhanced D-lactic acid production at low cost. *Process Biochemistry, 71*, 18–22.
27. Kelkar, S. T., & Mahanwar, P. A. (2015). Production of lactic acid from tamarind kernel by *Lactobacillus casei*. *International Journal of Technical and Educational Engineering, 3*, 25–31.
28. Montero-Zamora, J., Fernández-Fernández, S., Redondo-Solano, M., Mazón-Villegas, B., Mora-Villalobos, J. A., & Barboza, N. (2022). Assessment of different lactic acid bacteria isolated from agro-industrial residues: First report of the potential role of Weissella soli for lactic acid production from milk whey. *Applied Microbiology, 2*(3), 626–635.
29. Abdel-Rahman, M. A., Hassan, S. E., El-Din, M. N., Azab, M. S., El-Belely, E. F., Alrefaey, H. M., & Elsakhawy, T. (2020). One-factor-at-a-time and response surface statistical designs for improved lactic acid production from beet molasses by *Enterococcus hirae* ds10. *SN Applied Sciences, 2*(1), 1–14.
30. de Oliveira, R. A., Schneider, R., Rossell, C. E. V., Maciel Filho, R., & Venus, J. (2019). Polymer grade L-lactic acid production from sugarcane bagasse hemicellulosic hydrolysate using *Bacillus coagulans*. *Bioresource Technology Reports, 6*, 26–31.
31. Thakur, A., Panesar, P. S., & Saini, M. S. (2019). L (+)-Lactic acid production by immobilized Lactobacillus casei using low cost agro-industrial waste as carbon and nitrogen sources. *Waste and Biomass Valorization, 10*, 1119–1129.
32. Liu, T., Xu, X., Liu, Y., Li, J., Du, G., Lv, X., & Liu, L. (2022). Engineered microbial cell factories for sustainable production of L-lactic acid: A critical review. *Fermentation, 8*(6), 279.
33. Berlowska, J., Cieciura-Włoch, W., Kalinowska, H., Kregiel, D., Borowski, S., Pawlikowska, E., & Witonska, I. (2018). Enzymatic conversion of sugar beet pulp: A comparison of simultaneous saccharification and fermentation and separate hydrolysis and fermentation for lactic acid production. *Food Technology and Biotechnology, 56*(2), 188–197.
34. Aulitto, M., Fusco, S., Nickel, D. B., Bartolucci, S., Contursi, P., & Franzén, C. J. (2019). Seed culture pre-adaptation of *Bacillus coagulans* MA-13 improves lactic acid production in simultaneous saccharification and fermentation. *Biotechnology for Biofuels, 12*, 1–11.
35. Malacara-Becerra, A., Melchor-Martínez, E. M., Sosa-Hernández, J. E., Riquelme-Jiménez, L. M., Mansouri, S. S., Iqbal, H. M., & Parra-Saldívar, R. (2022). Bioconversion of corn crop residues: Lactic acid production through simultaneous saccharification and fermentation. *Sustainability, 14*(19), 11799.
36. Nguyen, C. M., Choi, G. J., Choi, Y. H., Jang, K. S., & Kim, J. C. (2013). D- and L-lactic acid production from fresh sweet potato through simultaneous saccharification and fermentation. *Biochemical Engineering Journal, 81*, 40–46.
37. Wang, Y., Wang, M., Cai, D., Wang, B., Wang, Z., Qin, P., & Tan, T. (2016). Efficient L-lactic acid production from sweet sorghum bagasse by open simultaneous saccharification and fermentation. *RSC Advances, 6*(42), 35771–35777.

38. Zhou, J., Ouyang, J., Xu, Q., & Zheng, Z. (2016). Cost-effective simultaneous saccharification and fermentation of l-lactic acid from bagasse sulfite pulp by *Bacillus coagulans* CC17. *Bioresource Technology, 222*, 431–438.
39. Pleissner, D., Demichelis, F., Mariano, S., Fiore, S., Gutiérrez, I. M. N., Schneider, R., & Venus, J. (2017). Direct production of lactic acid based on simultaneous saccharification and fermentation of mixed restaurant food waste. *Journal of Cleaner Production, 143*, 615–623.
40. Chen, H., Huo, W., Wang, B., Wang, Y., Wen, H., Cai, D., & Qin, P. (2019). L-lactic acid production by simultaneous saccharification and fermentation of dilute ethylediamine pre-treated rice straw. *Industrial Crops and Products, 141*, 111749.
41. Panesar, P. S., Kennedy, J. F., Gandhi, D. N., & Bunko, K. (2007). Bioutilisation of whey for lactic acid production. *Food Chemistry, 105*(1), 1–14.
42. Dosuky, A. S., Elsayed, T. R., Yousef, E. T., Barakat, O. S., & Nasr, N. F. (2022). Isolation, identification, and application of lactic acid-producing bacteria using salted cheese whey substrate and immobilized cells technology. *Journal of Genetic Engineering and Biotechnology, 20*(1), 1–16.
43. Eş, I., Khaneghah, A. M., Barba, F. J., Saraiva, J. A., Sant'Ana, A. S., & Hashemi, S. M. B. (2018). Recent advancements in lactic acid production—a review. *Food Research International, 107*, 763–770.
44. Esquivel-Hernández, D. A., García-Pérez, J. S., López-Pacheco, I. Y., Iqbal, H. M., & Parra-Saldívar, R. (2022). Resource recovery of lignocellulosic biomass waste into lactic acid—trends to sustain cleaner production. *Journal of Environmental Management, 301*, 113925.
45. Nwamba, M. C., Sun, F., Mukasekuru, M. R., Song, G., Harindintwali, J. D., Boyi, S. A., & Sun, H. (2021). Trends and hassles in the microbial production of lactic acid from lignocellulosic biomass. *Environmental Technology & Innovation, 21*, 101337.
46. Riofrio, A., Cornejo, M., & Baykara, H. (2022). Life cycle and environmental impact evaluation of polylactic acid (PLA) production in Ecuador. *International Journal of Life Cycle Assessment, 27*(6), 834–848.
47. Li, G., Zhao, M., Xu, F., Yang, B., Li, X., Meng, X., & Li, Y. (2020). Synthesis and biological application of polylactic acid. *Molecules, 25*(21), 5023.
48. Gan, I., & Chow, W. S. (2018). Antimicrobial poly (lactic acid)/cellulose bionanocomposite for food packaging application: A review. *Food Packaging and Shelf Life, 17*, 150–161.
49. Ncube, L. K., Ude, A. U., Ogunmuyiwa, E. N., Zulkifli, R., & Beas, I. N. (2020). Environmental impact of food packaging materials: A review of contemporary development from conventional plastics to polylactic acid based materials. *Materials, 13*(21), 4994.
50. Boey, J. Y., Mohamad, L., Khok, Y. S., Tay, G. S., & Baidurah, S. (2021). A review of the applications and biodegradation of polyhydroxyalkanoates and poly(lactic acid) and its composites. *Polymers, 13*(10), 1544.
51. Winnacker, M. (2019). Polyhydroxyalkanoates: Recent advances in their synthesis and applications. *European Journal of Lipid Science and Technology, 121*(11), 1900101.
52. Amaro, T. M., Rosa, D., Comi, G., & Iacumin, L. (2019). Prospects for the use of whey for polyhydroxyalkanoate (PHA) production. *Frontiers in Microbiology, 10*, 992.
53. Tripathi, A. D., Mishra, P. K., Darani, K. K., Agarwal, A., & Paul, V. (2022). Hydrothermal treatment of lignocellulose waste for the production of polyhydroxyalkanoates

copolymer with potential application in food packaging. *Trends in Food Science & Technology, 123*, 233–250.
54. Shevchenko, T., Ranjbari, M., Shams Esfandabadi, Z., Danko, Y., & Bliumska-Danko, K. (2022). Promising developments in bio-based products as alternatives to conventional plastics to enable circular economy in Ukraine. *Recycling, 7*(2), 20.
55. Reddy, V. U. N., Ramanaiah, S. V., Reddy, M. V., & Chang, Y. C. (2022). Review of the developments of bacterial medium-chain-length polyhydroxyalkanoates (mcl-PHAs). *Bioengineering, 9*(5), 225.
56. Tan, D., Wang, Y., Tong, Y., & Chen, G. Q. (2021). Grand challenges for industrializing polyhydroxyalkanoates (PHAs). *Trends in Biotechnology, 39*(9), 953–963.
57. Corchado-Lopo, C., Martínez-Avila, O., Marti, E., Llimós, J., Busquets, A. M., Kucera, D., & Ponsá, S. (2021). Brewer's spent grain as a no-cost substrate for polyhydroxyalkanoates production: Assessment of pretreatment strategies and different bacterial strains. *New Biotechnology, 62*, 60–67.
58. Kanavaki, I., Drakonaki, A., Geladas, E. D., Spyros, A., Xie, H., & Tsiotis, G. (2021). Polyhydroxyalkanoate (PHA) production in *Pseudomonas* sp. phDV1 strain grown on phenol as carbon sources. *Microorganisms, 9*(8), 1636.
59. Thomas, T., Sudesh, K., Bazire, A., Elain, A., Tan, H. T., Lim, H., & Bruzaud, S. (2020). PHA production and PHA synthases of the halophilic bacterium *Halomonas* sp. SF2003. *Bioengineering, 7*(1), 29.
60. Aljuraifani, A. A., Berekaa, M. M., & Ghazwani, A. A. (2019). Bacterial biopolymer (polyhydroxyalkanoate) production from low-cost sustainable sources. *MicrobiologyOpen, 8*(6), e00755.
61. Tyagi, P., & Sharma, A. (2021). Utilization of crude paper industry effluent for polyhydroxyalkanoate (PHA) production. *Environmental Technology & Innovation*, 23, 101692.
62. Muneer, F., Rasul, I., Qasim, M., Sajid, A., & Nadeem, H. (2022). Optimization, production and characterization of polyhydroxyalkanoate (PHA) from indigenously isolated novel bacteria. *Journal of Polymers and the Environment, 30*(8), 3523–3533.
63. Tan, G. Y. A., Chen, C. L., Ge, L., Li, L., Tan, S. N., & Wang, J. Y. (2015). Bioconversion of styrene to poly (hydroxyalkanoate)(PHA) by the new bacterial strain *Pseudomonas putida* NBUS12. *Microbes and Environments, 30*(1), 76–85.
64. Yasin, A. R., & Al-Mayaly, I. K. (2021). Study of the fermentation conditions of the *Bacillus cereus* strain ARY73 to produce polyhydroxyalkanoate (PHA) from glucose. *Journal of Ecological Engineering, 22*(8), 41–53.
65. de Paula, F. C., Kakazu, S., de Paula, C. B. C., de Almeida, A. F., Gomez, J. G. C., & Contiero, J. (2019). *Burkholderia glumae* MA13: A newly isolated bacterial strain suitable for polyhydroxyalkanoate production from crude glycerol. *Biocatalysis and Agricultural Biotechnology, 20*, 101268.
66. Kiselev, E. G., Demidenko, A. V., Zhila, N. O., Shishatskaya, E. I., & Volova, T. G. (2022). Sugar beet molasses as a potential C-substrate for PHA production by *Cupriavidus necator*. *Bioengineering, 9*(4), 154.
67. Vega-Vidaurri, J. A., Hernández-Rosas, F., Ríos-Corripio, M. A., Loeza-Corte, J. M., Rojas-López, M., & Hernández-Martínez, R. (2022). Coproduction of polyhydroxyalkanoates and exopolysaccharide by submerged fermentation using autochthonous bacterial strains. *Chemical Papers, 76*(4), 2419–2429.

68. Yin, F., Li, D., Ma, X., Li, J., & Qiu, Y. (2020). Poly (3-hydroxybutyrate-3-hydroxyvalerate) production from pretreated waste lignocellulosic hydrolysates and acetate co-substrate. *Bioresource Technology, 316*, 123911.
69. Lorini, L., Martinelli, A., Pavan, P., Majone, M., & Valentino, F. (2021). Downstream processing and characterization of polyhydroxyalkanoates (PHAs) produced by mixed microbial culture (MMC) and organic urban waste as substrate. *Biomass Conversion and Biorefinery*, *11*, 693–703.
70. Kumar, M., Rathour, R., Singh, R., Sun, Y., Pandey, A., Gnansounou, E., & Thakur, I. S. (2020). Bacterial polyhydroxyalkanoates: Opportunities, challenges, and prospects. *Journal of Cleaner Production*, *263*, 121500.
71. Sharma, V., Sehgal, R., & Gupta, R. (2021). Polyhydroxyalkanoate (PHA): Properties and modifications. *Polymer*, *212*, 123161.
72. Kourmentza, C., Plácido, J., Venetsaneas, N., Burniol-Figols, A., Varrone, C., Gavala, H. N., & Reis, M. A. (2017). Recent advances and challenges towards sustainable polyhydroxyalkanoate (PHA) production. *Bioengineering*, *4*(2), 55.
73. Khomlaem, C., Aloui, H., Singhvi, M., & Kim, B. S. (2023). Production of polyhydroxyalkanoates and astaxanthin from lignocellulosic biomass in high cell density membrane bioreactor. *Chemical Engineering Journal, 451*, 138641.
74. de Siqueira, E. C., & Houllou, L. M. (2022). Co-production of polyhydroxyalkanoates and levan by *Halomonas smyrnensis* AAD6T. *Research, Society and Development*, *11*(16), e180111637925-e180111637925.
75. Brojanigo, S., Alvarado-Morales, M., Basaglia, M., Casella, S., Favaro, L., & Angelidaki, I. (2022). Innovative co-production of polyhydroxyalkanoates and methane from broken rice. *Science of the Total Environment, 825*, 153931.
76. Aloui, H., Khomlaem, C., Torres, C. A., Freitas, F., Reis, M. A., & Kim, B. S. (2022). Enhanced co-production of medium-chain-length polyhydroxyalkanoates and phenazines from crude glycerol by high cell density cultivation of *Pseudomonas chlororaphis* in membrane bioreactor. *International Journal of Biological Macromolecules, 211*, 545–555.
77. Tariq, M. R., Raza, Z. A., Majeed, M. I., Ayub, A., & Khubaib, M. A. (2022). Bioreactor scale co-production of poly (hydroxyalkanoate) and rhamnolipids with distinct nitrogen sources. *Biologia, 77*(5), 1391–1404.
78. Romero Soto, L., Thabet, H., Maghembe, R., Gameiro, D., Van-Thuc, D., Dishisha, T., & Hatti-Kaul, R. (2021). Metabolic potential of the moderate halophile *Yangia* sp. ND199 for co-production of polyhydroxyalkanoates and exopolysaccharides. *MicrobiologyOpen*, *10*(1), e1160.
79. Khomlaem, C., Aloui, H., Oh, W. G., & Kim, B. S. (2021). High cell density culture of *Paracoccus* sp. LL1 in membrane bioreactor for enhanced co-production of polyhydroxyalkanoates and astaxanthin. *International Journal of Biological Macromolecules*, *192*, 289–297.
80. Kumar, V., Darnal, S., Kumar, S., Kumar, S., & Singh, D. (2021). Bioprocess for co-production of polyhydroxybutyrate and violacein using Himalayan bacterium *Iodobacter* sp. PCH194. *Bioresource Technology*, *319*, 124235.
81. Arumugam, A., & Shereen, M. F. (2020). Bioconversion of *Calophyllum inophyllum* oilcake for intensification of rhamnolipid and polyhydroxyalkanoates co-production by *Enterobacter aerogenes*. *Bioresource Technology, 296*, 122321.

82. Muhammad, M., Aloui, H., Khomlaem, C., Hou, C. T., & Kim, B. S. (2020). Production of polyhydroxyalkanoates and carotenoids through cultivation of different bacterial strains using brown algae hydrolysate as a carbon source. *Biocatalysis and Agricultural Biotechnology*, *30*, 101852.
83. Ma, H., Zhao, Y., Huang, W., Zhang, L., Wu, F., Ye, J., & Chen, G. Q. (2020). Rational flux-tuning of Halomonas bluephagenesis for co-production of bioplastic PHB and ectoine. *Nature Communications*, *11*(1), 3313.
84. Kumar, P., & Kim, B. S. (2019). *Paracoccus* sp. strain LL1 as a single cell factory for the conversion of waste cooking oil to polyhydroxyalkanoates and carotenoids. *Applied Food Biotechnology*, *6*(1), 53–60.
85. Kumar, P., Jun, H. B., & Kim, B. S. (2018). Co-production of polyhydroxyalkanoates and carotenoids through bioconversion of glycerol by *Paracoccus* sp. strain LL1. *International Journal of Biological Macromolecules*, *107*, 2552–2558.
86. Tohme, S., Hacıosmanoğlu, G. G., Eroğlu, M. S., Kasavi, C., Genç, S., Can, Z. S., & Oner, E. T. (2018). *Halomonas smyrnensis* as a cell factory for co-production of PHB and levan. *International Journal of Biological Macromolecules, 118*, 1238–1246.
87. Raza, Z. A., Abid, S., & Banat, I. M. (2018). Polyhydroxyalkanoates: Characteristics, production, recent developments and applications. *International Biodeterioration & Biodegradation*, *126*, 45–56.
88. de Souza Reis, G. A., Michels, M. H., Fajardo, G. L., Lamot, I., & de Best, J. H. (2020). Optimization of green extraction and purification of PHA produced by mixed microbial cultures from sludge. *Water, 12*(4), 1185.
89. Koller, M. (2020). Established and advanced approaches for recovery of microbial polyhydroxyalkanoate (PHA) biopolyesters from surrounding microbial biomass. *EuroBiotech Journal*, *4*(3), 113–126.
90. Kacanski, M., Pucher, L., Peral, C., & Dietrich, T. (2022). Cell retention as a viable strategy for PHA production from diluted VFAs with *Bacillus megaterium*. *Bioengineering, 9*(3), 122.
91. Macagnan, K. L., Alves, M. I., & Moreira, A. D. S. (2019). Approaches for enhancing extraction of bacterial polyhydroxyalkanoates for industrial applications. *Biotechnological Applications of Polyhydroxyalkanoates*, *1*, 389–408.
92. Alfano, S., Lorini, L., Majone, M., Sciubba, F., Valentino, F., & Martinelli, A. (2021). Ethylic esters as green solvents for the extraction of intracellular polyhydroxyalkanoates produced by mixed microbial culture. *Polymers, 13*(16), 2789.
93. Pérez-Rivero, C., López-Gómez, J. P., & Roy, I. S. (2019). Sustainable approach for the downstream processing of bacterial polyhydroxyalkanoates: State-of-the-art and latest developments. *Biochemical Engineering Journal*, *150*, 107283.
94. Yabueng, N., & Napathorn, S. C. (2018). Toward non-toxic and simple recovery process of poly (3-hydroxybutyrate) using the green solvent 1,3-dioxolane. *Process Biochemistry, 69*, 197–207.
95. Samorì, C., Basaglia, M., Casella, S., Favaro, L., Galletti, P., Giorgini, L., & Tagliavini, E. (2015). Dimethyl carbonate and switchable anionic surfactants: Two effective tools for the extraction of polyhydroxyalkanoates from microbial biomass. *Green Chemistry*, *17*(2), 1047–1056.
96. Kurian, N. S., & Das, B. (2021). Comparative analysis of various extraction processes based on economy, eco-friendly, purity and recovery of polyhydroxyalkanoate: A review. *International Journal of Biological Macromolecules*, *183*, 1881–1890.

97. Yasim-Anuar, T. A. T., Norrahim, M. N. F., Sapuan, S. M., Ilyas, R. A., Jenol, M. A., Razali, N. A. M., & Najmuddin, S. U. F. S. (2021). Polyhydroxyalkanoates for packaging application. In S. M. Sapuan & R. A. Ilyas (Eds.), *Bio-Based Packaging: Material, Environmental and Economic Aspects* (pp. 279–293). Wiley.
98. Shao, L., Xi, Y., & Weng, Y. (2022). Recent advances in PLA-based antibacterial food packaging and its applications. *Molecules*, *27*(18), 5953.
99. Sabapathy, P. C., Devaraj, S., Meixner, K., Anburajan, P., Kathirvel, P., Ravikumar, Y., & Qi, X. (2020). Recent developments in polyhydroxyalkanoates (PHAs) production–A review. *Bioresource Technology*, *306*, 123132.
100. Dilkes-Hoffman, L. S., Lane, J. L., Grant, T., Pratt, S., Lant, P. A., & Laycock, B. (2018). Environmental impact of biodegradable food packaging when considering food waste. *Journal of Cleaner Production, 180*, 325–334.
101. Flores-Albino, B., Arias, L., Gómez, J., Castillo, A., Gimeno, M., & Shirai, K. (2012). Chitin and L (+)-lactic acid production from crab (*Callinectes bellicosus*) wastes by fermentation of *Lactobacillus* sp. B2 using sugar cane molasses as carbon source. *Bioprocess and Biosystems Engineering, 35*, 1193–1200.
102. Gnanasekaran, D. (2019). The use of biopolymers in food packaging. In D. Gnanasekaran (Ed.), *Green Biopolymers and Their Nanocomposites; Materials Horizons: From Nature to Nanomaterials* (pp. 137–158). SpringerLink Books.
103. Kuddus, M., & Roohi. (2021). Microbial production of bioplastics: Current trends and future perspectives. In M. Kuddus & Roohi (Eds.), *Bioplastics for Sustainable Development* (pp. 1–60). Springer, Singapore.
104. Sapuan, S. M., & Ilyas, R. A. (2021). Polyhydroxyalkanoates for packaging application. In S. M. Sapuan & R. A. Ilyas (Eds.), *Bio-Based Packaging: Material, Environmental and Economic Aspects* (pp. 279–293). Wiley, New York.

CHAPTER 5

Proposal of a Biorefinery Design for the Utilization of Pineapple Stubble Using Aspen Plus

RODOLFO ANTONIO HERNÁNDEZ-CHAVERRI* and
DANIEL DÍAZ-RODRÍGUEZ

Bioprocess Engineering Laboratory, School of Exact and Natural Sciences, State Distance University, San José, Costa Rica

*Corresponding author.

ABSTRACT

The lignocellulosic biomass from pineapple stubble has several characteristics that make it an optimal raw material for the development of a biorefinery process. Therefore, a biorefinery process was developed using Aspen Plus V12 software, in which the milling process and its subsequent transformation into different products and by-products were simulated. For this purpose, a process flow diagram was elaborated, based on the saccharification of pineapple stubble. The different chemical compounds required by the chemical routes and physical processes of the equipment were entered into the software, and the "NRTL" thermodynamic model was selected as the thermodynamic model with the help of the software's tool, and the biorefinery process was designed. In addition, an investment analysis was conducted for the proposed process, where direct and indirect costs were incorporated, and the required investment capital was obtained. The results obtained indicate that when entering the simulator with a flow of 100 kg/h of wet biomass, a total of 3.27 kg/h of ethanol, 3.28 kg/h of activated carbon, and 8.29 kg/h of DAP, among others, are obtained. Similarly, the investment analysis shows direct costs associated with the process of $2,798,424, indirect costs of $6,99,606, and a capital investment of $4,309,573. Finally, it is concluded

Food Security, Safety, and Sustainability: Validated Innovative Approaches and Technological Advancements.
José Juan Buenrostro-Figueroa and Julio César Tafolla-Arellano (Eds.)

that the pineapple stubble biorefinery process presents various routes of use. Additionally, it is recommended that a more robust study be conducted on the income obtained from the possible commercialization of the products obtained from the stubble and the cost associated with production.

5.1 INTRODUCTION

Biorefineries play a key role in the circular bioeconomy. A biorefinery is a facility in which various conversion methods (e.g., biological, chemical, mechanical, and thermal) are used to process renewable biomass resources into multiple products, thus maximizing the value of biomass [1, 2].

To be truly sustainable, biorefineries must fully utilize biomass by reusing the residues and by-products generated by the conversion processes [3].

Biomass-based lignocellulosic waste is generated from agro-industrial process wastes and has gained interest from the biorefinery perspective [4]. The biorefinery concept is similar to that of a traditional petroleum refinery in which a substrate is fully utilized to obtain biofuels, bio-based chemicals, and other products derived from biomass [5, 6]. Due to its abundance, biomass is the primary lignocellulosic feedstock for sustainable bioeconomy. In addition to established uses such as co-firing with coal and biofuel production, there are also new uses and process applications [7]. Solid wastes have a composition rich in carbohydrates, lipids, proteins, minerals, and bioactive compounds, enabling their application in a wide range of bioprocesses, from the production of biofuels to the acquisition of commercially relevant biochemical compounds [8].

Disposing of waste from agro-industrial processes requires a significant capital investment, with transportation costs being the primary expense. Landfilling is one of the most frequently used methods of waste disposal; however, this method has issues, such as greenhouse gas emissions, as does leaving the waste in the field for decomposition [9]. Developing countries such as ours rarely have the capacity to treat the waste generated, and most of the time, waste is dumped into open dumps, causing severe environmental damage. Incineration is not a viable option for agro-industrial waste due to its high moisture content. Other problems associated with incineration include the formation and release of pollutants such as dioxins, furans, acid gases, and particulates [1, 7, 9].

The agro-industrial exploitation of pineapple (*Ananas comosus*) is no exception in the treatment of residues. Once harvested, the plant remains in the field (stubble) for mechanical integration into the soil, and fruit residues such as crowns, hulls, discarded fruit, and residues from industrial processing of the game are generally taken to open dumps or used in some cases as animal feed fodder. But pineapple plant residues contain relatively high levels of cellulose, hemicellulose, and soluble mono sugars, representing a potential feedstock for microbial conversion to produce fuels (ethanol, butanol) and value-added products (proteins, organic acids, and enzymes). Pineapple residues have been used to produce biogas and fertilizers, bromelain protease, fungal growth media, and fiber substrates, among others.

The objective of this research is to define a basic biorefinery process for the use of pineapple stubble as a lignocellulosic feedstock using ASPEN PLUS V12 simulation software as a tool.

5.2 METHODOLOGY

5.2.1 ASPEN PLUS V12

Aspen Plus V.12 was used to model a pilot biorefinery process using pineapple stubble lignocellulosic biomass as a feedstock source and to predict the composition of the output streams. Each process operation is represented in the block flow diagram, where material and energy flows are specified. This software includes an extensive database of chemical component properties that are used in the calculations; it also can incorporate custom-made Fortran or Excel subroutines when needed [10]. The flowchart of the simulation procedure used by the Aspen Plus® software is described in Figures 5.1 and 5.2. For the simulation performed, the "Non-Random Two Liquid Model" (NTRL) thermodynamic model was selected using the "Aspen Plus V12 methods wizard."

5.2.2 BIOMASS CHARACTERISTIC

Biomass and ash are considered non-conventional components. They are defined as attributes of the components in Aspen Plus called PROXANAL, ULTANAL, and SULFANAL using proximate and ultimate analysis [10]. Table 5.1 shows the characteristics of the proximate analysis [11] (moisture, fixed carbon, ash, volatiles), and particle size distribution. In the same table, the ULTANAL and SULFANAL biomass composition (ash, carbon, hydrogen, oxygen, nitrogen, and sulfur) can be found, as well as biomass characterization such as the percentage of acetates, cellulose, hemicellulose, among other components present in the dry biomass fraction.

TABLE 5.1 Biomass Characterization of Pineapple Stubble

Feature	Value	References
Average size	(900–1,000) mm	[12–14]
Humidity	85.58% m/m	
Fixed coal	14.42% m/m	
Ashes	11.36% m/m	
Volatiles	74.22% m/m	
Carbon	37.71% m/m	[12–14]
Hydrogen	4.97% m/m	
Nitrogen	1.13% m/m	
Oxygen	44.58% m/m	
Sulfur	0.25% m/m	
Chloride	0.00% m/m	
Humidity	85.58% m/m	[12]
Dry matter	14.42% m/m	
Extracts	2.12% m/m	[6, 14, 15]
Lignin	0.58% m/m	
Sucralose	0.12% m/m	
Acetate	0.26% m/m	
Ash	1.77% m/m	[6, 14]
Cellulose	4.99% m/m	
Hemicellulose	2.92% m/m	
Protein	1.66% m/m	

5.2.3 SIMULATION PROCESS

5.2.3.1 SIMULATION PROCESS ASSUMPTIONS

In the simulation, an operating time of 24 hours per 330 working days (7,920 h) was established. A mass flow of 100 kg/g of wet biomass entering the system was taken as a basis. For the cellulose and hemicellulose, Davis et al. [15] propose to take cellulose as glucan since it's composed of β-glucose chains, and hemicellulose as Xylan since it is made up of xylose units. Other considerations for the systems are shown in the following list [10, 16].

The following temperature conditions are considered in the milling process: biomass enters the knife mill at 25C (or ambient temperature), enters hammer mill 1 at 30°C, and enters hammer mill 2 at 35°C.

1. The process takes place under steady-state conditions.

2. The ash is inert without catalytic activity and does not play a role during the gasification process (activated carbon production).
3. The charcoal composition consists only of carbon (organic solid) and ash (inorganic solid).
4. All reactions reach equilibrium conditions, and kinetics are not considered.
5. Drying and pyrolysis take place simultaneously, and the volatile products consist of H_2, CO, CO_2, CH_4, and H_2O.
6. Tar formation is not considered during the process.
7. The ammonium sulfate stream is obtained in a liquid state and the separation process is not indicated.
8. The simulation uses cellulase enzyme as the enzyme responsible for the hydrolysis process and *Zymomonas mobilis* as the ethanologenic bacteria.
9. Recirculation streams are not considered in the process.
10. The simulation process was divided into two parts: the grinding process, and the pretreatment process, hydrolysis, fermentation, activated carbon production, and other products.

5.2.3.2 SIMULATION OF THE GRINDING PROCESS

The simulation for the milling process consists of a preliminary step that involves defining the type of biomass used in the Aspen Plus V12 software. The characteristics from Table 5.1 were utilized for this purpose. Subsequently, the entry conditions into the system were established and are detailed in Table 5.2. For the grinding process, a combination of three pieces of equipment in the flow sequence was proposed, as illustrated in Figure 5.1. This equipment was labeled as "EQ1," "EQ2," and "EQ3" to achieve a particle size close to 2 mm [15].

EQ1 corresponds to a knife mill for the initial size reduction, while EQ2 and EQ3 are hammer mills chosen for their high size reduction rate, easy adjustment of the particle size range, and ability to maintain a constant cubic shape of the particle [17, 18]. Additionally, the American Society of Agricultural and Biological Engineers (ASABE) has conducted various studies recommending the use of a knife mill followed by a hammer mill to achieve an average particle size close to 2 mm [19].

TABLE 5.2 Working Conditions for Grinding Simulation in Aspen Plus V12

FEATURE	VALUE
Temperature	25°C
Pressure	1 atm
Mass flow	100 kg/h
Annual operating time	7,920 h
Initial biomass size	9,000 mm
Target value of ground biomass	2 mm

Figure 5.1 shows a macro-scale of the milling proposal where the wet pineapple stubble biomass enters, is milled and passes through the following process. The desired particle size is 2 mm so that the following processes are as efficient as possible [15]. The simulation output is presented in Figure 5.3, where the proposed mill is the integration of three pieces of equipment; these are called "EQ1," "EQ2" and "EQ3." EQ1 corresponds to a knife mill for an initial size reduction; EQ2 and EQ3 to hammer mills, as these provide a high size reduction rate, easy adjustment of the particle size range, and maintain a constant cubic shape of the particles [17, 18]. In addition, the ASABE has conducted several studies in which it recommends the use of knife mills and later hammer mills to obtain an average particle size of approximately 2 mm [19].

For the grinding simulation, the US Bureau of Mines model was selected, since only the maximum outlet diameter was required to be specified. For the stress number representing the contacts with the media, the probability of the particle being trapped and stressed by the contacting media, and the number of feed particles [20], a stress number of 2 was selected, which is also related to the feed flow rate and the final particle size. Additionally, the Austin model is a kinetic model that relates to the size reduction process [19]. The conditions for each of these equipment is presented in Table 5.3.

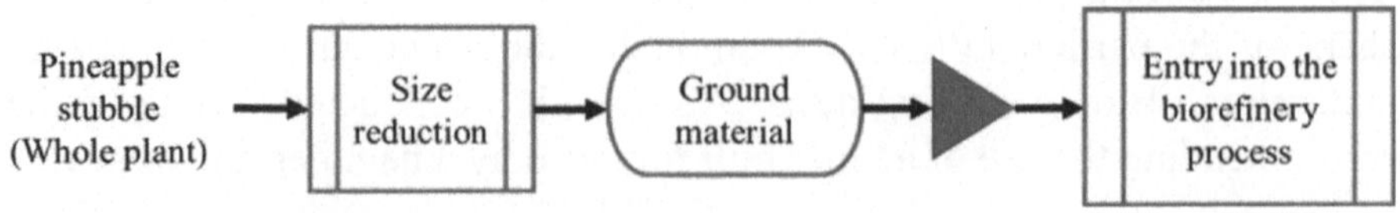

FIGURE 5.1 Block diagram for pineapple stubble grinding.

TABLE 5.3 Operating Conditions for the Knife Mill and Hammer Mills Used in the Grinding Process Simulation

Parameter	Blade Mill/(EQ1)	Hammer Mill/ (EQ2)	Hammer Mill/ (EQ3)	References
Type of mill	Rotating	Hammer	Hammer	–
Function	US Bureau of Mines	Austin	Austin	–
Maximum particle diameter	20 mm	N/A	N/A	19
Operating model	Primary	N/A	N/A	–
The ratio of cut size and outlet diameter	1.7	N/A	N/A	Aspen Plus V12
Number of stresses	None	2	2	–
S1, d1 y a	N/A	1, 1.18 mm y 1	1, 1.18 mm y 1	22 y Aspen Plus V12
γ, β y φ	N/A	0.7, 0.85 y 3.35	0.7, 0.85 y 3.35	–

5.2.3.3 SIMULATION OF THE PROCESS FOR OBTAINING MAIN PRODUCTS

The block diagram in Figure 5.2 shows the simulation of the pretreatment process, hydrolysis, fermentation, production of activated carbon, and other products. The Aspen Plus simulation is presented in detail in Figure 5.4.

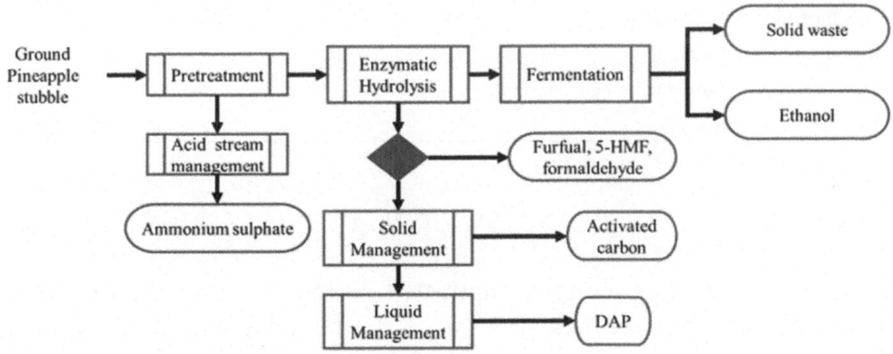

FIGURE 5.2 Block diagram for the process of pretreatment, hydrolysis, fermentation, production of activated carbon, and other products from pineapple stubble.

Table 5.4 shows the input streams to the simulation process for the production of primary and secondary products on an annualized basis. Thus, stream (1) is the flow of ground stubble biomass; (2) sulfuric acid of concentration 0.5 M; (4) water vapor; (8) cooling water; (14) gaseous ammonia for ammonium sulfate production; (18) enzymes; (23) 30% v/v phosphoric acid solution; (26) activated carbon wash water; (32) nutrients for fermentation and *Z. mobilis*; and (35) gaseous ammonia for the di-ammonium phosphate (DAP) production process.

TABLE 5.4 Conditions and Composition of Process Input Streams for Primary and Secondary Product Simulation

Feature	Units	1	2	4	8	14	18	23	26	32	35
Temperature	°C	40	25	300	25	100	25	25	25	25	25
Pressure	Atm	1	1	5	1	1	1	1	1	1	1
Mass flow	ton/year	792.0	4048.1	4950	198	277.2	8.7	105.9	118.8	41.6	792
Water	ton/year	677.8	3853.9	4950	198	198.0	7.9	57.2	118.8	39.6	594
Extracts	ton/year	16.8	–	–	–	–	–	–	–	–	–
Lignin	ton/year	4.6	–	–	–	–	–	–	–	–	–
Sucrose	ton/year	0.9	–	–	–	–	–	–	–	–	–
Acetate	ton/year	2.1	–	–	–	–	–	–	–	–	–
Ashes	ton/year	14.0	–	–	–	–	–	–	–	–	–
Cellulose	ton/year	39.5	–	–	–	–	–	–	–	–	–
Hemicellulose	ton/year	23.2	–	–	–	–	–	–	–	–	–
Protein	ton/year	13.2	–	–	–	–	–	–	–	0.99	–
Sulfuric acid	ton/year	–	194.2	–	–	–	–	–	–	–	–
Phosphoric acid	ton/year	–	–	–	–	–	–	48.7	–	–	–
Ammonia	ton/year	–	–	–	–	79.2	–	–	–	–	198
Enzyme	ton/year	–	–	–	–	–	0.78	–	–	–	–
DAP	ton/year	–	–	–	–	–	–	–	–	0.99	–

5.2.3.4 REACTIONS IMPLICATED DURING THE SIMULATION PROCESS

Table 5.5 shows the reactions implicated during the simulation process. It includes the pretreatment process of the biomass with sulfuric acid, the reactions for the enzymatic hydrolysis process using cellulases, the reactions for the fermentation process with *Z. mobilis* as a yeast, as well as the expected conversion percentage to the product. This information is entered into the Aspen software to carry out the simulation of the respective reactor [23].

TABLE 5.5 Conversions Associated with Pretreatment Reactions [15, 22]

Reaction for Pretreatment Process Using Sulfuric Acid	Reactant	Conversion to Product (%)
$(Glucan)_n + n*H_2O \rightarrow n*Glucose$	Glucan	9.9
$(Glucan)_n \rightarrow 2n*H_2O + n*HMF$	Glucan	0.3
$Sucrose \rightarrow Glucosa + 2H_2O + HMF$	Sucrose	100
$(Xilan)_n + n*H_2O \rightarrow n*Xylose$	Xylan	90.0
$(Xilan)_n \rightarrow n*Furfural\, n + 2n*H_2O$	Xylan	5.0
$(Lignin)_n \rightarrow n*soluble\, lignin$	Lignin	5.0
$(Glucan)_n + 2n*H_2O \rightarrow n*Glucose\, Oligomer$	Glucan	0.3
$(Xilano)_n + n*H_2O \rightarrow n*Xilose\, Oligomer$	Xylan	2.4
Reaction enzymatic hydrolysis	Reactant	Conversion to product (%)
$(Glucan)_n + 2n*H_2O \rightarrow n*Glucose\, Oligomer$	Glucan	4.0
$(Glucan)_n + \frac{1}{2}n*H_2O \rightarrow \frac{1}{2}n*Celullose$	Glucan	1.2
$(Glucan)_n + n*H_2O \rightarrow n*Glucose$	Sucrose	90
$Cellulose + H_2O \rightarrow 2*Glucose$	Xylan	100
Reaction fermentation process	Reactant	Conversion to product (%)
$Glucose \rightarrow 2*Ethanol + 2*CO_2$	Glucose	90.0
$Glucose + 2n*H_2O \rightarrow 2*Glycerol + O_2$	Glucose	0.4
$Glucose + 2*CO_2 \rightarrow 2*Succinic\ Acid + O_2$	Glucose	0.6
$3*Xylose + 5*H_2O \rightarrow 5*Glycerol + 2.5*O_2$	Xylose	0.3
$3*Xylose + 5*CO_2 \rightarrow 5*Succinic\ Acid + 2.5*O_2$	Xylose	0.9
$3*Xylose + 5*CO_2 \rightarrow 5*Succinic\ Acid + 2.5*O_2$	Xylose	80.0
Inhibition Reactions (Contamination)		
$3*Xylose \rightarrow 5*Ethanol + 5*CO_2$	Glucose	3.0
$3\ Xiloses \rightarrow 5*Lactic\ acid$	Xylose	3.0

5.2.4 ECONOMIC ESTIMATE

A preliminary economic estimate was made for the proposed process for a pineapple stubble mass flow rate of 100 kg/h. This includes the necessary investment of the simulated biorefinery process: directly related costs such as equipment, installation, instrumentation, pipes and piping, electrical equipment, insulation (paints or insulators), civil works, utility installations, land improvements, and land costs. In addition, indirect costs are included, which correspond to engineering and supervision, legal expenses, construction costs, contractor's fees, and eventual [23–25].

5.3 RESULTS

5.3.1 GRINDING SIMULATION RESULTS

Figure 5.3 shows the output diagram of the simulation of the milling process in Aspen Plus C12. Table 5.6 summarizes the results of the integration of the three pieces of equipment in the milling process to obtain the highest percentage of material with a diameter less than or equal to 5 mm depending on the inlet or outlet stream [15, 16, 26].

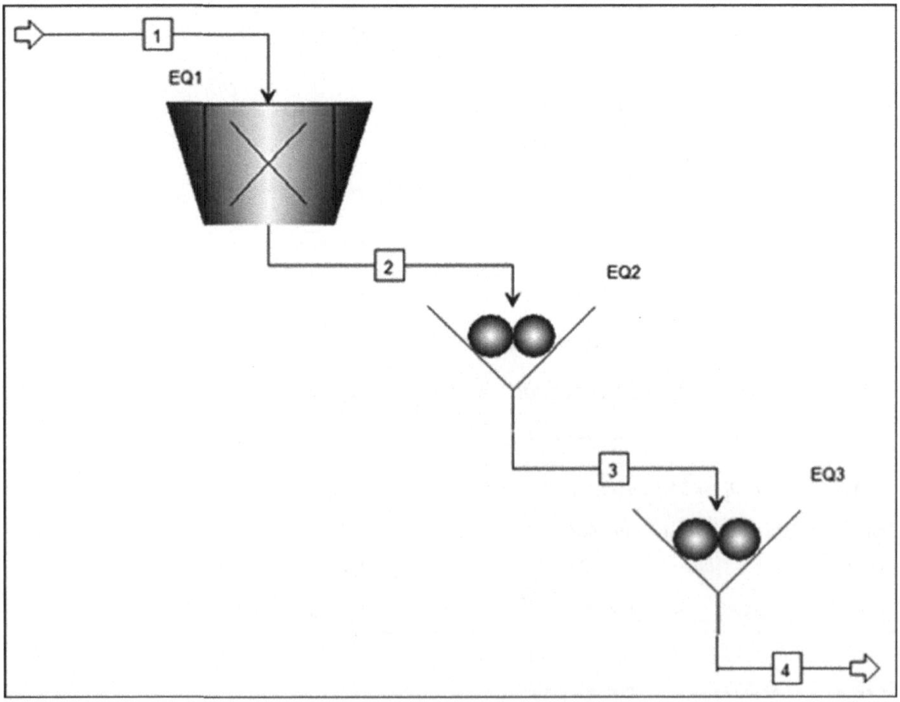

FIGURE 5.3 Process flow diagram for pineapple stubble milling. Output from ASPENV12 program.

The average diameter of the biomass obtained after the knife mill is between 0 mm to 35 mm. Subsequently, hammer mills 1 and 2 grind the biomass to a fraction of 0.93 on the 0 mm to 5 mm sieve. This indicates that the milling operation for the biorefinery process requires a minimum of three mills to achieve the desired diameter for the production process of the different products and by-products [3, 15]. In addition, a 5°C increase in temperature is assumed due to the friction process of grinding in each equipment [3].

TABLE 5.6 Percent Particle Size Fraction Obtained for the Inlet and Outlet Streams of the Milling Process in Aspen Plus V12

Parameter	Stream 1/(1)	Stream 2/(2)	Stream 3/(3)	Stream 4/(4)
Temperature/(°C)	25	30	35	40
Pressure/(atm)	1	1	1	1
Mass flow/(kg/h)	100	100	100	100
		Particle Size Distribution		
0–5/(mm)	0	0.37	0.68	0.93
5–10/(mm)	0	0.16	0.19	0.06
10–15/(mm)	0	0.14	0.08	0.009
15–20/(mm)	0	0.12	0.03	0.001
20–25/(mm)	0	0.11	0.009	4.37e-05
25–30/(mm)	0	0.10	0.0005	0
30–35/(mm)	0	0.010	0	0
900–950/(mm)	0.5	0	0	0
950–1,000/(mm)	0.5	0	0	0

Note: In the particle size distribution, only those with a percentage of solid present in them are included.

In the case of the equipment, the values described in Table 5.7, where the characteristics of the milling process are included, were obtained. The Aspen Plus V12 software presents the results of the Sauter diameter, which represents, through a volume-surface radius relationship, the average diameter of the particles [27]. The Sauter diameter obtained after the knife mill is 5.11 mm, for the case of hammer mill 1 it is 3.24 mm, and for hammer mill 2 it is 2.64 mm. Based on the above, it is observed that after the hammer milling process, an average diameter of 2.62 mm for the 100 kg/h of biomass is obtained, very close to the 2 mm recommended by Davis et al. [15]. Another mill is not included because it is considered that the process would become more expensive with one more milling unit, and the distribution obtained is the one that is entered into Aspen Plus V12 for the following simulation process, which will proceed to the process of obtaining products and by-products.

TABLE 5.7 Results Obtained for the Equipment Used in the Simulation of the Milling Process in Aspen Plus V12

Parameter	Blade Mill/(EQ1)	Hammer Mill/(EQ2)	Hammer Mill/(EQ3)
Particle diameter greater than 80% at inlet.	0.99 m	20.96 mm	7.97 mm
Particle diameter greater than 80% of output.	20.96 mm	7.97 mm	4.28 mm
Particle diameter greater than 50% at inlet.	0.99 m	9.15 mm	3.65 mm
Particle diameter greater than 50% of output.	9.15 mm	3.65 mm	2.67 mm
Average Sauter input distribution.	0.99 m	5.11 mm	3.24 mm
Average Sauter output distribution.	5.11 mm	3.24 mm	2.62 mm

5.3.2 RESULTS OF THE SIMULATION OF THE PRODUCT-OBTAINING PROCESS

From the high-level block diagram (Figure 5.2), it is transformed into the process flow diagram in the simulation, as shown in Figure 5.4. This allows a visual division of the process into three parts: the first one corresponding to the biomass pretreatment process (equipments 1, 2, 3, 4, 5, and 6), the second to the hydrolysis process, biomass treatment, and ammonium sulfate formation (equipment 7, 9, 10, and 12), and finally the production process of activated carbon, DAP, and alcoholic fermentation (equipments 11, 13, 14, 15, 16, 17, and 18).

The process enters a separation equipment (equipment 3) as shown in Figure 5.4. This equipment exemplifies a pressure relief of the system with the purpose of releasing the steam that previously entered the pretreatment process. After the steam is released, the hot biomass enters a mixing tank (equipment 4), where a water flow corresponding to the stream (8) is incorporated, and its conditions are specified in Table 5.5. This process allows for lowering the temperature of the biomass in the filter (equipment 5). The filter allows dividing the solid section in the stream (10) by the pre-treated material (11).

The last equipment of this process is a FLASH (equipment 7) where the pretreated material enters and is separated into the stream (12), which corresponds to the by-products of furfural, hydroxymethyl furfural (HMF), and extracts (formaldehyde); and stream (13) the solid pretreated material with a low concentration of contaminants (inhibitors) of the enzymatic hydrolysis process such as furfural and HMF. The parameters used for the equipment

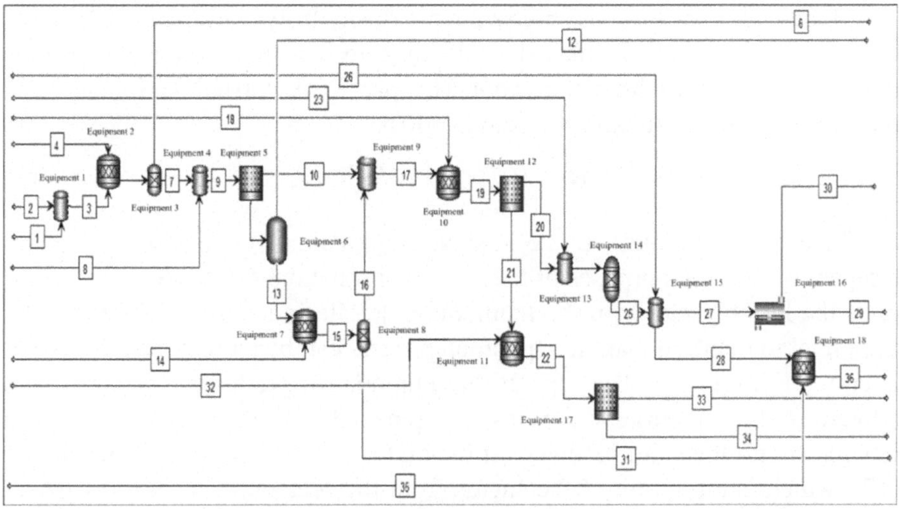

FIGURE 5.4 Process flow diagram for obtaining primary and secondary products.

specifications were based on the studies of the ethanol production process and similar to the National Renewable Energy Laboratory of the U.S. Department of Energy [15, 22, 28]. These conditions are summarized in Table 5.8.

TABLE 5.8 Conditions Employed in the Equipment Used for Biomass Pretreatment, for the Simulation Process [15, 22, 28]

Parameter	Units	Equipment Number				
		1	2	3	4	5
Equipment		Mixing Tank	Reactor	Separator	Mixing Tank	Filter
Temperature defined	°C	27.33	121	–	90	–
Defined pressure	atm	1.00	2	–	1	–
Caloric work obtained	J/s	–	91458.28	52250.07	–1,62,414	–
Fraction of liquid in the liquid stream	–	–	–	–	–	0.98
The solid fraction in the solid stream	–	–	–	–	–	0.98
Separated vapor fraction	–	–	–	1	–	–

Once the pretreatment stage of the material is completed, the process stream (12) is obtained from the process. This stream corresponds to the first process byproduct, where the following amounts of components are obtained per year: 181 kg/year of HMF, 81.9 kg/year of furfural, and 16.45 tons/year of formaldehyde (the process of separation of these products is not included in the simulation, the potential for obtaining them is proposed).

At the end of the biomass pretreatment, the hydrolysate stream (13) enters a reactor (equipment 7) where the sulfuric acid is neutralized using gaseous ammonia entrained with water vapor. This process is carried out by means of the following reaction, and its yield is 100%:

$$2*NH_3 + H_2SO_4 \rightarrow (NH_4)_2\,SO_4 \quad (1)$$

Once the ammonium sulfate in solution (stream 15) is obtained, it enters a separator. For the purposes of the current simulation, the process only separates the hydrolysate from the ammonium sulfate in solution; however, it is important to mention that the separation process can be more robust than what is currently proposed. The separator (equipment 8) produces a sugary liquid (stream 16) that enters a mixing tank (equipment 9), which mixes the liquid rich in sugars with the solid biomass. At the outlet of the equipment is the stream (17), which is a mixture of the sugars from the pretreatment, most of which are xylose monomers, and the remaining solid biomass. This stream enters a reactor (equipment 10) where enzymatic hydrolysis takes place through the addition of enzymes corresponding to cellulases [23] (stream 18) [2].

The by-product obtained from this stage of the process is the formation of ammonium sulfate in solution, for an amount of 256.02 tons per year, which is used as fertilizer. The conditions used for the simulation of the above process are specified in Table 5.9.

At the end of the second stage, the hydrolysate stream (19) is separated from the remaining solid material (stream 20) by means of a filter (equipment 12), which separates the hydrolysate (stream 21) from the remaining solid material (stream 20).

TABLE 5.9 Conditions Employed in the Equipment Used for Stage 2 of the Pineapple Stubble Biorefinery Process

Parameter	Units	Equipment Number				
		7	8	9	10	12
Equipment		Reactor	Separator	Mixing Tank	Reactor	Filter
Temperature defined	°C	100	–	73.	48	–
Defined pressure	atm	1.00	–	1	1	–
Caloric work obtained	J/s	–18401.3	–11267.56	–	–1126.83	–
Fraction of liquid in the liquid stream	–	–	–	–	–	0.98
The solid fraction in the solid stream	–	–	–	–	–	0.98
Separated ammonium sulfate fraction	–	–	–	1	–	–

Once the previous stage is completed, the third stage of the process, which corresponds to the formation of the main products, is carried out. This begins with the entry of the hydrolysate (stream 21) into a fermentation reactor, which simultaneously receives a stream of nutrients necessary for the activation of the ethanologenous bacterium *Z. mobilis*. From the fermentation reactor, it obtains an output stream (22) which enters a filter (equipment 17), which separates the solids from the fermentation process (33) from the fermented liquid (34). Both streams correspond to product streams from the pineapple stubble biorefinery process. The ethanol obtained from the process corresponds to 25.86 tons/year; glycerol (0.22 tons/year) and succinic acid (0.54 tons/year) are also obtained from the process. The conditions for the equipment used in this stage are summarized in Table 5.10.

TABLE 5.10 Conditions Employed in the Equipment Used for Fermentation and Fermentate Separation

Parameter	Units	Equipment Number	
		11	17
Equipment		Reactor	Filter
Temperature defined	°C	32	–
Defined pressure	atm	1	–
Caloric work obtained	J/s	–1328.3	–
Fraction of liquid in the liquid stream	–	–	0.98
The solid fraction in the solid stream	–	–	0.98

The current (20), which corresponds to the biomass remaining from the enzymatic hydrolysis process, is introduced into a mixing tank (equipment 13) where a 30% v/v solution of phosphoric acid is added, with the objective of chemically activating the lignocellulosic material to produce activated carbon. Phosphoric acid is used since it is environmentally better than other activating agents such as zinc chloride, in addition to being less costly [29, 30]. As an output of the previous process, stream (24) is obtained, which enters a muffle (equipment 14) to carry out the carbonization of the lignocellulosic material and obtain activated carbon (stream 25). Once the activated carbon is obtained, it requires a series of washes to remove the remaining phosphoric acid, so it enters a mixing tank (equipment 15) where the washes are carried out and the streams (27) and (28) are obtained. Stream (27) corresponds to the washed activated carbon obtained, which then enters a drying oven (equipment 16), from which production of 25.96 tons/year is obtained. On the other hand, stream (28) contains the phosphoric acid and the water from the washings; this stream enters a reactor (equipment 18) where ammonia entrained with water vapor (stream 35) is simultaneously introduced to form di-ammonium phosphate (DAP) (stream 36). The chemical reaction involved in the DAP formation process is exemplified below:

$$H_3PO_4 + 2 * NH_3 \rightarrow (NH_4)_2HPO_4 \quad (2)$$

The conditions of the equipment used in this third stage are summarized in Table 5.11.

TABLE 5.11 Conditions Employed in the Equipment Used for Fermentation and Fermentation Separation

Parameter	Units	Equipment Number				
		13	14	15	16	18
Equipment		Mixing Tank	Mufla	Washing Tank	Drying Oven	Reactor
Temperature defined	°C	29.29	600	102.5	105	25
Defined pressure	atm	1	1	1	1	1
Caloric work obtained	J/s	–	14945.9	–	22.2093	20048.7

The simulated biorefinery process for pineapple stubble has the following by-products and products as outputs which are summarized in Table 5.12.

5.3.3 ESTIMATED INVESTMENT FOR BIOREFINERY PILOT PROCESS

The results obtained for the estimation of the capital investment for the pineapple stubble biorefinery pilot process are presented below. Table 5.13 shows the direct costs associated with the capital investment such as equipment, installation, instrumentation, controls, piping, and electrical equipment, among others. Each determination is a percentage of the equipment investment [23–25].

Indirect costs are obtained from the total direct costs, for which percentage estimates are made and presented in Table 5.14 [24–26].

The initial working capital was estimated at 10% of the fixed capital investment, which corresponds to the sum of direct and indirect costs, for the biorefinery pilot project using 100 kg/h of wet pineapple stubble, without considering the distillation processes. It is approximately $4,309,573 and is shown in Table 5.15.

TABLE 5.12 Product Streams from the Pineapple Stubble Biorefinery Simulation Process

Feature	Units	Process Streams							
		6	12	29	30	31	33	34	36
Temperature	C	100.3	130	105	105	100	32.0	32.0	25.0
Pressure	atm	1	0.5	1	1	1	1	1	1
Mass flow	ton/year	4989.6	4635.1	26.0	130.1	279.4	359.1	8.7	904.3
Water	ton/year	4989.6	4585.7	–	119.1	39.6	292.0	6.0	656.2
Sulfuric acid	ton/year	–	29.7	–	1.57E-05	–	3.7	7.61E-02	0.08
Glucose	ton/year	–	8.08E-04	–	3.64E-08	–	2.3	4.73E-02	8.04E-01
Cellulose	ton/year	–	–	–	3.21E-13	–	4.30E-01	8.78E-03	8.96E-03
Xylose	ton/year	–	3.0	–	7.53E-05	–	2.22E+00	4.54E-02	4.13E-01
HMF	ton/year	–	1.81E-01	–	5.45E-06	–	2.29E-01	4.67E-03	–
Furfural	ton/year	–	8.19E-02	–	3.92E-05	–	2.30E-03	4.69E-05	–
Acetate	ton/year	–	–	–	–	–	8.22E-04	4.03E-02	–
Ash	ton/year	–	–	–	–	–	5.61E-03	2.75E-01	–
Glucose oligomer	ton/year	–	1.98E-05	–	1.53E-09	–	1.6	3.31E-02	3.38E-02
Extracts	ton/year	–	16.5	–	6.59E-03	1.44E-03	3.23E-01	5.88E-03	1.22E-04
Carbon	ton/year	–	–	26.0	–	–	–	–	–
Carbon dioxide	ton/year	–	–	–	5.5	–	24.8	2.04E-01	1.17E-02
Carbon monoxide	ton/year	–	–	–	5.5	–	–	–	–
Ethanol	ton/year	–	–	–	–	–	25.9	0.5	–
Glycerol	ton/year	–	–	–	–	–	0.2	4.46E-03	–
Succinic acid	ton/year	–	–	–	–	–	0.5	1.10E-02	–
Oxygen	ton/year	–	–	–	–	–	2.12E-01	2.47E-04	–
Ammonia	ton/year	–	–	–	–	23.4	–	–	–
Enzyme	ton/year	–	–	–	4.53E-18	–	0.8	1.53E-02	1.56E-02
Protein	ton/year	–	–	–	–	–	2.51E-02	1.23	–
Ammonia sulfate	ton/year	–	–	–	–	216.4	–	–	–
DAP	ton/year	–	–	–	–	–	–	–	65.7

TABLE 5.13 Direct Costs Associated with Capital Investment for Pineapple Stubble Harvesting

Parameter	Value
Equipment acquired	$511,757
Installation (painting and insulation)	$179,115
Instrumentation and controls installed	$204,703
Piping and piping installed	$255,878
Electrical equipment installed	$204,703
Works	$307,054
Services and improvements	$409,406
Terrain	$725,806
Total	$ 2,798,424

Note: The cost of the land is based on the mapping of homogeneous zones of the Ministry of Finance of Costa Rica and has an extension of 3,000 m^2 for the North zone of Costa Rica.

TABLE 5.14 Indirect Costs Associated with Capital Investment for Pineapple Stubble Harvesting

Parameter	Value
Engineering and supervision	$2,79,842
Legal expenses	$69,960
Construction costs and contractor	$4,19,763
Eventual	$3,49,803
Total	$6,99,606

TABLE 5.15 Capital Investment Required to Implement the Pineapple Stubble Harvesting Process

Parameters	Value
Fixed capital investment	$3,917,794
Working capital	$3,91,779
Capital investment	$4,309,573

5.4 CONCLUSION

The simulation with Aspen Plus V12 shows that it is possible to obtain a series of products of commercial interest from pineapple stubble such as ethanol, furfural, HMF, formaldehyde, activated carbon, and minimize waste streams by producing ammonium sulfate and di-ammonium phosphate (DAP) as by-products that can be integrated into the cultivation process as fertilizers. In addition, the use of Aspen Plus V12 allows continuing to add processes or alternatives to generate new products, but it is essential to have information on the characterization of the biomass under study, as in this case, the pineapple stubble. Subsequently, these results can be compared with laboratory-scale results and progressively scaled up. It is clear that in order to advance in the integration of biorefinery processes at a national level, it is necessary to make an economic investment to have a pilot biorefinery model, to which, through investment, operational modules can be added to obtain new value-added products from lignocellulosic biomass, in this particular case, pineapple stubble. In addition, it is recommended to incorporate a more robust study on the income obtained from the possible commercialization of the products obtained from the stubble and the cost associated with production.

ACKNOWLEDGMENTS

AMI-UNED Institutional Improvement Program and to the Institutional Scholarship Council for the allocation of scholarships for doctoral studies, to the UNED Vice Rector's Office for Research for continued support.

KEYWORDS

- **activated carbon**
- **ammonium sulfate**
- **Aspen Plus V12**
- **biomass**
- **biorefinery model**
- **formaldehyde**
- **operational modules**

REFERENCES

1. Tursi, A. (2019). A Review on biomass: Importance, chemistry, classification, and conversion. *Biofuel Research Journal, 6*(2), 962–979. https://doi.org/10.18331/BRJ2019.6.2.3.
2. Ubando, A. T., Felix, C. B., & Chen, W. H. (2020). Biorefineries in circular bioeconomy: A comprehensive review. *Bioresource Technology, 299*, 122585. https://doi.org/10.1016/j.biortech.2019.122585.

3. McCance, K. R., Suarez, A., McAlexander, S. L., Davis, G., Blanchard, M. R., & Venditti, R. A. (2021). Modeling a biorefinery: Converting pineapple waste to bioproducts and biofuel. *J. Chem. Educ.*, *98*(6), 2047–2054. https://doi.org/10.1021/acs.jchemed.1c00020.
4. Banerjee, S., Patti, A. F., Ranganathan, V., & Arora, A. (2019). Hemicellulose based biorefinery from pineapple peel waste: Xylan extraction and its conversion into xylooligosaccharides. *Food and Bioproducts Processing, 117*, 38–50. https://doi.org/10.1016/j.fbp.2019.06.012.
5. Cherubini, F. (2010). The biorefinery concept: Using biomass instead of oil for producing energy and chemicals. *Energy Conversion and Management, 51*(7), 1412–1421. https://doi.org/10.1016/j.enconman.2010.01.015.
6. Hernández-Chaverri, R. A., Buenrostro-Figueroa, J. J., & Prado-Barragán, L. A. (2021). Biomass: Biorefinery as a model to boost the bioeconomy in *Costa Rica*, a review. *Agronomía Mesoamericana, 32*(3), 1047–1070. https://doi.org/10.15517/am.v32i3.43736.
7. Dimian, A. C. (2015). Biorefinery, the future of chemical process industries. *Bulletin of the Romanian Chemical Society, 22*(1), 15–44.
8. Vieira, I. M. M., Santos, B. L. P., Santos, C. V. M., Ruzene, D. S., & Silva, D. P. (2022). Valorization of pineapple waste: A review on how the fruit's potential can reduce residue generation. *Bioenerg. Res. 15*, 924–934 https://doi.org/10.1007/s12155-021-10318-9.
9. Banerjee, S., Ranganathan, V., Patti, A., & Arora, A. (2018). Valorisation of pineapple wastes for food and therapeutic applications. *Trends in Food Science & Technology, 82*(March), 60–70. https://doi.org/10.1016/j.tifs.2018.09.024.
10. González-Vázquez, M. P., Rubiera, F., Pevida, C., Pio, D. T., & Tarelho, L. A. C. (2021). Thermodynamic analysis of biomass gasification using aspen plus: Comparison of stoichiometric and non-stoichiometric models. *Energies*, *14*(1), 189-206. https://doi.org/10.3390/en14010189.
11. Guo, H., Chang, Y., & Lee, D. J. (2018). Enzymatic saccharification of lignocellulosic biorefinery: research focuses. *Bioresource Technology, 252*, 198–215. https://doi.org/10.1016/j.biortech.2017.12.062.
12. Hernández-Chaverri, R. A., & Prado-Barragán, L. A. (2018). Impacto y oportunidades de biorrefinería de los desechos agrícolas del cultivo de piña (*Ananas Comosus*) En Costa Rica. *UNED Research Journal, 10*(2), 455–468. https://doi.org/10.22458/urj.v10i2.2059.
13. Montenegro Quesada, D., Montero Rambla, N., Hernández Chaverri, R. A., & Méndez Arias, J. (2021). Production and characterization of activated carbon from pineapple stubble for removal of methylene blue and surfactants. *Journal of Applied Research and Technology, 19*(5), 521–539. https://doi.org/10.22201/icat.24486736e.2021.19.5.1480.
14. Vargas-Vargas, M. A., Hernández-Chaverri, R. A., & Jiménez-Silva, A. (2019). Caracterización fisicoquímica de la biomasa del rastrojo de piña (*Ananas comosus* L. Merr.) y Su Valoración Preliminar Como Sustrato En La Propagación Micelial Del Hongo Shiitake (*Lentinula edodes* (Berk) Pegler). *Yulök Revista de Innovación Académica, 3*(1), 13–27.
15. Davis, R., Tao, L., Tan, E. C. D., Biddy, M. J., Beckham, G. T., Scarlata, C., Jacobson, J., Cafferty, K., Ross, J., Lukas, J., Knorr, D., Schoen, P. (2013). *Process design and economics for the conversion of lignocellulosic biomass to hydrocarbons: Dilute-Acid and enzymatic deconstruction of biomass to sugars and biological conversion of sugars*

to hydrocarbons. Golden, CO, (Unites States): National Renewable Energy Laboratory. Technical Report NREL/TP-5100-60223. https://doi.org/10.2172/1107470.

16. Ryan Davis, Nicholas Grundl, Ling Tao, Mary J. Biddy, Eric C.D. Tan, Gregg T. Beckham, David Humbird, David N. Thompson, & Mohammad S. Roni. (2018). *Process Design and Economics for the Conversion of Lignocellulosic Biomass to Hydrocarbon Fuels and Coproducts: 2018 Biochemical Design Case Update: Biochemical Deconstruction and Conversion of Biomass to Fuels and Products via Integrated Biorefinery Pathways*. Golden, CO, (Unites States): National Renewable Energy Laboratory. NREL/TP-5100-71949. https://doi.org/10.2172/1483234.
17. Kratky, L., & Jirout, T. (2011). Biomass size reduction machines for enhancing biogas production. *Chemical Engineering & Technology, 34*(3), 391–399. https://doi.org/10.1002/ceat.201000357.
18. Lyu, F., Thomas, M., Hendriks, W. H., & van der Poel, A. F. B. (2020). Size reduction in feed technology and methods for determining, expressing and predicting particle size: a review. *Animal Feed Science and Technology, 261*, 114347. https://doi.org/10.1016/j.anifeedsci.2019.114347.
19. Naimi, L. J., Sokhansanj, S., Bi, X., Lim, C. J., Mani, S., Hoque, M., Lam, P. S., & Womac, A. R. (2007). Modeling and characterization of biomass size reduction. *2007 ASABE Annual International Meeting, Technical Papers, 076047*, 1–17. https://doi.org/10.13031/2013.23432.
20. Kwade, A., & Schwedes, J. (2002). Breaking characteristics of different materials and their effect on stress intensity and stress number in stirred media mills. *Powder Technology, 122*(2–3), 109–121. https://doi.org/10.1016/S0032-5910(01)00406-5.
21. Kwon, J., Cho, H., Lee, D., & Kim, R. (2014). Investigation of breakage characteristics of low rank coals in a laboratory swing hammer mill. *Powder Technology, 256*, 377–384. https://doi.org/10.1016/j.powtec.2014.01.047.
22. Humbird, D., Davis, R., Tao, L., Kinchin, C., Hsu, D., Aden, A., Schoen, P., Lukas, J., Olthof, B., Worley, M., Sexton, D., & Dudgeon, D. (2011). *Process design and economics for biochemical conversion of lignocellulosic biomass to ethanol: Dilute-acid pretreatment and enzymatic hydrolysis of corn stover*. Golden, CO, (Unites States): National Renewable Energy Laboratory. NREL/TP-5100-47764. https://doi.org/10.2172/1013269.
23. Peters, M. S., Timmerhaus, K. D., & West, R. E. (2003). Plant design and economics for chemical engineers. *The Engineering Economist, 5*(1), 27–30. https://doi.org/10.1080/00137915908965075.
24. Sinnott, R., & Towler, G. (2013). *Chemical engineering design: Principles, practice and economics of plant and process design* (2nd ed.). Oxford, UK: Elsevier. 307–354. https://doi.org/10.1016/B978-0-08-096659-5.00007-9.
25. Ulrich, G. D. (1986). *Procesos de ingeniería química* (1st ed.). México: McGraw-Hill. 296–358.
26. Decker, S. R., Sykes, R. W., Turner, G. B., Lupoi, J. S., Doepkke, C., Tucker, M. P., Schuster, L. A., Mazza, K., Himmel, M. E., Davis, M. F., & Gjersing, E. (2015). High-throughput Screening of Recalcitrance Variations in Lignocellulosic Biomass: Total Lignin, Lignin Monomers, and Enzymatic Sugar Release. *J. Vis. Exp. 103*, e53163, https://doi.org/10.3791/53163.

27. Kowalczuk, P. B., & Drzymala, J. (2016). Physical meaning of the sauter mean diameter of spherical particulate matter. *Particulate Science and Technology, 34*(6), 645–647. https://doi.org/10.1080/02726351.2015.1099582.
28. Wooley, R., Ruth, M., Sheehan, J., Ibsen, K., Majdeski, H., & Galvez, A. (1999). *Lignocellulosic biomass to ethanol process design and economics utilizing co-current dilute acid prehydrolysis and enzymatic hydrolysis current and futuristic scenarios.* Golden, CO, (Unites States): National Renewable Energy Laboratory. NREL/TP-580-26157. https://doi.org/10.2172/12150.
29. Hijab, M., Saleem, J., Parthasarathy, P., Mackey, H. R., & McKay, G. (2020). Two-stage optimisation for malachite green removal using activated date pits. *Biomass Conv. Bioref. 11*, 727–740. https://doi.org/10.1007/s13399-020-00813-y.
30. Saleem, J., Shahid, U. Bin, Hijab, M., Mackey, H., & McKay, G. (2019). Production and applications of activated carbons as adsorbents from olive stones. *Biomass Conv. Bioref. 9,* 775–802. https://doi.org/10.1007/s13399-019-00473-7.

CHAPTER 6

Valorization of Apple Residues: A Potential Biorefinery

LETICIA XÓCHITL LÓPEZ-MARTÍNEZ,[1] RAMIRO BAEZA-JIMÉNEZ,[2] HILDA KARINA SAENZ-HIDALGO,[2] and EDWIN ROJO-GUTIÉRREZ[2*]

[1]*Laboratorio de Antioxidantes y Alimentos Funcionales, CONACYT–Centro de Investigación en Alimentación y Desarrollo, A.C. Carr. Gustavo Enrique Astiazarán Rosas 46, Col. La Victoria, CP., Hermosillo, Sonora, Mexico*

[2]*Laboratorio de Biotecnología y Bioingeniería, Centro de Investigación en Alimentación y Desarrollo, A.C. Av. Cuarta Sur, Fracc. Vencedores del Desierto, CP., Delicias, Chihuahua, México*

*Corresponding author.

ABSTRACT

Apple (*Malus domestica*) is a very healthy fruit because of its chemical composition and bioactive content. It is mainly consumed fresh, and when processed, products such as juices, cider, nectars, yogurt, and vinegar are obtained and commercialized. During such processes, residues including peel, pomace, and seeds are generated. Even though they could be a good feedstock for fermentation, they are often disposed of in land or landfills. These residues are rich in bioactive compounds such as phenolic compounds, flavonoids, and oil. If all the material flows from these processes are combined to utilize all resources, an apple-based biorefinery can be integrated. This review describes the main characteristics of apples and the potential of their residues for designing a biorefinery.

6.1 INTRODUCTION

Our society demands food, environmental preservation, and the sustainable obtaining of commodities. If a complete transformation of our knowledge

Food Security, Safety, and Sustainability: Validated Innovative Approaches and Technological Advancements.
José Juan Buenrostro-Figueroa and Julio César Tafolla-Arellano (Eds.)

is achieved in order to design more eco-efficient and competitive processes, our dependence on non-renewable resources will be minimized, and bioeconomy will constitute a new model. Hodson de Jaramillo [1] points out the potential benefits of that transition: (i) reduction of greenhouse gas emissions; (ii) food sovereignty and security; (iii) job creation; and (iv) revalorization of by-products and agroindustry residues. The latter is a potential research area for the food industry because such by-products and residues are rich in carbohydrates, proteins, and lipids, and they are sustainable feedstocks for the biotechnological production of bioactive compounds, biofuels, enzymes, and prebiotics, among others [2]. If the complete reuse and recycling of all those biomass components is achieved, a biorefinery can be integrated, which is consistent with the fundamentals of bioeconomy [3].

The sustainability of food residue biorefinery is related to the amount of biomass generated and its composition (both chemical and functional). Cereals, dairy, eggs, fish, fruit, meat, oil crops, sugar crops, and tubers, and vegetables are the main groups of foods for conducting techno-economic and environmental assessments for designing efficient and economically profitable biorefineries [4]. For instance, fruit residues such as seeds, peels, skin, or rind can contain phenolic compounds, flavonoids, carotenoids, essential oils, and pectin, as well as a relevant chemical composition for animal feeding.

As an example, an apple-based biorefinery can be defined. Apple is a fruit of a firm and fleshy structure, of worldwide production and consumption. The present review describes the main characteristics of apples (production and composition) and the valorization of their residues for designing a biorefinery.

6.2 APPLE

Fruit crops are a major part of agricultural production. Apples are the third most important fruit crop worldwide [5].

Apple (*Malus domestica*) is a very healthy fruit due to its chemical composition (*see* Table 6.1). The apple tree is native to Western Asia, and it has been cultivated for thousands of years in Asia and Europe. In the case of America, it was introduced by colonizers. Different varieties are cultivated and harvested; however, the most recognized are Golden Delicious, Granny Smith, and Red Starking.

Depending on the variety, apples differ in color, size, and flavor. Apples are mainly consumed fresh and cooked in many desserts, and at an industrial level, they are used to prepare several products: beverages (juice, cider, concentrate, vinegar), sauce, and dried fruit, among others [6, 7]. Apples grow in humid climates, sandy-loam soils, and a temperature range of –10 to 30°C (*see* Figure 6.1).

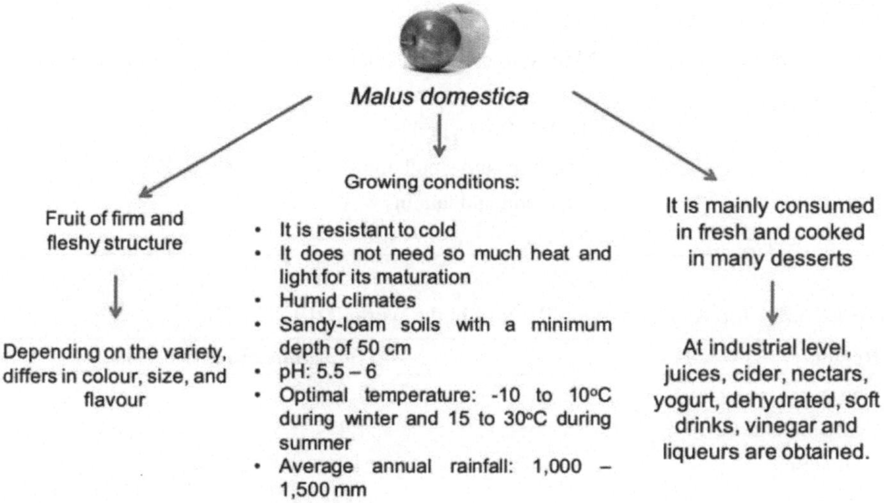

FIGURE 6.1 Apple: Characteristics, growing conditions, and uses.

Source: Adapted from Ref. [67]

During the 2019/2020 period, China was the main producing country, followed by the European Union (*see* Table 6.2). In Mexico, 10 varieties (Golden Delicious, Red Delicious, and Manzana Criolla are the most important) are harvested, and in 2020, 7,46,521 tons were produced [8].

According to the composition listed in Table 6.1 and the industrial uses indicated in Figure 6.1, during apple processing, several residues are generated (25–30% of the apple mass). The disposal of such residues compromises the environmental and economic sustainability of the agro-industrial market; significant environmental pollution is produced as a consequence of enzymatic and microbial deterioration, and the loss of commercial and biological values [9]. Those residues include seeds, peels, and pomace, which are rich in carbohydrates, proteins, and lipids, and then, they are sustainable feedstocks for the biotechnological production of beverages, prebiotics, and functional foods. The utilization of all the available biomass components will integrate an apple-based biorefinery, which is consistent with the fundamentals of bioeconomy.

In the next section, the valorization of apple residues is described in order to design an efficient and economically profitable biorefinery.

TABLE 6.1 Chemical and Functional Composition of Apple

Component	Value
Water	85%
Carbohydrates	Mainly fructose and small amounts of glucose and sucrose
Fiber	–
Vitamins	B1, B2, B6, C, and E
Minerals	Mainly K and small amounts of Zn, Mn, S, F, I, Br, or Se
Flavonoids	Quercetin and tannins
Energy supply	80 kcal

TABLE 6.2 Top Apple Growing Regions in the World 2019/2020 [68]

Region	Thousands of Metric Tons
China	41,000
European Union	11,477
United States	4,821
Turkey	3,000
India	2,370
Iran	2,097
Russia	1,714
Brazil	1,301
Chile	1,144
Ukraine	1,076

6.3 VALORIZATION: BIOREFINERY CONCEPT

Waste and residue management from the food industry has become a key factor for environmental protection and the exploitation of high-value commercial bioactive compounds. Nowadays, the popularity of nutraceuticals has increased since consumers are more concerned about seeking health-promoting compounds and preventing chronic illness [10]. Moreover, the extraction of bioactive compounds from apple by-products and their implementation in different industrial sectors could minimize significant monetary losses, transitioning to a bioeconomy.

Three main residues are obtained during apple processing (*see* Figure 6.2), which constitute the potential feedstocks for an apple-based biorefinery and are described as follows.

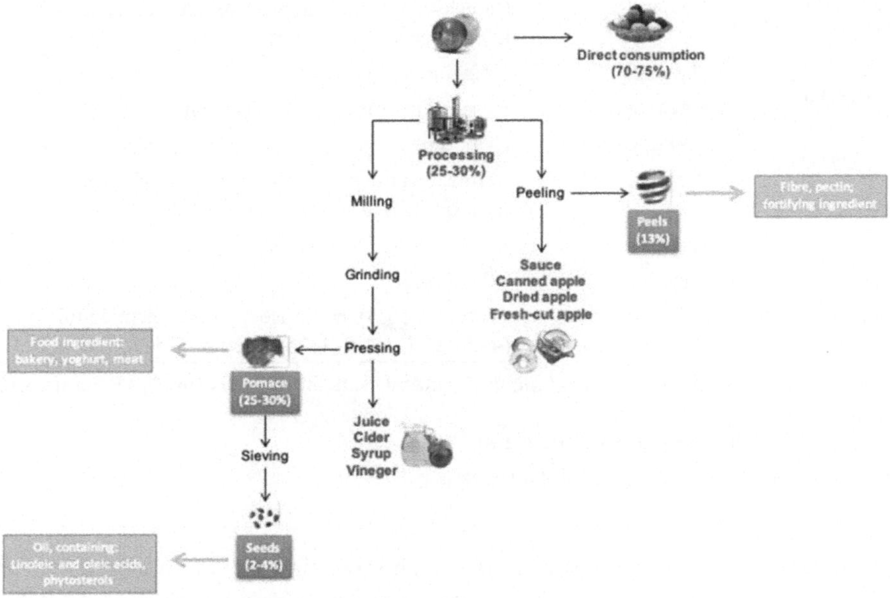

FIGURE 6.2 Apple-based biorefinery.

6.3.1 *APPLE POMACE (AP)*

Apple processing residues are typically known as apple pomace (AP), and it is constituted by skin/flesh (95%), seeds (2–4%), and stems (1%) [6, 11]. AP is rich in polysaccharides and phenolic compounds (Table 6.3). It retains higher amounts of bioactive substances than apple juice since the phenolic compounds are mainly concentrated in peels [12, 13]. AP has been used as a food ingredient to prepare a wide variety of products, where some of the most important are listed in Table 6.3.

6.3.1.1 BAKERY PRODUCTS

Several successful studies have been conducted using AP as a fortifying agent in baked food products. Jung et al. [14] used two different amounts (15 and 20%) of AP (wet and dried), which were incorporated in wheat flour for

TABLE 6.3 Main Phenolic Compounds and Carbohydrates Contained in Industrial Apple Pomace [69–71]

Components	Subclasses	Examples
Phenolic compounds (5 g/kg dm)	Dihydrochalcones	Phloridzin
	Flavanols	Catechin, epicatechin, procyanidin, (epi)catechin trimer, (epi)catechin tetramer.
TPC = 9 g GAE/kg	Flavones	Apigenin, chrysoeriol, luteolin.
	Flavanones	Hesperidin, eriodictyol, naringenin
	Anthocyanins	Cyanidin
	Phenolic acids	Protocatechuic acid, gallic acid, salicylic acid, ferulic acid, caffeic acid, and chlorogenic acid.
	Tannins	–
	Ellagitannins	Ellagic acid
Carbohydrates (720 g/kg dm)	Polysaccharides (622 g/kg dm)	Pectin (5–10%), cellulose (7–40%), hemicellulose (4–25%), lignins (4–25%), gums (15–25%)

Note: TPC: Total phenolic content; dm: dry matter; DS: defatted seeds; and GAE: gallic acid equivalents.

* Peels (TPC 500–589 mg GAE/100 g dm).

* Seeds (TPC 2861–5141 mg GAE/kg DS flour).

the production of cookies and muffins. The fortified baked goods presented higher amounts of fibers and the partial substitution of wheat flour with AP did not produce significant effects on the physicochemical and textural properties in comparison with their controls. Zlatanović et al. [15] studied the effect of three wheat substitution concentrations (25, 50, and 75%) of fine and coarse AP (0.16 and 0.50 mm, respectively) for the industrial production of cookies. Coarse AP flour (50%) cookies preserved the best sensorial properties (texture, flavor, odor) and were acceptable for consumers according to the 9-point hedonic test. Fortified cookies contained 21 g of dietary fiber/100 g of cookies. The total phenolic and flavonoid content was several times higher than control cookies, as well as the antioxidant activity. Lu et al. [16] added into wheat flour 3, 6, and 9% of AP for bread production. Fortified wheat flour with 9% AP increased approximately three times the dietary fiber content with respect to its control sample. However, their sensory analysis of bread (a fuzzy mathematical model) indicated that the optimum mixing concentration of AP was 3%. They stated that adding an optimum amount of AP could improve the dietary fiber content as well as the sensory properties of the fortified bread. Mir et al. [17] developed gluten-free crackers with different formulations of brown rice flour (k-332 and Khosar) and AP flour

(0, 3, 6, and 9%). They observed significantly higher total phenolic and total flavonoid content, antioxidant properties (evaluated by reducing power and DPPH methods), total dietary fiber, and mineral contents (chlorine, potassium, phosphorus, sulfur) by increasing the concentration of AP in the flour formulations. Although the incorporation of AP resulted in color and texture variations, the sensory evaluation of rice crackers (9-hedonic scale) showed that the products were acceptable for consumers.

6.3.1.2 YOGHURT

Sharma et al. [18] developed a fiber-enriched acidophilus yogurt. Four different concentrations of AP were added (2.5, 5, 7.5, and 10%) to whole milk, followed by inoculation with a combined culture (1%) of *Bifidobacterium longum* and *Lactobacillus acidophilus*. Results indicated that fiber-enriched yogurt exhibited lower acidity and fat contents, while minerals and fiber concentration increased compared to control samples. The highest fiber concentration achieved was 9.94%, implying approximately a 10-fold increment. After sensory evaluation, treatment with 5% AP concentration showed the best acceptability (including control). Like fortified baked products, the addition of an adequate AP concentration enhances fiber content and grants better sensorial properties. Wang et al. [19] incorporated AP as a source of phytochemicals and dietary fiber and evaluated its effect on the physicochemical properties of set-type yogurt. They concluded that AP could be potentially used as a natural texturizer and stabilizer in yogurt fermentation. AP addition shortened fermentation time by increasing the gelation pH. Consequently, a more consistent and firmer yogurt during cold storage was obtained. From the treatments applied (0.1, 0.5, and 1% AP addition), the yogurt with 0.5% AP showed the most stable structure after the storage period (28 days). Wang et al. [20] conducted another study where the stabilized attributes of AP were once again observed in the fortification of stirred-type and drinkable yogurt. They reported that AP addition could simultaneously confer attractive structural changes to the yogurt matrix and increase dietary fiber and antioxidant concentrations. AP addition caused a significant decrease in syneresis, increasing the viscosity, cohesiveness, and firmness of the yogurt matrix during cold storage. Jovanović et al. [21] fortified a probiotic yogurt with AP flour (1%, 3%, and 5%). The AP flour was immediately added at the end of the inoculation step process. *Bifidobacterium bifidum, Streptococcus thermophilus,* and *Lactobacillus acidophilus* were used as culture strains. According to previously described studies, the syneresis of the fortified yogurt was reduced up to 1.8 times compared to control samples. Probiotic strain viability remained within the indicated range (>log 7 CFU/g). Supernatants prepared from yogurts showed an increased value within 1.4–2.3-fold, 1.3–17.7-fold, and 3.3–8.0-fold of total phenolic content, reducing activity, and radical scavenging capacity, respectively, in comparison to its controls. The sensorial properties (firmness, cohesiveness, viscosity index) were enhanced by the AP, where the 3% concentration of AP presented the highest scores for color and taste overall.

6.3.1.3 MEAT

Consumption of meat is growing every year, and it has substantial effects on world population health. It has been demonstrated that high consumption of meat can produce harmful health effects such as colorectal cancer and cardiovascular disease [22]. In that matter, AP could be a potential alternative to conventional meat preservatives (nitrates, smoking) since it is a rich source of bioactive compounds and has functional properties (stabilizer, emulsifier) [23]. Jung et al. [14] also prepared fortified meat products in addition to bakery products in the same study. Around 10 and 20% of wet AP were incorporated in chicken patties and beef jerky. Antioxidant activity and dietary fiber were significantly increased, with the latter being up to 9 times higher than that of the control. A reduction in the firmness of the meat products was detected, probably because the addition of dietary fiber from wet AP dissociated the food matrix. Conversely, in the study by Younis & Ahmad [23], AP powder improved the textural properties (firmness, toughness, cohesiveness) of buffalo meat patties. This could be because AP significantly increased the water and fat holding capacity, as well as enhanced emulsion stability, diameter, and thickness. Formulations consisted of the addition of 2, 4, 6, and 8% (w/w) of AP. Moreover, the cooking yield was increased, as well as emulsion stability, water holding capacity, diameter, and thickness. These results are in agreement with those of Rather et al. [24]. They utilized AP powder as an antioxidant fat replacer in meat emulsions and analyzed the effect on emulsion stability, crude fiber, and oxidative stability. Partial substitution of fat (10%) with AP powder (1–5%) resulted in significantly higher emulsion stability and a crude fiber content (up to 15 times) than control samples. On the other hand, lipid oxidation was significantly reduced from 0.370 to 0.221 mg MDA/kg of the sample (treatment with 5% AP).

6.3.1.4 FERMENTATION SUBSTRATE

Since apple pomace contains valuable compounds such as free, insoluble, and soluble sugars, structural carbohydrates such as hemicellulose and cellulose, vitamins, and minerals, fermentation is an excellent tool to exploit all its components for the obtaining of different products of interest, including enzymes, antioxidants, organic acids, biogas, ethanol, and butanol.

Madrera et al. [25] enhanced the nutritional and bioactive properties of apple pomace using solid-state fermentation. Three autochthonous cider yeast strains (*Saccharomyces cerevisiae*, *Saccharomyces bayanus,* and *Hanseniaspora uvarum*) were used as substrate for the fermentable sugars from apple pomace, incubated for 7 days at 25°C. Protein, fat, and dietary fiber values significantly increased up to 49, 39, and 41%, respectively. Also, a significant increase in phenolic compounds was detected. A higher percentage value of quercetin and phloretin derivatives and fatty acids, mainly oleic and linoleic acids were obtained. Piwowareck et al. [26] used apple pomace as an alternative source of nitrogen to conventional yeasts or peptones. Appropriate control of fermentation parameters such as maintaining pH within 5.5–7.0, increasing the temperature from 30 to 37°C, and the addition of biotin (0.2 mg/L) to the medium, led to a higher propionic acid production (0.38 g/g) from an apple pomace extract in comparison to conventional sources (0.29 g/g). *Propionibacterium freudenreichii* T82 was the strain used in their study for the production of propionic acid. They concluded that future works can still improve the fermentation of apple pomace by using a more efficient strain like *Acidipropionibacterium acidipropionici*, a larger scale experiment, and a medium containing only waste materials.

Vashisht et al. [27] effectively produced acetic acid from AP employing a bacterial isolate *Acetobacter pasteurianus* SKYAA25. The acetic acid was obtained via bio-ethanol production using commercial brewing yeasts (Red Star dry yeast and Lalvin yeast EC-1118). AP slurry was incubated with the Red Star yeast at 30°C for 4 days. Bioethanol from the fermented supernatant was inoculated with *A. pasteurianus* SKYAA25 strain and incubated at 37°C with 180 rpm to finally obtain acetic acid. Bioconversion from bioethanol yielded an acetic acid production of 52.4 g/100 g of dry matter (dm).

Singh et al. [28] analyzed the AP capacity for inulinase production by *Mucor circinelloides* under solid-state fermentation. A central composite rotatable matrix design of 15 runs was conducted to obtain the optimal conditions of fermentation time, pH, and moisture and its effects on inulinase production. Experimental optimal conditions fermentation time (5.8 days), pH (6.4), and moisture (83.5%) allowed an inulinase production of 411.3 IU/gds, which is promising.

On the other hand, Ricci et al. [29] used the AP as an alcoholic aromatizer by lacto-fermentation of *Lactobacillus casei* 224, *Lactobacillus rhamnosus* 1019, and *Lactobacillus rhamnosus* 1473. The results showed that *L. rhamnosus* 1473 was the strain that conferred the best complex aroma profile in beer with the highest concentration of total volatile compounds, constituted by ketones, esters, norisoprenoids, terpenes, terpene derivatives, and alcohols. It is worth noting that alcohols were highly improved since statistically higher concentrations of hexanol and 2-hexanol were detected.

Jin et al. [30] obtained water-soluble and hydrolyzed sugars with alkali and acid from AP through hot water and the enzymatic hydrolysis process. Sugars alone or in combination were used as substrates for *Clostridium beijerinckii* P260 to generate acetone-butanol-ethanol (ABE). They found that combining water-soluble sugars and hydrolyzed sugars enhanced ABE production yields, obtaining up to 260.1 and 262.2 g for acid and alkali of ABE/kg of dry AP, respectively.

Kumar et al. [31] used isolated microorganisms (IM) from soil combined as co-culture with *Saccharomyces cerevisiae* to produce bioethanol. IM potential to produce xylanase, cellulase, and ethanol was analyzed to achieve the highest ethanol yield production. Isolate APW-12 (an *Actinomyces*) co-culture with *S. cerevisiae* resulted in the highest ethanol production (49.64 g/L).

6.3.1.5 EMULSIFIER

Lu et al. [32] developed a novel micronized AP emulsifier for food oil/water-based Pickering emulsion. Wet AP was subjected to 8 hours of milling until they achieved a particle size of 550 nm. Altering the wet AP structure by reducing its particle size conferred gel-like properties, enhancing the emulsion's physicochemical and antioxidant attributes. Pickering emulsions stabilized with milled AP showed strong stability and oxidation resistance. Their results are in accordance with those reported by Huc-Mathis et al. [33]. Micronized AP powder was dispersed in water and different oils (rapeseed, jojoba, and myritol) to obtain an emulsion of 50/50 (w/w) oil in water. Dynamic rheological measurements (backscattered light, droplet size distribution, light, and confocal microscopy) demonstrated that the AP exhibited excellent emulsifying properties. The authors stated that insoluble and soluble fibers from AP complement each other by forming the emulsion since insoluble fibers stabilize emulsions through a Pickering mechanism, enhancing stronger network construction in the continuous phase, while the soluble fraction stabilizes proteins and pectin. These produce the gel emulsifying behavior, which lasted for two weeks. Micronized AP powder is then a bioresource that needs minimal processing to benefit from its emulsifying properties. Lu et al. [34] made a novel cholesterol-free mayonnaise using a Pickering emulsion stabilized by apple pomace. The rheological properties of the novel cholesterol-free mayonnaise were acceptable compared to the commercial control, as well as its appearance. After 210 days of storage at 25°C, the cholesterol-free mayonnaise showed excellent stability, as no oil-water separation was observed.

6.3.1.6 SOURCE OF BIOACTIVE COMPOUNDS

Ferrentino et al. [35] conducted a supercritical fluid extraction of phenolic compounds from AP (fresh, oven-dried, and freeze-dried). The extraction was carried out under 20 and 30 MPa and 45 and 55°C in the absence and presence of the co-solvent ethanol. They found that the extracts from freeze-dried AP obtained by supercritical extraction at 30 MPa and 45°C for 2 hours with the co-solvent presented the maximum antioxidant activity (5.63 mg TEA/g of extract) in comparison to the conventional techniques of extraction Soxhlet and boiling water maceration (2.05 and 1.14 mg TEA/g of extract), respectively. Zardo et al. [36] obtained acetone, ethanol, and methanol phenolic extracts from AP to further evaluate their antibacterial activity against *Salmonella enterica, Staphylococcus aureus,* and *Escherichia coli.* Acetone exhibited significantly ($p < 0.01$) the highest extraction of phenolic compounds and percentage of inhibition. *S. aureus* was the most inhibited bacteria (65.51%), followed by *E. coli* (36.34%) and *S. enterica* (23.17%). Pearson products indicated a high correlation between total phenols and inhibition values. Chen & Lahaye [37] proposed a novel green pectin extraction process from AP. It consisted of employing eutectic solvent mixtures (choline chloride, lactic acid, oxalic acid, urea) in a 1:2 ratio to AP as pre-treatment of a pectin hot water extraction. Times and temperatures of the pre-treatment were within the ranges of 0.5–2 h and 40–80°C, respectively. AP pre-treated with a chloride-lactic acid (1:2) solvent mixture at 40°C for 1 h led to a more efficient extraction amount of pectin than the conventional method. Besides, they detected the recovery of a high concentration of uronic acid (33.1–56.1%).

6.3.1.7 BIOFILMS

Biofilms are an excellent eco-friendly alternative to synthetic plastics due to the actual issues of pollution and greenhouse gas emissions derived from their production process and imprudent disposal. AP may be a potential alternative feedstock since studies have shown that carbohydrates in AP (free sugars, cellulose, hemicellulose, lignin) can be used to develop completely functional biofilms; consequently, mitigating issues mentioned above. Gustafsson et al. [38] could successfully produce AP biofilms. For the production of the biofilms from AP, glycerol and naturally occurring sugars were used as plasticizers. The latter granted the best mechanical properties to the AP films, presenting a tensile strength and elongation of 3.27 MPa and 55.41%, respectively. Carpes et al. [39] prepared biofilms from cassava starch with different percentages of AP concentrations. Total phenolic content, mechanical properties, and antioxidant and antimicrobial activities of the prepared bio-based films were then evaluated. The addition of 8% of AP to cassava starch enhanced its total phenolic content (3.32 mg GAE/g) and antioxidant activity (2.78 mmol Trolox/g), approximately 4.7 and 2.7 times higher in comparison to the control film, respectively. All bacteria strains (*S. aureus*, *Salmonella typhimurium*, and *E. coli*) in contact with films containing AP showed an inhibitory concentration above 12.5 mg/mL. All variables analyzed were influenced by the amount of AP incorporated into films. Their results showed that using AP as an ingredient for the preparation of bio-based films had good mechanical properties and antioxidant and antibacterial properties. Similarly, Lan et al. [40] used AP as an ingredient for developing a multifunctional bioactive organic/inorganic-based film. Their film preparation consisted of adding 5% of an AP extract (w/w of chitosan) in an aqueous solution (100 mL) containing 2.5 g of chitosan. The mixture was then reinforced with 10% of nano-TiO_2 (w/w of chitosan). The combination of nano-TiO_2 and AP extract with chitosan for the development of a biofilm strongly enhanced ($p < 0.05$) its mechanical properties (tensile strength, elongation), antioxidant properties (around 66% of scavenging ability), and antibacterial properties (>8 and 7 mm of inhibition against *S. aureus* and *E. coli*, respectively) in comparison to the control chitosan-based film. Furthermore, the film was color-sensitive to pH variations, functioning as an intelligent packaging with the ability to monitor salmon fillets' freshness.

Stojanović et al. [41] explore the preparation of a novel edible biofilm using a pea protein isolate (8% w/w) with the addition of AP extract (15% w/w) as a substrate for printed edible electronics for medical purposes. Edible electronics are a promising alternative to the invasive conventional medical sensors used for health monitoring and diagnostics. Once the edible biofilms were prepared, they could print silver ink (5 μm of silver was employed, which represents 1.4% of the safe daily dose) into them to fabricate inductor capacitors, successfully manufacturing an edible electronic device. Since a sintering treatment is necessary after the ink printing, biofilms were submitted to microstructural and morphological analysis by nanoindentation. Four temperatures in a range of 27–100°C were evaluated, where the increasing temperature had a negative effect on the film's mechanical properties. Nevertheless, their results indicated that it is possible to obtain a functional electronic device using a biofilm as a substrate.

6.3.2 APPLE SEEDS

Apple seeds contain a considerable amount of oil that can range from 15% to 29.4% [42] on average. Oil variations can be attributed to agroclimatic, variety, and genetic factors [43]. Apple seed oil (ASO) is mainly composed of linoleic and oleic acids, but it is also a rich source of bioactive compounds such as hydroxycinnamic acids, flavan-3-ols, dihydrochalcones, especially phytosterols, tocopherols, and phloridzin [44, 45]. ASO also has significant potential for use in the food, pharmaceutical, biotechnological, and cosmetic industries. Pieszka et al. [46] evaluated different oil seeds, including apple seeds, as a source of fatty acids, tocopherols, tocotrienols, and phytosterols. Their fatty acid profile showed high levels of linoleic (55.5%) and oleic (29.4%) acids. It also contained α- (41.7 mg/100 g), β- (62.7 mg/100 g), γ- (13.6 mg/100 g), and δ- (21.8 mg/100 g) tocopherol isomers. Tocotrienols and phytosterols values were 4.21 mg/100 g and 3,460 μg/g, respectively. Results indicated that the high amounts of essential unsaturated fatty acids and bioactive compounds might have strong nutraceutical effects that could be applied in the cosmetic, pharmaceutical, and food industries. A similar study was conducted by Górnaś [47], who determined tocopherol composition from dessert and crab ASO by RP-HPLC/FLD, extracted from residual seeds recovered from the fruit industry. Once again, all homologs were identified (α:β:γ:δ), obtaining an average of 70.36:46.60:26.27:16.99 mg/100 g oil in crab ASO, respectively. With respect to dessert ASO samples, the tocopherol profile was 103.11:97.57:40.84:39.00 mg/100 g oil, respectively. It is worth noting that a single dilution of ASO in 2-propanol was needed for the DPPH assay, which also enhanced the separation of tocopherol homologs by a pentafluoro phenyl column, incorporated into an RP-HPLC. Later, Górnaś & Rudzińnska [48] analyzed the composition of fatty acids and phytosterols of oils extracted from different residual seeds (watermelon, honeydew melon, sea buckthorn, red currant, pomegranate, Japanese quince, grape, gooseberry, and apple) derived from the fruit industry to evaluate their utility as non-conventional feedstocks for biodiesel production and cosmetic and pharmaceutical sectors. All the evaluated oils presented a unique and varied composition and concentration of fatty acids, phytosterols, and squalene. An oil yield of approximately 22% (w/w) from apple seeds was obtained. Linoleic and oleic acids were the major fatty acids found in ASO (61.57 and 26.69%, respectively). In general, the ASO fatty acid composition and physicochemical properties were suitable for biodiesel production. The composition of phytosterols comprises β-sitosterol, cycloartenol, campesterol, stigmasterol, avenasterol, and Δ7-stigmasterol in ASO. β-sitosterol presented the highest concentration (2.46 mg/g of oil) of all. Squalene determination resulted in 0.26 mg/g oil. Their conclusions highlighted the potential utility of ASO in many industrial sectors, such as biodiesel production.

Montañés et al. [49] explored the extraction of ASO using supercritical carbon dioxide at pressures within 300–1,300 bar since this extraction technique preserves more of the bioactive compounds contained in the oil. An extraction yield of 21–22% was obtained. The highest solubility achieved was approximately 191 g extract/kg CO_2 under 1,300 bar and 336 K, which was 19 times higher than the value obtained when a typical extraction pressure was employed (300 bar). Although polyphenols were not recovered during the oil extraction, lipid composition remained unaltered, and tocopherols could be exhaustively extracted (up to 809 mg/kg) under moderate extraction conditions ($\geq$ 750 bar and $\leq$ 326 K). ASO tocopherol content resulted in being between 3 and 6 times higher than in the seeds. Ferrentino et al. [35] carried out a comparison between supercritical fluid and conventional Soxhlet extraction methods for obtaining ASO. The supercritical fluid extraction (SFE) process was optimized (24 MPa, 40°C, 1 L/h CO_2, 140 min), and a maximum yield was 20.5% (w/w), slightly lower than that obtained by Soxhlet (22.5%, w/w). Interestingly, the SFE fatty acid profile exhibited a higher linoleic acid concentration (63.76 g/100 g oil) than Soxhlet (49.03 g/100 g oil). Consequently, SFE oil exhibited a higher unsaturated fatty acid content. Surprisingly, ASO obtained through SFE was still significantly more stable against oxidation (21.4 h) than the oil extracted by the Soxhlet method. They attributed this to the fact that ASO extracted by SFE presented a greatly higher total phenolic content and antioxidant activity (2 and 2–3 times higher than Soxhlet extractions, respectively), where the predominant phenolic compound was phloridzin (2.96 μg/g seed). Data demonstrated that SFE is a promising alternative method to obtain ASO with high phenolic content and antioxidant activity. Khalil & Mustafa [50] isolated four novel simple coumarins: *8-(tert-butyl) officinalin* (R2), *8-Hydroxy officinalin* (R3), *Officinalin-8-acetic acid* (R4), and 8-(2'-hydroxypropan-2'-yl) *officinalin* (R5), from Granny Smith apples. Officinalin (R1) was also detected but already previously identified. Water, methanol, chloroform, and n-hexane were used as extract solvents. The effect of kinetic maceration, ultrasound, and microwave extraction methods on the antioxidant and antitumor activities in the apple seed extracts was investigated. DPPH and hydroxyl free radicals' tests were conducted to determine the antioxidant activity of apple seed extracts. For the determination of antitumor activity, an MTT assay against HeLa and MCF-7 cancer cell lines was carried out. Successful structure characterization of the isolated coumarins was achieved by employing FTIR, H-NMR, and C-NMR spectra analysis. It was found that the ultrasound extraction method with chloroform as a solvent was the most suitable for the isolation of coumarins. Their findings also showed that all the novel isolated coumarins presented important antioxidant and antitumor activity with a significant correlation. R4 presented the highest scavenging activities (57.14 and 61.39 [SC_{50}, μM] for DPPH and OH radicals, respectively) and antitumor activities (52.47 and 45.70 [IC_{50}, μM] against HeLa and MCF-7, respectively). According to these results, new bioactive agents could be obtained. This work was extrapolated to isolate coumarins from Red Delicious apple seeds by Mohammed & Mustafa [51]. Four novel furanocoumarins [(*E*)-*12-(2'-Chlorovinyl) bergapten,* E1; *12-(1,'1'-dihydroxyethyl) bergapten*, E2; *12-(2'-chloroppropan-2'-yl)*-8-hydroxybergapten, E3; *12-Hydroxy-11*-chloromethylbergapten, E4] were isolated and structurally characterized. However, in this study, only an antimicrobial activity analysis was approached. Bacterial strains (*Pseudomonas aeruginosa, Klebsiella pneumonia, Haemophilus influenzae*, and *E. coli*) and fungi strains (*Candida albicans* and *Aspergillus niger*) were used for the antimicrobial test, using Ciprofloxacin and Nystatin as a reference, respectively. Once again, all isolated furanocoumarins exerted important inhibition activity against bacterial and fungi strains, where E1 presented the highest antibacterial and antifungal activities.

Mustafa et al. [52] obtained methanolic extracts from Red Delicious and Granny Smith apples' seeds. Similar methodologies from their previous work were employed [50]. For the determination of antitumor activity,

AMN3, and SKG cancer cell lines were also considered. The methanolic extracts showed a remarkable capacity to neutralize DPPH (8.57–37.91% of antiradical activity) and hydroxyl (5.84–36.83% of antiradical activity) free radicals. Results from the antitumor assessment suggested good biological activity. The percentage ranges of growth inhibition obtained from methanolic extracts were 8.48–44.30%, 7.20–54.71%, 9.02–56.76%, 8.19–42.45% for AMN3, HeLa, MCF-7, and SKG, respectively. Overall findings indicated that Red Delicious apple seed extract possessed higher antioxidant and antitumor (except for AMN3 cancer cell line) activity, and the ultrasound technique allowed for achieving the highest antioxidant and antitumor result values.

Opyd et al. [53] evaluated the possibility of using defatted apple seeds (DAS) as a source of protein and fiber. Rats were fed with DAS to provide them with dietary fiber and protein since it is contained in the endosperm of apple seeds. Supplementation with DAS exerted beneficial health effects on the intestinal tract, blood lipid profile, and antioxidant status of the rats. Despite the protein value being to some extent low, amygdalin contained in apple seeds increased the protein digestibility and nitrogen retention without compromising the health of the rats since amygdalin is slightly toxic. They also observed that DAS increased the satiety of rats since a decrease in their body weights was recorded. Senica et al. [54] investigated the phenolics and cyanogenic glycosides contained in apple seeds (Golden Delicious) from four countries (Azerbaijan, Russia, Serbia, and Slovenia). Apple seeds from Russia presented the highest levels of phenolics (6.2 mg/g). They also highlighted that the introduction of apple seeds in human diets would not compromise health because it would be necessary to consume high amounts of cyanogenic glycosides (20 μg/kg of body weight), and the highest dose found was 2.2 mg/g (Azerbaijan apple seeds).

6.3.3 APPLE PEEL

Apple peels are highly rich in phenolic compounds since they contain up to 46% of the polyphenolics found in the apple [55]. For this reason, the applications of the apple peel are very similar to those of the AP.

Ahmad et al. [56] evaluated the influence of polyphenols extracted from apple peels on the physicochemical and microbiological properties of probiotic yogurt. Five fortifying apple peel polyphenol extracts (APPEs) were used (1–5%). The total phenolic contents of probiotic yogurt significantly increased, achieving up to 1.35 (APPE 5%). Also, an increase in the sensorial attributes (taste, flavor, texture, color) of the fortified probiotic yogurt was observed, where the addition of 3 and 4% of APPE was the most accepted by the panelists. The fortification of probiotic yogurt using apple peels was accomplished since its viable counts of probiotics and antioxidant activity were enhanced. Ranjha et al. [57] also employed APPEs for the development of functional date bars. Ethanolic extracts of apple peels were obtained by ultrasound-assisted extraction. After preliminary tests, the fortification values of 2 and 3% of APPE were the most suitable for the fortification of the date bars. Data showed that the total phenolic content was 5-fold (136.14 mg GAE/g date bar) and 7-fold (192.47 mg GAE/g date bar) higher in comparison to the control sample. Their sensorial analysis resulted in selecting 2% APPE as the most suitable fortification value since it was the most preferred among the treatment and control bars. Nevertheless, data indicated that the panelists accepted all bars.

Rahman et al. [58] used a mix of banana, carrot, and apple peels as partial substitutes for wheat flour to prepare a high-fiber biscuit. Bananas, carrots, and apple peels were dried at 60°C for 12 h to grind them subsequently. Each feedstock was ground separately into fine powders and then mixed in equal amounts. The partial substitution of wheat flour with mixed peel powders triplicated the fiber content of the control biscuit. They also observed that Mn and K content increased approximately two times the concentration of the minerals of the control sample. The sensory evaluation indicated that the mixed peel powders improved the aroma, sweetness, flavor, and overall acceptance. Nakov et al. [59] enriched bread wheat cookies with different formulations (0, 4, 8, 16, 24, and 32%) of apple peel powder (APW). Chemical compounds such as moisture, ash, lipid, fiber, and polyphenols content significantly increased as higher amounts of APW were used. The maximum total phenolic content achieved was 622.12 μg GAE/g (32% APW). The inclusion of higher amounts of APW also strongly influenced the antioxidant activity, reaching a scavenging activity of 126.7 μg TE/g dm and 2280.9 μg TE/g of dm against DPPH and FRAP radicals, respectively. Sensorial attributes of the cookies also benefited from the addition of APW since an improvement in their aroma, taste, texture, appearance, and internal structure was detected by the panelists. Based on the sensorial results and analysis as mentioned above, an addition of 24% of APW showed the best overall quality. In that matter, the latter was selected as the optimum concentration for the enrichment of cookies.

Matta et al. [60] produced bioactive edible methylcellulose films with ethanolic APPEs (Granny Smith). For the film preparation, they used 10, 20, and 25% (v/v) of APPEs concentrations and mixed with 1.5 g of methylcellulose dispersed in 50 mL, until 20 g of film-forming solutions were obtained. The films exhibited a total phenolic content within 18–21 mg chlorogenic acid/g film and inhibition value within 10–21% of DPPH radicals, increasing according to the percentage of APPE concentration used. They also reported that the functionalized films presented good thermal and mechanical properties. In that matter, the prepared edible bioactive films of methylcellulose-APPE could potentially be used in the food industry.

Riaz et al. [61] investigated the efficiency of edible chitosan-apple peel polyphenols (APP) coatings as a strategy to improve the storage quality of strawberries. The coatings applied to the strawberries contained chitosan and different concentrations of APP (0, 0.25, 0.50, 0.75, and 1.0%). The use of chitosan-APP coatings significantly improved the preservation of strawberries after 6 days in weight loss, decay, firmness, bioactive compounds, titratable acidity, and total soluble solids. The coating of chitosan with 0.50% of APP presented the highest total phenolic content (1.3 mg/g) on day 6 of storage. Overall, they considered that 0.50% of APP was the most suitable concentration to prolong the shelf life of the strawberries, preserving their quality attributes. In another study, Riaz et al. [62] used apple peels as feedstock to manufacture apple peel nanoparticles, which were later used to prepare chitosan/gelatin-based bioactive food packaging. About 5, 10, 15, and 20 g of APPE-containing nanoparticles were employed. The addition of apple nanoparticles significantly influenced the physical properties of the films positively, such as thickness, swelling ratio, solubility, and water vapor permeability. The films showed statistically higher scavenging activity against DPPH and ABTS free radicals compared to the control sample and between each concentration of APPE employed for the preparation of the films. The chitosan/gel-based film with 20 g of APPE exhibited the highest scavenging activity obtained (approximately 66 and 90% for DPPH and ABTS, respectively). Data suggest that nanoparticles from natural sources are potential candidates for the production of bioactive food packaging, like the chitosan/gel-APPE films developed.

Alpaslan et al. [63] analyzed the potential of a novel hydrogel intelligent food packaging. A random redox polymerization was performed for the synthesis and preparation of hydrogels from N, N-dimethylacrylamide (DMAAm), gelatin, and citric acid (CA) – red apple peel extract (RAPE). According to the investigators, the addition of RAPE conferred to the intelligent package more thermal stability (up to 565°C), an excellent total phenolic

(1,120 mg/g) and anthocyanin concentrations (0.4 mg/g), in comparison to previous studies, as well as antioxidant activity (111 μmol/g). Moreover, it was observed that the RAPE contained color-sensitive compounds (anthocyanin glucosides, phenolic acids, and flavonoids, among others) that act as an indicator of pH variations. Antimicrobial activity tests indicated high antimicrobial properties against *E. coli, Bacillus subtilis*, and *S. aureus* (minimal bactericidal concentration of hydrogels was > 200 mg/mL against all strains). Experimental assays with whole milk and cheese demonstrated that the intelligent, active packaging presented convincingly antimicrobial activity and a color indicator for quality control. They stated that this poly-DMAAm/CA-RAPE packaging material could not only be used in the food industry but also the medical sector.

Hoseinifar et al. [64] supplemented carp fish with dietary pectin derived from apple peels. Four formulations with 0, 0.5, 1, and 2% of apple peel-derived pectin were prepared to feed common carp for eight weeks to evaluate its effect on the growth, antioxidant enzyme activity, and immune parameters in their blood. Carp-fed diets supplemented with dietary pectin from apple peels significantly increased their growth, feed conversion ratio, serum total immune globulin, and lysozyme, catalase, and superoxide dismutase activity values. An antibacterial activity evaluation demonstrated that dietary pectin significantly enhanced the antibacterial activity of the skin mucus and serum against 10 different bacterial strains. Results showed that the dietary supplementation of dietary pectin derived from apple peel potentiated the growth and immune parameters of carp fish.

Susilowati et al. [65] proposed the use of APPE as an alternative to the conventional simvastatin drug to treat hyperlipidemic rats; for groups, rats were fed with different doses of APPE (0, 57, and 114 mg/kg body weight) and simvastatin (3.6 mg/kg body weight), where the latter was established as the control group. Results showed that quercetin, chlorogenic acid, epicatechin, and catechins presented higher inhibition activities than simvastatin against 3-hydroxy-3-methylglutaryl coenzyme A (HMG-CoA) reductase, which is a regulatory enzyme involved in the cholesterol biosynthetic pathway. Nevertheless, quercetin presented the most similar bonding mechanism to amino acid residues as simvastatin. The optimal APPE concentration that significantly decreased low-density lipoprotein cholesterol and increased high-density lipoprotein cholesterol serum to the greatest extent was 114 mg/kg BW. Results showed that APPE may be a more effective product to treat hyperlipidemia than simvastatin since a higher inhibition of the cholesterol biosynthesis pathway was observed.

Singh et al. [66] immobilized apple peel in sodium alginate beads to produce a biosorbent for heavy metals that could be used in a drinking water purification system. For this purpose, a solution containing seven heavy metals (arsenic, cadmium, chromium, copper, mercury, lead, nickel) potentially toxic to human health was prepared. Results showed that the pH of the medium greatly influenced the biosorption of the heavy metals (except arsenic and chromium). An apple peel bead (APB) and a control bead were put separately in contact with a metal solution for 72 hours; however, sampling started at 15 minutes. The biosorption levels of the heavy metals were in the following decreasing order (pH = 7): Cd (90%), Cu (80%), Pb (73%), Ni (70%), Hg (10%), and As (2%). Cu was the metal that required more time (48 hours) to reach equilibrium, while Pb only required 6 hours. The apple peel bead showed a significantly higher biosorption capacity than the control bead throughout the entire assay period. Despite As, Cr, and Hg not being purified by the APB, the concentrations of the other heavy metals in the solution were significantly reduced. Moreover, the efficiency of biosorption can be enhanced by adding more APBs.

As it has been described, those three apple residues exhibit a biotechnological potential to integrate a biorefinery (*see* Figure 6.2). During apple processing, conventional products can be obtained, and, on the other hand, the amounts of pomace, seed, and peel generated are suitable for further bioconversion, namely, apple seed oil, fortified (containing prebiotics) and functional foods (containing phenolic compounds).

The proposed biorefinery from apple residues meets the bioeconomy principles and combining the biochemical engineering concept, new and novel processes can be developed for apple processing and residues revalorization, contributing to minimizing the environmental impact, creation of jobs, extraction and recovery of bioactive compounds, design of more efficient and sustainable processes, and transition to cleaner industries.

6.4 CONCLUSION

The concept of bioeconomy is still in the early stages and further understanding and development are needed. The transition to such a new model implies complete biomass utilization during food processing. In this review, the commercial products obtained from apple processing were mentioned; however, the aim of this contribution is to highlight the biotechnological potential of apple residues in order to integrate them into an apple-based biorefinery. The revalorization of such residues will reduce the environmental impact of their disposal and promote their use for preparing fortified foods containing bioactive compounds (such as phenolic compounds and fiber), as well as apple seed oil rich in linoleic and oleic acids. This review intends to emphasize biochemical engineering for its importance in the development, diversification, and sustainability of apple residues for their complete integration and utilization during apple processing.

KEYWORDS

- **apple**
- **biochemical engineering**
- **bioeconomy**
- **biomass utilization**
- **biorefinery**
- **residues**
- **sustainability**
- **valorization**

REFERENCES

1. Hodson de Jaramillo, E. (2018). Bioeconomía: el futuro sostenible. *Revista Académica Colombiana de Ciencias Exactas, Físicas y Naturales, 42*, 188–201.
2. Ng, H. S., Kee, P. E., Yim, H. S., Chen, P. T., Wei, Y. H., & Lan, J. C. W. (2020). Recent advances on the sustainable approaches for conversion and reutilization of food wastes to valuable bioproducts. *Bioresource Technology, 302*, 122889.
3. Cherubini, F. (2010). The biorefinery concept: Using biomass instead of oil for producing energy and chemicals. *Energy Conversion and Management, 51*, 1412–1421.
4. Caldeira, C., Vlysidis, A., Fiore, G., De Laurentis, V., Vignali, G., & Sala, S. (2020). Sustainability of food waste biorefinery: A review on valorisation pathways, techno-economic constraints, and environmental assessment. *Bioresource Technology, 312*, 123575.
5. Statista. (n.d.). Global fruit production in 2022, by selected variety (in million metric tons). https://www.statista.com/statistics/264001/worldwide-production-of-fruit-by-variety/ (accessed on May 26, 2024)..
6. Rabetafika, H. N., Bchir, B., Blecker, C., & Richel, A. (2014). Fractionation of apple by-products as source of new ingredients: Current situation and perspectives. *Trends in Food Science & Technology, 40*, 99–114.

7. Rodríguez-Madrera, R., & Suárez-Valles, B. (2018). Characterization of apple seeds and their oils from the cider-making industry. *European Food Research and Technology, 244*, 1821–1827.
8. Secretaría de Agricultura y Desarrollo Rural. Dulce sabor de la manzana. https://www.gob.mx/agricultura/articulos/dulce-sabor-de-la-manzana?idiom=es (accessed May 27, 2024).
9. Gómez, M., & Martinez, M. M. (2017). Fruit and vegetable by-products as novel ingredients to improve the nutritional quality of baked goods. *Critical Reviews in Food Science and Nutrition, 58*(13), 2119–2135.
10. Birch, C. S., & Bonwick, G. A. (2019). Ensuring the future of functional foods. *International Journal of Food Science & Technology, 54*, 1467–1485.
11. Antonic, B., Jancikova, S., Dordevic, D., & Tremlova, B. (2020). Apple pomace as food fortification ingredient: A systematic review and meta-analysis. *Journal of Food Science, 85*(10), 2977–2985.
12. Krawitzky, M., Arias, E., Peiro, J. M., Negueruela, A. I., Val, J., & Oria, R. (2014). Determination of colour, antioxidant activity, and phenolic profile of different fruit tissue of Spanish "Verde Doncella" apple cultivar. *International Journal of Food Properties, 17*(10), 2298–2311.
13. Kruczek, M., Gumul, D., Kačániová, M., Ivanišhová, E., Marecek, J., & Gambuś, H. (2017). Industrial apple pomace by-products as a potential source of pro-health compounds in functional food. *Journal of Microbiology, Biotechnology and Food Sciences, 7*, 22–26.
14. Jung, J., Cavender, G., & Zhao, Y. (2014). Impingement drying for preparing dried apple pomace flour and its fortification in bakery and meat products. *International Journal of Food Science & Technology, 52*(9), 5568–5578.
15. Zlatavonič, S., Kaluševič, A., Micić, D., Laličić-Petronijević, J., Tomić, N., Ostojić, S., & Gorjanović, S. (2019). Funcionality and storability of cookies fortified at the industrial scale with up to 75% of apple pomace flour produced by dehydration. *Food, 8*, 561.
16. Lu, Q., Liu, H., Wang, Q., & Liu, J. (2017). Sensory and physical quality characteristics of bread fortified with apple pomace using fuzzy mathematical model. *International Journal of Food Science & Technology, 52*(5), 1092–1100.
17. Mir, S. A., Bosco, S. J. D., Shah, M. A., Santhalakshmy, S., & Mir, M. M. (2017). Effect of apple pomace on quality characteristics of brown rice-based cracker. *Journal of Saudi Society of Agricultural Sciences, 16*(1), 25–32.
18. Sharma, P. C., Gupta, A., & Issar, K. (2016). Effect of packaging and storage and dried apple pomace and fiber extracted from pomace. *Journal of Food Processing and Preservation, 41*(3), e12913.
19. Wang, X., Kristo, E., & LaPointe, G. (2019). The effect of apple pomace on the texture, rheology and microstructure of set type yogurt. *Food Hydrocolloids, 91*, 83–91.
20. Wang, X., Kristo, E., & LaPointe, G. (2019). Adding apple pomace as a functional ingredient in stirred-type yogurt and yogurt drinks. *Food Hydrocolloids, 100*, 105453.
21. Jovanović, M., Petrović, M., Miočinović, J., Zlatanović, S., Petronijević, J. L., Mitić-Ćulafić, D., & Gorjanović, S. (2020). Bioactivity and sensory properties of probiotic yogurt fortified with apple pomace flour. *Foods, 9*(6), 763.

22. Godfray, H. C. J., Aveyard, P., Garnett, T., Hall, J. W., Key, T. J., Lorimer, J., & Jebb, S. A. (2018). Meat consumption, health, and the environment. *Science, 361*(6399), eaam5324.
23. Younis, K., & Ahmad, S. (2018). Quality evaluation of buffalo meat patties incorporated with apple pomace powder. *Buffalo Bulletin, 37*(3), 389–401. Retrieved from: https://158.108.80.26/kuojs/index.php/BufBu/article/view/592 (accessed on 27 May 2024).
24. Rather, S. A., Masoodi, F. A., Akhter, R., & Gani, A. (2018). Utilization of apple pomace powder as an antioxidant fat replacer in meat emulsions. In S. M. Wani, A. Gani, I. A. Rather, M. A. Shah, T. A. Wani, & M. Ahmad (Eds.), *National Conference on Recent Advances in Understanding the Role of Phytochemicals in Human Health* (pp. 69–76). Excel India Publishers.
25. Madrera, R. R., Bedriñana, R. P., & Valles, B. S. (2017). Enhancement of the nutritional properties of apple pomace by fermentation with autochthonous yeasts. *LWT, 79*, 27–33.
26. Piwowareck, K., Lipińska, E., Hać-Szymańczuk, E., Rudziak, A., & Kieliszek, M. (2019). Optimization of propionic acid production in apple pomace extract with *Propionibacterium freudenreichii. Preparative Biochemistry & Biotechnology, 49*(19), 974–986.
27. Vashisht, A., Thakur, K., Kauldhar, B. S., Kumar, V., & Yadav, S. K. (2019). Waste valorization: Identification of an ethanol tolerant bacterium *Acetobacter pasteurianus* SKYAA25 for acetic acid production from apple pomace. *Science of the Total Environment, 690*, 956–964.
28. Singh, R. S., Chauhan, K., Kaur, K., & Pandey, A. (2019). Statistical optimization of solid-state fermentation for the production of fungal inulinase from apple pomace. *Bioresource Technology Reports, 9*, 100364.
29. Ricci, A., Cirlini, M., Guido, A., Liberatore, C. M., Ganino, T., Lazzi, C., & Chiancone, B. (2019). From byproduct to resource: Fermented apple pomace as beer flavouring. *Foods, 8*(8), 309.
30. Jin, Q., Qureshi, N., Wang, H., & Huang, H. (2019). Acetone-butanol-ethanol (ABE) fermentation of soluble and hydrolysed sugars in apple pomace by *Clostridium Beijerinckii* P260. *Fuel, 244*, 536–544.
31. Kumar, D., Surya, K., & Verma, R. (2020). Bioethanol production from apple pomace using co-cultures with *Saccharomyces cerevisiae* in solid-state fermentation. *Journal of Microbiology, Biotechnology and Food Sciences, 9*(4), 742–745.
32. Lu, Z., Ye, F., Zhou, G., Gao, R., Qin, D., & Zhao, G. (2020). Micronized apple pomace as a novel emulsifier for food O/W Pickering emulsion. *Food Chemistry, 330*, 127325.
33. Huc-Mathis, D., Journet, C., Fayolle, N., & Bosc, V. (2019). Emulsifying properties of food by-products: valorising apple pomace and oat bran. *Colloids and Surfaces A: Physicochemical and Engineering Aspects, 568*(1), 84–91.
34. Lu, Z., Zhoua, S., Ye, F., Zhou, G., Gao, R., Qin, D., & Zhao, G. (2021). A novel cholesterol-free mayonnaise made from Pickering emulsion stabilized by apple pomace particles. *Food Chemistry, 353*, 129418.
35. Ferrentino, G., Giampiccolo, S., Morozova, K., Haman, N., Spilimbergo, S., & Scampicchio, M. (2020). Supercritical fluid extraction of oils from apple seeds: Process optimization, chemical characterization and comparison with a conventional solvent extraction. *Innovative Food Science & Emerging Technologies, 64*, 102428.

36. Zardo, D. M., Alberti, A., & Zielinski, A. A. F. (2020). Influence of solvents in the extraction of phenolic compounds with antibacterial activity from apple pomace. *Separation Science and Technology, 56*(5), 903–911.
37. Chen, M., & Lahaye, M. (2021). Natural deep eutectic solvents pretreatment as an aid for pectin extraction form apple pomace. *Food Hydrocolloids, 115*, 06601.
38. Gustafsson, J., Landberg, M., Bátori, V., Åkesson, D., Taherzadeh, M. J., & Zamani, A. (2019). Development of bio-based films and 3D objects from apple pomace. *Polymers, 11*(2), 289.
39. Carpes, S. T., Certotto, C., Bilck, A. P., Yamashita, F., Anjos, O., Siddique, M. A. B., & Harrison, S. M. Brunton, N. P. (2021). Bio-based films prepared with apple pomace: Volatiles compound composition and mechanical, antioxidant and antibacterial properties. *LWT, 144*, 111241.
40. Lan, W., Wang, S., Zhang, Z., Liang, X., Liu, X., & Zhang, J. (2021). Development of red apple pomace extract/chitosan-based films reinforced by TiO2 nanoparticles as a multifunctional packaging material. *International Journal of Biological Macromolecules, 168*, 105–115.
41. Stojanović, G., Pojić, M., Kojić, S., Mišan, A., & Vasilijević, D. (2019). Mechanical properties of edible biofilm as a substrate for printed electronics. *Applied Physics A, 125*(8), 576.
42. Fromm, M., Bayha, S., Carle, R., & Kammerer, D. R. (2012). Comparison of fatty acid profiles and contents of seed oils recovered from dessert and cider apples and further Rosaceous plants. *European Food Research and Technology, 234*, 1033–1041.
43. Yukui, R., Wenya, W., Rashid, F., & Qing, L. (2009). Fatty acids composition of apple and pear seed oils. *International Journal of Food Properties, 12*(4), 774–779.
44. Fromm, M., Loos, H. M., Bayha, S., Carle, R., & Kammerer, D. R. (2013). Recovery and characterisation of coloured phenolic preparations from apple seeds. *Food Chemistry, 136*(3–4), 1277–1287.
45. Abbas, A., Anwar, F., & Ahmad, N. (2019). Apple (*Malus pumila*) seed oil. In M. Ramadan (Ed.), *Fruit Oils: Chemistry and Functionality* (pp. 495–504). Springer.
46. Pieszka, M., Migdał, W., Gąsior, R., Rudzińska, M., Bederska-Łojewska, D., Pieszka, M., & Szczurek, P. (2015). Native Oils from Apple, Blackcurrant, Raspberry, and Strawberry Seeds as a Source of Polyenoic Fatty Acids, Tocochromanols, and Phytosterols: A Health Implication. *Journal of Chemistry*, *2015*(1), 659541. https://doi.org/https://doi.org/10.1155/2015/659541.
47. Górnaś, P. (2015). Unique variability of tocopherol composition in various seed oils recovered from by-products of apple industry: Rapid and simple determination of all four homologues (α, β, γ and δ) by RP-HPLC/FLD. *Food Chemistry, 172*, 129–134.
48. Górnaś, P., & Rudzińska, M. (2016). Seeds recovered from industry by-products of nine fruit species with a high potential utility as a source of unconventional oil for biodiesel and cosmetic and pharmaceutical sectors. *Industrial Crops and Products, 83*, 329–338.
49. Montañés, F., Catchpole, O. J., Tallon, S., Mitchell, K. A., Scott, D., & Webby, R. F. (2018). Extraction of apple seed oil by supercritical carbon dioxide at pressures up to 1300 bar. *The Journal of Supercritical Fluids, 141*, 128–136.
50. Khalil, R. R., & Mustafa, Y. F. (2020). Phytochemical, antioxidant and antitumor studies of coumarins extracted from Granny Smith apple seeds by different methods. *Systematic Reviews in Pharmacy, 11*(2), 57–63.

51. Mohammed, E. T., & Mustafa, Y. F. (2020). Coumarins from Red Delicious Apple Seeds: Extraction, phytochemical analysis, and evaluation as antimicrobial agents. *Systematic Reviews in Pharmacy, 11*(2), 64–70.
52. Mustafa, Y. F., Mohammed, E. T., & Khalil, R. (2020). Antioxidant and antitumor activities of methanolic extracts obtained from Red Delicious and Granny Smith apples' seeds. *Systematic Reviews in Pharmacy, 11*(4), 570–576.
53. Opyd, P., Jurgoński, A., Juśkiewicz, J., Milala, J., Zduńzyk, Z., & Kroi, B. (2017). Nutritional and health-related effects of a diet containing apple seed meal in rats: The case of amygdalin. *Nutrients, 9*(10), 1091.
54. Senica, M., Stampar, F., Veberic, R., & Mikulic-Petkovsek, M. (2019). Cyanogenic glycosides and phenolics in apple seeds and their changes during long-term storage. *Scientia Horticulturae, 255*, 30–36.
55. McGhie, T. K., Hunt, M., & Barnett, L. E. (2005). Cultivar and growing region determine the antioxidant polyphenolic concentration and composition of apple grown in New Zealand. *Journal of Agricultural and Food Chemistry, 53*, 3065–3070.
56. Ahmad, I., Khalique, A., Shahid, M. Q., Rashid, A. A., Faiz, F., Ikram, M. A., & Rasool, B. (2020). Studying the influence of apple peel polyphenol extract fortification on the characteristics of probiotic yoghurt. *Plants, 9*(1), 77.
57. Ranjha, M. M. A. N., Amjad, S., Ashraf, S., Khawar, L., Safdar, M. N., Jabbar, S., & Murtaza, M. A. (2020). Extraction of polyphenols from apple and pomegranate peels employing different extraction techniques for the development of functional date bars. *International Journal of Fruit Science, 20*(sup3), S1201–S1221.
58. Rahman, N., Uddin, Md. B., Quader, Md. F. B., & Bakar, M. A. (2020). Optimization of mixed peels from banana, carrot and apple to develop high fiber biscuit. *International Journal of Natural and Social Sciences, 7*(1), 21–25.
59. Nakov, G., Brandolini, A., Hidalgo, A., Ivanova, N., Jukić, M., Komlenić, D. K., & Lukinac, J. (2020). Influence of apple peel powder addition on the physico-chemical characteristics and nutritional quality of bread wheat cookies. *Food Science and Technology International, 26*(7), 574–582.
60. Matta, E., Tavera-Quiroz, M. J., & Bertola, N. (2019). Active edible films of methylcellulose with extracts of green apple (Granny Smith) skin. *International Journal of Biological Macromolecules, 124*, 1292–1298.
61. Riaz, A., Aadil, R. M., Amoussa, A. M. O., Bashari, M., Abid, M., & Hashim, M. M. (2020). Application of chitosan-based apple peel polyphenols edible coating on the preservation of strawberry (*Fragaria ananassa* cv Hongyan) fruit. *Journal of Food Processing and Preservation, 45*, e15018.
62. Riaz, A., Lagnika, C., Abdin, M., Hashim, M., Malik, M., & Ahmed, W. (2020). Preparation and characterization of chitosan/gelatin-based active food packaging films containing apple peel nanoparticles. *Journal of Polymers and the Environment, 28*, 411–420.
63. Alpaslan, D., Dudu, T. E., & Aktas, N. (2020). Synthesis of smart food packaging from poly(gelatin-co-dimethyl acrylamide)/citric acid-red apple peel extract. *Soft Materials, 19*(1), 64–77.
64. Hoseinifar, S. H., Jahazi, M. A., Mohensi, R., Yousefi, M., Bayani, M., Mazandarni, M., & El-Haroun, E. (2021). Dietary apple peel-derived pectin improved growth performance, antioxidant enzymes and immune response in common carp, *Cyprinus carpio* (Linnaeus, 1758). *Aquaculture, 535*, 736311.

65. Susilowati, R., Jannah, J., Maghfuroh, Z., & Kusuma, M. T. (2020). Antihyperlipidemic effects of apple peel extract in high-fat diet-induced hyperlipidemic rats. *Journal of Advanced Pharmaceutical Technology & Research, 11*(3), 128–133.
66. Singh, R. J., Martin, C. E., Barr, D., & Rosengren, R. J. (2019). Immobilised apple peel biosorbent for the simultaneous removal of heavy metals from a cocktail solution. *Cogent Environmental Science, 5*(1), 1673116.
67. SAGARPA (2016). Planeación Agrícola Nacional 2017 – 2030. https://www.gob.mx/cms/uploads/attachment/file/255625/Planeaci_n_Agr_cola_Nacional_2017-2030-_parte_dos.pdf.
68. Statista. Países líderes en producción de manzanas a nivel mundial en 2022/2023. https://es.statista.com/estadisticas/636019/principales-paises-productores-de-manzanas-del-mundo/#statisticContainer (accessed May 22, 2024).
69. Barreira, J. C. M., Arraibi, A. A., & Ferreira, I. C. F. R. (2019). Bioactive and functional compounds in apple pomace from juice and cider manufacturing: potential use in dermal formulations. *Trends in Food Science & Technology, 90*, 76–87.
70. Coman, V., Teleky, B. E., Mitrea, L., Martău, G. A., Szabo, K., Călinoiu, L. F., & Vodnar, D. C. (2019). Bioactive potential of fruit and vegetable wastes. *Advances in Food and Nutrition Research, 91*, 157–225.
71. Fernandes, P. A. R., Ferreira, S. S., Bastos, R., Ferreira, I., Cruz, M. T., Pinto, A., & Wessel, D. F. (2019). Apple pomace extract as a sustainable food ingredient. *Antioxidants, 8*(6), 189.

CHAPTER 7

Insights into Nuclear Magnetic Resonance Spectroscopy as a Tool for Metabolic Markers in Pre- and Post-Harvest Sustainable Production of Food Crops

JOSÉ CARLOS ESPINOZA HICKS,
ALEJANDRO ALBERTO CAMACHO DÁVILA, and
DANIELA SÁNCHEZ-ALDANA VILLARRUEL*

Facultad de Ciencias Químicas, Universidad Autónoma de Chihuahua, Circuito Universitario s/n, Campus UACH II, Chihuahua, Chih., México

*Corresponding author.

ABSTRACT

One of the challenges in agricultural production is improving the quality of fruits and vegetables leading to enhancing their biological properties, but with the goal to reduce food waste or even achieve zero food waste because everything can be used. New technological trends in food product analysis include the area of metabolomics, with the use of nuclear magnetic resonance (NMR), where it is possible to relate the changes that occur at the molecular level in a food product, with the natural or technological processes applied in food processing. The metabolomic analysis can help in the identification of characteristic chemical markers that are used to understand and explain phenomena observed at first sight such as changes in color, firmness, rotting, adulterations, and evidence of deterioration of food, among others. This technology presents superior advantages over the use of other analysis techniques such as HPLC or GS-MS given the ease of sample preparation, the detection and identification of multiple analytes per sample, non-invasive

Food Security, Safety, and Sustainability: Validated Innovative Approaches and Technological Advancements.
José Juan Buenrostro-Figueroa and Julio César Tafolla-Arellano (Eds.)

methods, and the relatively short analysis time. Also, NMR is a potential method for food quality and authenticity purposes. In this way, this review aims to comprehensively explain and discuss the use of NMR spectroscopy as an analytical tool that allows the detection of the optimal state of maturity for harvest. Also, metabolomic markers can be used to correlate the physical, chemical, and microbiological parameters with storage conditions so that the food products can be consumed when the nutrients are in the optimal presentation. Therefore, it will help ensure the quality and safety of food products before they reach the end of their shelf life.

7.1 INTRODUCTION

Crops that are grown for human consumption are called food crops, where fruits, vegetables, cereals, oilseeds, and tubers are included. The data recently published by the Statistics Division of the Food and Agriculture Organization of the United Nations (FAO) extends until 2020. World production of primary crops has increased by 52% from 2000 to 2020. Sugarcane, rice, corn, and wheat are the main primary crops representing half of the primary crops produced in 2020. In this period, the world production of fruit increased by 55% and that of vegetables increased by 65% [1]. While vast amounts of crops are grown and harvested around the world, greenhouse gas emissions from agricultural practices have remained constant over the past 30 years, yet agricultural practices contribute significantly to climate change, with 20% of the total emissions produced by human activities. In 2019, global emissions from agriculture were almost 11 billion tons of CO_2 equivalent (CO_2eq) [1]. At the same time, agriculture turns out to be one of the practices most affected by climate change, considerably affecting crop yields, causing losses, and destruction of crops; extreme variations in temperatures cause pests and diseases to spread rapidly with devastating consequences in the agricultural sector [2].

In this sense, massive agricultural practices for food production generate a considerable environmental impact mainly due to the use of soil and water [3], coupled with the subsequent processing of food for its conservation, storage, and distribution, as well as the implementation of analytical techniques to produce foods that meet increasingly demanding quality parameters that ensure their safety.

Both the food industry, suppliers of raw materials, additives, and other components, researchers, and other professionals involved in the food production chain must have full knowledge of the composition of food products, as well as their physical and chemical properties to implement conservation processes, the establishment of optimal storage and distribution conditions, this to ensure safety and quality. In food analysis, the type of food sample, the time required for analysis, the precision, and the accuracy, directly influence the choice of analytical methods. In this sense, different approaches and procedures have been created and approved, as well as official methodologies for food analysis that are practiced in a standardized manner worldwide [4]. However, the use of analytical techniques is increasingly demanding due to the awareness of consumers about the composition of the food they consume and their selection, due to sensory aspects that are appreciated at first contact such as color, smell, and texture, among others, these properties being the main referents of food quality. In addition to this, the determination of properties that are not directly perceived by the consumer as toxic, microbiological components, and anti-nutritional factors that are mainly responsible for the safety and innocuousness of food [5].

This has led to the development and implementation of new analytical techniques for the control and analysis of crops, for the determination of the optimal harvest time, for post-harvest processing and conservation, to guarantee quality, to determine the origin, and to identify possible food adulterations. Various instrumental techniques have successfully replaced chemical experimental methods in various analytical applications, and it has been recognized that they tend to be more precise [5]. One of the main trends is the implementation of non-invasive and non-destructive analytical techniques to carry out direct analysis of food products and obtain information in a short time, with lower costs, and reducing food waste. Within these analytical tools, metabolomics can be highlighted, which is closely linked to the phenotype or characteristics of food [6].

Metabolomics is the study of metabolites, which are small molecules that are produced during metabolism in living organisms, including plants, animals, and humans. Metabolites can be amino acids, lipids, sugars, phosphates, nitrogen-containing compounds, polyols, and other low molecular weight (<1,500 Da) chemicals that are produced during metabolic processes. Such molecules can be derived from primary or secondary metabolism. The primary metabolites are related to the maintenance of physiological processes that usually focus on obtaining energy, and the secondary ones are associated with late growth phases where cell signaling, defense, and communication play an important role in the development of the organism. In both cases, the metabolites synthesized during crop growth and those that develop during the ripening and/or postharvest storage phases can be identified and quantified through various tools, which in turn can be correlated with the behavior of the food during its storage, the state of maturity, the nutritional composition, the quality, and its shelf life [7].

The metabolomic analysis techniques have been carried out through the combination of various analytical techniques such as high-performance liquid chromatography (HPLC) [8, 9], mass spectrometry (MS) [10, 11], infrared spectroscopy (IR) [12] and nuclear magnetic resonance (NMR) [13–15]. These techniques have made it possible to carry out increasingly detailed and exhaustive analyses of the physical and chemical changes suffered by different types of food [16–19].

In this chapter, we will focus on the analysis of metabolites using Nuclear Magnetic Resonance (NMR). This is a spectroscopic technique that allows the analysis of nuclei (atoms) contained in organic structures. Being the analysis of proton (^1H) the nucleus is most used for metabolic analysis due to the great abundance of this atom in organic structures [15]. This technique uses intense magnetic fields that allow the aligning of the spins contained in a sample, which are subsequently excited by radiofrequency pulses to generate movements in the spin rotation orientation, which are detected and analyzed to generate spectra containing specific signal patterns (Figure 7.1). Within the spectra obtained in a sample, three main parameters can be identified: chemical shift, coupling constants, and signal integral. Together, the interpretation of all these parameters leads to the elucidation or description of the chemical structure of a molecule [13, 20].

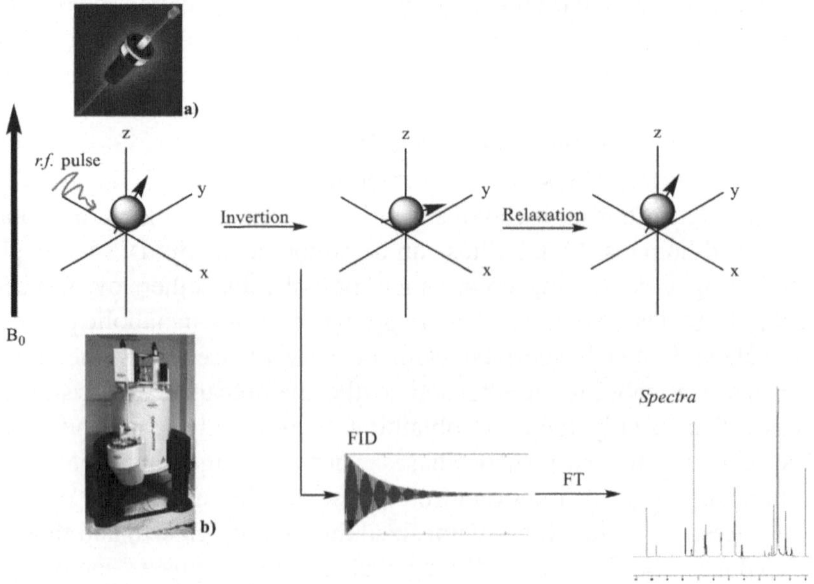

FIGURE 7.1 Steps involving an NMR experiment – (a) NMR sample tube and spinner; and (b) 400 MHz NMR instrument.

This type of analysis was initially used to determine the structure of pure molecules, mainly in organic chemistry; however, in recent years the use of this technique has diversified, mainly for the analysis of mixtures from various sources, which currently represents one of the main applications of Nuclear Magnetic Resonance (NMR). In addition to this, the development of two-dimensional techniques (2D NMR) has allowed the addition of a greater number of analyses that help to determine the various molecules contained in a mixture as well as their identification. This type of analysis allows for correlations between existing nuclei. In the structure, there are experiments of the homonuclear type (COSY, TOCSY, *J*-resolved), heteronuclear (HSQC, HMBC) as well as experiments where spatial relationships between nuclei are determined (NOESY, ROESY) [14, 21].

Together, the ^1H NMR analysis mixed with statistical analysis techniques allows metabolomic studies to be carried out on mixtures from a wide range of organisms. Within the metabolomic analysis, non-directed studies can be carried out (where comparisons are made between complete metabolic profiles from each sample analyzed) and directed (in which a metabolite or group of metabolites is monitored or quantified). For performing each of the analyses separately, the signal profile obtained by ^1H NMR is subjected to binning (where the areas of the signals are transformed into a kind of histogram that contains sums of the areas of each signal contained in a given space, generally 0.04 ppm). This data series is subsequently subjected to multiparameter analysis such as principal component analysis (PCA) among others, which helps to find similarities or differences between the different types of samples. With these results, it is possible to determine the main zones or regions responsible for the variation between the samples (which generally correspond to a particular type of metabolite) [15].

Also, with this type of analysis, it is possible to search for the main metabolites observed in the sample. There are currently a large number of databases such as the Human Metabolome Database [22], Food Database [23], Birmingham Metabolite Library [24], and Biological Magnetic Resonance Bank [25], among others, which help us to carry out the targeted search and identification of metabolites within the samples from the comparison of chemical shifts and coupling constants of each molecule. In conjunction with the two-dimensional studies, it is possible to establish within the sample the presence of low-weight molecules, mainly belonging to the primary type of metabolism in living organisms.

The benefits of the implementation of this analytical tool are many since the treatment of samples for this type of analysis generally corresponds to extractions with polar or nonpolar solvents (which will be selected depending on the type of metabolites to be analyzed) combined with sonication and centrifugation techniques to increase the quality of the extraction and eliminate the presence of insoluble solids in the sample. It is also important to mention that the use of deuterated solvents (necessary for the determination of the spectra) is generally used in samples soluble in water at 10% of the sample volume to be analyzed, which is generally found in a buffer solution of phosphates adjusted to a pH of 7.4, which considerably reduces the costs of the determination. Using pre-saturation pulse sequences, the signal corresponding to water in the spectrum can be significantly decreased, allowing a clearer analysis of the metabolites contained in the mixture (food or organism analyzed). Another advantage that can be highlighted of nuclear magnetic resonance is that in the case of food analysis, it is not necessary to carry out thermal, acid, or alkaline denaturation processes. The foregoing contributes to the sample being less manipulated and oxidation or decomposition reactions are not generated in the food; therefore, the results present fewer alterations or distortions and greater precision [26]. Due to the aforementioned, the determination of metabolic fingerprints using Nuclear Magnetic Resonance spectroscopy is presented as a versatile tool with wide application in various areas of science, including food science [27].

7.2 METABOLOMIC ANALYSIS FOR IDENTIFICATION OF CHARACTERISTIC CHEMICAL MARKERS

Metabolomic analysis has proven to be a useful tool in determining changes in the metabolic profile of various samples with applications in the areas of medicinal chemistry, biotechnology, botanical classification, and environmental and food science. The great relevance of this type of study is demonstrated by the growing number of scientific articles published in the last five years. Figure 7.2 shows the continuous growth of publications in the area in the period from 2001 to 2022. The graph shows that the articles related to metabonomic studies present an apparent decrease, which is aimed at measuring metabolic changes in living systems before physiological stimuli or genetic modification, while metabolomic studies continuously experience a gradual increase in the same period. The latter represents more general publications since all studies involving metabolites are integrated, and metabonomic studies can be included [28].

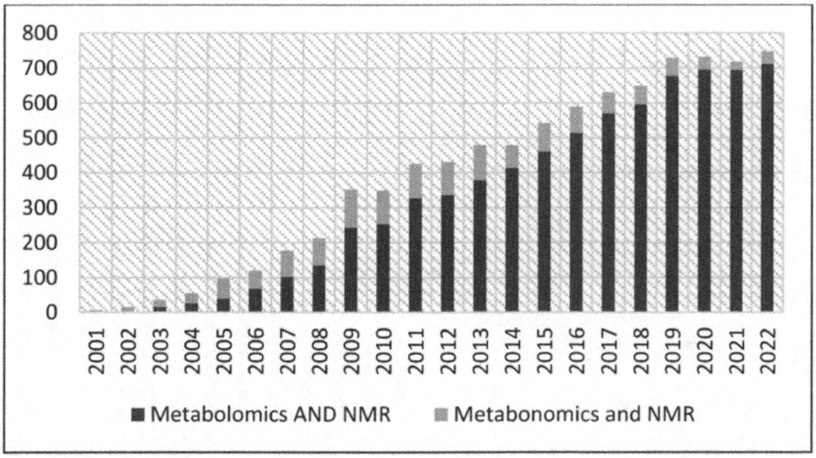

FIGURE 7.2 Number of scientific publications involving "metabonomic" and "metabolomics" studies.

Source: Data obtained from www.scopus.com, using the word metabolomics or metabolomics located in the title, abstract, and keywords.

In the area of food science, metabolomic studies using the Nuclear Magnetic Resonance technique show a similar trend to the previous one. Figure 7.3 shows the data obtained from published scientific articles related to the words food chemistry (orange), processed (gray), food microbiology (yellow), crop metabolomics (blue), and harvest metabolomics (green) linked to the NMR technique. In all cases, a gradual increase in the number of related publications is observed; however, it is important to highlight the increase in publications related to metabolomics, food cultivation, and harvesting, which makes this area of knowledge an opportunity for the development of related scientific research [28].

Within metabolomic studies, one of the main approaches is the search for biomarkers, defined as those metabolites derived from primary or secondary metabolism that are distinctive of the analyzed sample. The identification of these can be associated with many metabolic responses that explain phenomena or differentiate organisms. In food chemistry, many studies have revealed metabolic markers associated with differentiating quality, authenticity, designation of origin, and adulteration, placing this analytical technique in a privileged position in science. These metabolic markers are mainly associated with primary metabolism since they are low molecular weight compounds that can be easily distinguished within the mixtures analyzed by metabolomics [7, 29].

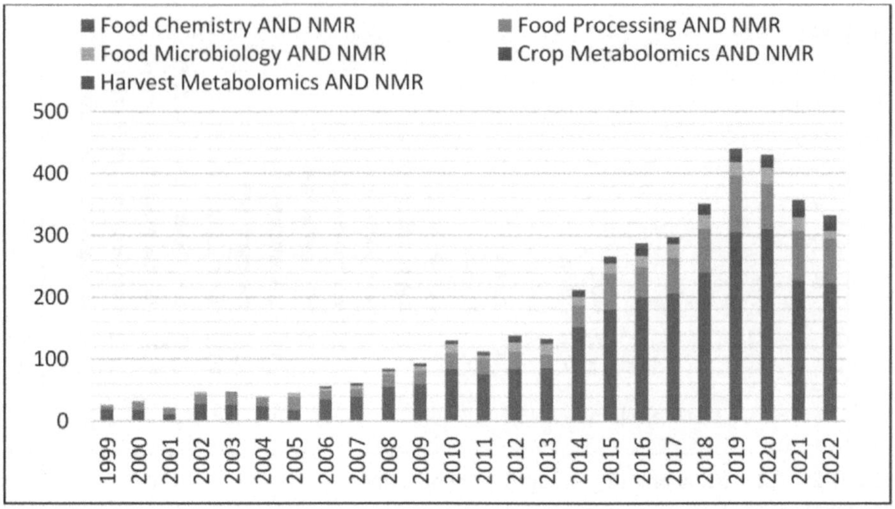

FIGURE 7.3 Number of scientific papers published related to food chemistry (orange), food processing (gray), food microbiology (yellow), crop metabolomics (blue), and harvest metabolomics (green) related to NMR analysis.

Source: Data obtained from www.scopus.com, the search was realized using the words anteriorly mentioned located in the title, abstract, or keywords.

Although the metabolite comes mainly from primary metabolism, it is possible to associate increases or decreases in concentration with secondary metabolism processes, which can lead to finding biochemical and physiological responses to the changes that are occurring in the food.

Initially, metabolic studies were aimed at determining the designation of origin, mainly for beverages such as wine [30, 31], coffee [32], tea [33], and various citrus juices [34, 35]. This provided information to consider this type of analysis also in the identification of adulterations in a wide variety of products such as honey, milk, meat products, juices, and sugary drinks, among others [36–38].

Currently, many international organizations are beginning to consider and associate this type of analysis as quality control techniques on a routine basis to guarantee the authenticity, designation of origin, and quality of food products.

7.3 NUCLEAR MAGNETIC RESONANCE SPECTROSCOPY-BASED METABOLITE PROFILING ON FOOD CROPS

Next, some examples of the metabolomic analysis carried out by ^{1}H NMR will be discussed, focused on the determination of metabolites involved in the development of the cultivation of various food products, where harvest times are related to nutritional quality and different fertilization treatments on cultivation, environmental parameters with the composition of the crop, as well as the varieties of the products associated with the presence or variation of concentration in metabolites with a focus on the most recent publications.

The cultivation of green tea (*Camellia sinensis*) has been the subject of metabolomic studies using ^{1}H HR-MAS NMR (high-resolution magic angle spinning) and ^{1}H-NMR. The studies have been directed at the relationship between environmental parameters such as relative humidity, maximum and minimum temperatures reported, solar radiation, and precipitation, and their correlation with changes in metabolic components at different harvest times. Studies have succeeded in identifying linoleic acid, theanine, caffeine, epigallocatechin gallate, epigallocatechin, epicatechin gallate, and epicatechin as main metabolic markers, as well as sugars typically found in tea such as α-glucose, β-glucose, and sucrose.

The main metabolite that could be correlated with environmental parameters was theanine, where it was found that high solar radiation causes a drop in the concentration of the metabolite. Another of the metabolites related to high solar radiation was epicatechin, where the highest concentrations were found. Finally, caffeine, which is considered together with theanine and epicatechin as a quality indicator, showed its highest concentration in December without finding a clear relationship with an environmental parameter [39].

Another example of the use of production and cultivation metabolomics is the one carried out on purslane in different areas of Mexico (Xochimilco, Mixquic, and Cuautla), which were cultivated under hydroponic conditions and were harvested at different times (32, 39, and 46 days). The analysis was able to identify 39 main metabolites of the aerial part of purslane, the principal component analysis (PCA), and the discriminant analysis of orthogonal partial least squares (OPLS-DA), which separate the crops into three main groups. Interestingly, the most notable changes in the metabolite content were observed at longer harvest times, where it was determined that the compounds that differentiate this culture time are glucose, fructose, galactose, pyruvate, choline, and 2-hydroxybutyrate. With the data generated in the study, the purslane harvest time can be correlated with its nutritional quality, which was observed 46 days after the appearance of purslane [40].

Another use of metabolomic analysis has been reported for lettuce cultivation, where three different types of biofertilization treatments were evaluated. One used compost from aerobic fermentation of organic matter, a second treatment involved lyophilized *Spirulina* (*Arthrospira platensis*) being added, and a third treatment involved the *Spirulina* microalga being exposed to fragmented DNA of the pathogen *F. oxysporum* (a common pathogen for this type of crop) applied foliar. The PCA chemometric analysis found a significant reduction in the presence of formic acid, fumaric acid, aromatic amino acids, and nucleosides, while an increase in ethanolamine concentration was observed. In treatment number three, where *Spirulina* exposed to fragmented DNA material was added, it was possible to determine a reduction in the concentration of phenolic acids and an increase in the concentration of citric acid and glutamine. In the culture, it was found that the treatment with *Spirulina* in contact with fragmented DNA presented the lowest frequency of infection by the pathogen *F. oxysporum*. This study relates to the concentration of phenolic acids, citric acid, and glutamine in the control of infection by *F. oxysporum*. However, additional studies are needed to describe the mechanism by which the fragmented DNA of the pathogen exposed to the microalga *Spirulina* helps to generate resistance to infection in the lettuce crop [41].

In tomato (*Solanum lycopersicum* L.), metabolomic studies have been carried out aimed at the search for metabolic biomarkers involved in the mechanisms of resistance to infection by the pathogen *Ralstonia solanacearum*. Using two crops resistant to this infection (Permata and Hawaii, 7996), the metabolomic profile of the different crops subjected to infection by the pathogen was determined and compared with a highly susceptible crop (GM2). It was found that leucine and valine are the metabolites that presented the main variations after the inoculation of the pathogen. The resistant cultivars showed a lower concentration of valine and an increase in the concentration of leucine compared to the susceptible variety. The study describes that the high concentration of leucine may be related to the formation of a leucine-rich receptor, which plays an important role in the plant's immune system against bacterial infection [42].

Sorghum (*Sorghum bicolor*) has also been the target of metabolomic analysis, where 12 crop fields in the Emilia-Romagna region of Italy were sampled at different stages of growth and maturation of the sorghum crop. In this analysis, it was found that Dhurrin, a cyanogenic glycoside, was one of the main biomarkers related to the quality and development of the sorghum crop. This analysis found that *p*-glucozyloxy-2-hydroxyphenylacetic acid was identified as the metabolite that mainly accumulates during the ripening period of sorghum. Interestingly, this study also found that chlorogenic acid was related to attacks by predators. With these results, it is possible to associate the concentration of Dhurrin with the quality of the sorghum crop, which will lead to improving agricultural practices [43].

Red beetroot has also been the target of metabolomic analysis, where an attempt was made to associate the metabolomic profile of beetroot juice harvested at 3 different months for three consecutive years in Italy. It was possible to determine that in the years 2016 and 2017 where the harvest was carried out in 3 months, an increase in the content of choline and betaine was found, while in 2018 a higher content of amino acids and dopamine was found. It was determined that 2018 was the year where the lowest temperatures were recorded, which can be related to the difference observed in the metabolic profile of this year. With these data and due to the uses of red beetroot, the production of metabolites can be associated with the modulation between the use of beetroot as a source of dyes or for its use as food. This is related to the betaine content, which is considered one of the main precursors of dyes extracted from beetroot [44].

With a focus on the circular economy, some crops are also analyzed to revalue the quality of the nutrients they contain and their future application as biostimulants or food supplements. An example of this is the study carried out on cauliflower and artichoke where the leaves, stems, and florets in the case of cauliflower, to determine the profile of metabolites contained in this type of crops, using a combined study of aqueous and organic phases. In general, the study manages to identify metabolites corresponding to free amino acids, organic acids, sugars, polyols, polyphenols, amines, glucosinolates, fatty acids, phosphor- and galacto-lipids, sterols, and sesquiterpene lactones in both crops. It is possible to differentiate in the case of artichokes, metabolites with reports of health benefits such as *chiro*-inositol, *scyllo*-inositol, sesquiterpene lactones (cynaropicrin, grosheimin, dehydrocynaropicrin), inulins, and chlorogenic acid. In the specific case of cauliflower, it was possible to identify bioactive compounds that contain sulfur in their structure, such as *S*-methyl-*L*-cysteine *S*-oxide and glucosinolates. Regarding specifically the type of tissue used for the analysis, it was possible to distinguish that the largest variation corresponds to the type of tissue, generally finding that the leaves are richer in amino acids and amines and poor in sugars concerning the stems. With these results, the study demonstrates that cauliflower leaves and stems can be subjected to extraction strategies that promote the reuse of these tissues (which are commonly discarded) and can be incorporated as functional foods or food supplements [45].

7.4 METABOLIC CHARACTERIZATION OF FOOD CROPS DURING PREHARVEST DEVELOPMENT, RIPENING, AND POSTHARVEST SHELF-LIFE

In this section, metabolic studies carried out on fruits will be addressed, mainly focusing on the parameters of pre and postharvest, maturation, and lifetime after harvesting the fruit related to the variation in the main metabolites described.

Postharvest metabolic analyses in citrus have been described, specifically in oranges and mandarins (using albedo, flavedo, and juice). The focus was to analyze the physiological changes that occur in the fruit by comparing the fresh fruit against fruit storage at 4°C for 10 days and the fruit treated with imazalil sulfate (IMZ), which is a fungicide frequently used to prevent infection by pathogens.

Principal component analysis manages to differentiate the samples for each treatment, mainly showing an effect of both storages at 4°C (where choline, leucine, malate, citrate, isoleucine, sucrose, phenylalanine, valine, arginine, and trigonelline) as well as treatment with imazalil sulfate (alanine, valine, succinate, proline, phenylalanine, leucine, arginine, adenosine being altered), mainly in the albedo and flavedo tissues. In the case of juice, only 4 altered metabolites corresponding to adenosine, malonate, and proline were detected. These results mainly highlight that the fruits undergo physiological alterations that affect the metabolic components, which are reflected in quality, in addition to proposing a whole series of metabolite indicators of said changes linked to each treatment [46].

Metabolomic studies focused on harvest periods have also been described. An example of this is the use of cane sugar collected in a period from November to January in Japan, where it is reported that the main variations found in these periods correspond to acid *trans*-aconitic, threonine, and alanine. These results propose these metabolites as possible candidates to indicate the best times to harvest cane sugar; however, it is indicated that a more robust study is needed to find better conditions [47].

Postharvest metabolomic studies on hazelnuts have also been reported, where different storage conditions were analyzed, varying temperatures (40, 25, 19, and 10°C), humidity percentages (75, 60, 50, and 25%), and storage time (24 weeks to 18 months) using a polar extract (phosphate buffer) and a non-polar extract (chloroform), making it possible to distinguish the analysis by principal components between each type of storage. The study found that inadequate storage conditions are related to an increase in the concentration of fumaric and lactic acid, glucose, fructose, and choline, and a decrease in the concentration of acetic acid. With this study, the previously described metabolites are proposed as possible markers of poor storage in hazelnuts [48].

The differentiation of citrus varieties and their relationship with consumer acceptability has also been carried out using metabolomic approaches. In this case, the pulp from 8 varieties of mandarin cultivated in Uruguay was described, and the analysis of the aqueous extract of the mandarin varieties achieved by chemometric techniques managed to metabolically differentiate each citrus variety, where the signals of sugars such as sucrose, glucose, and citric acid show the main differences. This study also describes the metabolic stability of mandarins in consecutive harvest times. The PCA analysis did not show significant differences in the metabolic composition of mandarins harvested in two periods (2015 and 2016), considering that environmental conditions could vary in the two seasons. Focusing on the maturation period (4 months after the optimal consumption period), the OPLS-DA analysis describes that the main changes occur at the sugar level, mainly in the sucrose content, and an increase in glucose and fructose is observed, while the signals corresponding to organic acids remain without notable changes. These changes are associated with the metabolism of sugars, which are notable in the maturation processes. With these results, it is possible to highlight the great sensitivity of the technique to associate different varieties of the fruit, such as its ripening stages, which in the future can be associated with the optimization of harvest times and fruit consumption [49].

Another interesting approach in the use of metabolomics by ^{1}H NMR is a study carried out on fruits such as grapes, apples, and peaches, where the metabolites contained in these types of fruit were evaluated when using them with the peel or peeling the fruits (pulp) for consumption. The objective of this study is differences in the presence of metabolites in the pulp and peel of these fruits after harvesting and maturation, with a focus on determining the nutritional quality of the use of fruits in healthy eating habits. This study managed to find several metabolites with health benefits which were found mainly in peels instead of the pulp of the fruit. Gallocatechin, epigallocatechin, and epicatechin are found in apples and peaches, quercetin which was only found in apples, chlorogenic acid in apples and grapes, neochlorogenic acid in apples and peaches, *p*-coumaric acid in grapes, and catechin and phloridzin in apples were identified. All the metabolites described have antioxidant capacity and participate in glucose metabolism. With these results, it is concluded that the consumption of the whole fruit entails greater health benefits, due to the presence of a greater number of metabolites that have a beneficial impact on nutrition [50].

It is known that during storage, food undergoes changes that influence its quality and shelf life. The effect of storage on fruits has also been described using metabolomic analysis by ^{1}H NMR. An example of this is the study carried out on melons, of which it is known that high temperatures at the time of harvest and storage significantly increase the metabolism of the fruit, leading to a reduction in the shelf life in the fruit market. One of the treatments used when harvesting melons is to apply hydrocooling at 4°C at different periods to significantly reduce the internal temperature of the melon, resulting in longer storage and longer useful life of the fruit. By combining methods to extend the useful shelf life of the melon and focusing on the changes in its metabolic composition, it has been possible to find that when studying the rind and pulp of the fruit after storing for 4 weeks and applying hydro-cooling treatments, each cluster was perfectly differentiated by principal component analysis (PCA) based on the type of hydro-cooling applied and the weeks of storage. This study showed that the treatment with hydrocooling, where an internal temperature of 14.8°C is reached, was the one that showed fewer variations in the first 2 weeks of melon storage. Also, the changes in aromatic amino acids (phenylalanine, tryptophan, and tyrosine) were lower during this treatment and time. In the case of phenolic compounds, hydrocooling showed an increase in the concentration of phenolic compounds present in the peel and pulp. With these results, as can be seen, it is possible to identify optimal storage times and temperatures that guarantee a high nutritional value of the fruit, and it is possible in the future to associate the metabolites found as storage time biomarkers [51].

As evidenced by NMR metabolomic studies, it is possible to obtain a broader profile of the changes that foods undergo after being harvested, stored, or during the fruit's ripening processes. What is interesting here is that these studies contain interesting proposals about the metabolites responsible for or associated with changes related to the processes. This sets the standard for the development of targeted analysis techniques where postharvest, harvest, ripening, or storage states can be associated with different crops or fruits to increase shelf life, as well as the quality of the food.

7.5 CONCLUSIONS AND FUTURE PERSPECTIVE

As has been addressed in recent years, the number of studies carried out at all stages that involve the growth and processing of food has increased greatly. It is important to highlight that studies on harvest times, maturation, and food processing have shown results that may be of interest to apply to sustainable agriculture and aimed at obtaining products of high nutritional value. Although this type of study is increasing, the direct application of these techniques in the field is still under development. As the metabolomics technique was analyzed using the ^{1}H-NMR analysis, it is presented as an excellent option to determine and associate metabolic components with quality parameters in the products, thus allowing to determine the best harvest times, the optimal ripening points, and add nutritional value to the product. In addition, the distribution and storage processes can be improved by relating metabolites to the degree or evolution in the maturation of food crops. Another of the great advantages of the technique is the great versatility that it demonstrates due to the little limitation in the type of samples to be analyzed, as well as the great variety of extracts with different components that can be analyzed, thus allowing more robust analyses with high precision.

The association of biomarkers with cultivation, harvesting, and fruit ripening processes represents one of the greatest advantages.

With the above, it has been shown that metabolomics can be useful for many of the main future objectives of agriculture and food production, such as sustainability, food quality, safety, storage, and nutrition.

KEYWORDS

- **cultivation**
- **food quality**
- **harvest**
- **nutrition**
- **resonance magnetic nuclear spectroscopy**
- **storage**
- **sustainability**
- **sustainable food crops**

REFERENCES

1. FAO (2021). Emissions from agriculture and forest land. Global, regional, and country trends 1990–2019. FAOSTAT Analytical Brief Series No 25. Rome 1-5pp. ISSN 2709-0078 [Online] Retrieved from https://openknowledge.fao.org/server/api/core/bitstreams/faa7344e-b2b8-42b8-ab0a-43f6523edebd/content.

2. Arteaga, L. E., & Burbano, J. E. (2018). Efectos del cambio climático: Una mirada al Campo. *Revista de Ciencias Agrícolas, 35*(2), 79–91.
3. Ritchie, H., Rosado, P. and Roser, M. (2022) *Environmental Impacts of Food Production*. OurWorldInData.org. Retrieved May 27, 2024 from, https://ourworldindata.org/environmental-impacts-of-food
4. Khan, M. K. I. (Ed.). (2019). *Advances in Noninvasive Food Analysis*. CRC Press. New York, NY.
5. Maan, A. A., & Hafeez, M. A. (2019). Introduction. In *Advances in Noninvasive Food Analysis* (pp. 1–8). CRC Press. New York, NY.
6. Jacobs, D. M., Van Den Berg, M. A., & Hall, R. D. (2021). Towards superior plant-based foods using metabolomics. *Current Opinion in Biotechnology, 70*, 23–28.
7. Hatzakis, E. (2019). Nuclear magnetic resonance (NMR) spectroscopy in food science: A comprehensive review. *Comprehensive Reviews in Food Science and Food Safety, 18*(1), 189–220.
8. Esteki, M., Simal-Gandara, J., Shahsavari, Z., Zandbaaf, S., Dashtaki, E., & Vander Heyden, Y. (2018). A review of the application of chromatographic methods, coupled to chemometrics, for food authentication. *Food Control, 93*, 165–182.
9. Wilson, I. D., Plumb, R., Granger, J., Major, H., Williams, R., & Lenz, E. M. (2005). HPLC-MS-based methods for the study of metabonomics. *Journal of Chromatography B, 817*(1), 67–76.
10. Liebal, U. W., Phan, A. N., Sudhakar, M., Raman, K., & Blank, L. M. (2020). Machine learning applications for mass spectrometry-based metabolomics. *Metabolites, 10*(6), 243.
11. Ren, J. L., Zhang, A. H., Kong, L., & Wang, X. J. (2018). Advances in mass spectrometry-based metabolomics for investigation of metabolites. *RSC Advances, 8*(40), 22335–22350.
12. Martens, J., van Outersterp, R. E., Vreeken, R. J., Cuyckens, F., Coene, K. L., Engelke, U. F., & Oomens, J. (2020). Infrared ion spectroscopy: New opportunities for small-molecule identification in mass spectrometry – A tutorial perspective. *Analytica Chimica Acta, 1093*, 1–15.
13. Emwas, A. H., Roy, R., McKay, R. T., Tenori, L., Saccenti, E., Gowda, G. N., & Wishart, D. S. (2019). NMR spectroscopy for metabolomics research. *Metabolites, 9*(7), 123.
14. Wishart, D. S. (2019). NMR metabolomics: A look ahead. *Journal of Magnetic Resonance, 306*, 155–161.
15. Vignoli, A., Ghini, V., Meoni, G., Licari, C., Takis, P. G., Tenori, L., & Luchinat, C. (2019). High-throughput metabolomics by 1D NMR. *Angewandte Chemie International Edition, 58*(4), 968–994.
16. Fraga-Corral, M., Carpena, M., Garcia-Oliveira, P., Pereira, A. G., Prieto, M. A., & Simal-Gandara, J. (2022). Analytical metabolomics and applications in health, environmental, and food science. *Critical Reviews in Analytical Chemistry, 52*(4), 712–734.
17. Gao, Y., Hou, L., Gao, J., Li, D., Tian, Z., Fan, B., & Li, S. (2021). Metabolomics approaches for the comprehensive evaluation of fermented foods: A review. *Foods, 10*(10), 2294.
18. Zhang, A., Sun, H., Wang, P., Han, Y., & Wang, X. (2012). Modern analytical techniques in metabolomics analysis. *Analyst, 137*(2), 293–300.

19. Griffiths, W. J., Koal, T., Wang, Y., Kohl, M., Enot, D. P., & Deigner, H. P. (2010). Targeted metabolomics for biomarker discovery. *Angewandte Chemie International Edition, 49*(32), 5426–5445.
20. Wishart, D. S., Cheng, L. L., Copié, V., Edison, A. S., Eghbalnia, H. R., Hoch, J. C., & Uchimiya, M. (2022). NMR and metabolomics – A roadmap for the future. *Metabolites, 12*(8), 678.
21. Lhoste, C., Lorandel, B., Praud, C., Marchand, A., Mishra, R., Dey, A., Bernard, A., Dumez, J-N., & Giraudeau, P. (2022). Ultrafast 2D NMR for the analysis of complex mixtures. *Progress in Nuclear Magnetic Resonance Spectroscopy, 130*, 1–46.
22. HMDB. (2023). *Human Metabolome Database*. https://hmdb.ca (accessed on 27 May 2024).
23. FooDB. (2023). *Food Database*. https://foodb.ca (accessed on 27 May 2024).
24. BML-NMR. (2023). *Birmingham Metabolite Library*. http://www.bml-nmr.org (accessed on 27 May 2024).
25. BMRB. (2023). *Biological Magnetic Resonance Bank*. https://bmrb.io (accessed on 27 May 2024).
26. Dumez, J. N. (2022). NMR methods for the analysis of mixtures. *Chemical Communications, 58*(100), 13855–13872.
27. Razzaq, A., Sadia, B., Raza, A., Khalid Hameed, M., & Saleem, F. (2019). Metabolomics: A way forward for crop improvement. *Metabolites, 9*(12), 303.
28. Scopus. (2023). *Elsevier Scopus Citation Database*. www.scopus.com (accessed on 27 May 2024).
29. Consonni, R., & Cagliani, L. R. (2019). The potentiality of NMR-based metabolomics in food science and food authentication assessment. *Magnetic Resonance in Chemistry, 57*(9), 558–578.
30. Viskić, M., Bandić, L. M., Korenika, A. M. J., & Jeromel, A. (2021). NMR in the service of wine differentiation. *Foods, 10*(1), 120.
31. Le Mao, I., Martin-Pernier, J., Bautista, C., Lacampagne, S., Richard, T., & Da Costa, G. (2021). 1H-NMR metabolomics as a tool for winemaking monitoring. *Molecules, 26*(22), 6771.
32. Coqueiro, A., Baqueta, M. R., Mandrone, M., & Poli, F. (2021). Coffee assessment using 1H NMR spectroscopy and multivariate data analysis: A review. *Applications of NMR Spectroscopy: Volume 9, 9*, 35.
33. Jiang, C. K., Ma, J. Q., Apostolides, Z., & Chen, L. (2019). Metabolomics for a millenniums-old crop: Tea plant (*Camellia sinensis*). *Journal of Agricultural and Food Chemistry, 67*(23), 6445–6457.
34. Xu, L., Xu, Z., & Liao, X. (2022). A review of fruit juice authenticity assessments: Targeted and untargeted analyses. *Critical Reviews in Food Science and Nutrition, 62*(22), 6081–6102.
35. Dasenaki, M. E., & Thomaidis, N. S. (2019). Quality and authenticity control of fruit juices review. *Molecules, 24*(6), 1014.
36. Yong, C. H., Muhammad, S. A., Nasir, F. I., Mustafa, M. Z., Ibrahim, B., Kelly, S. D., & Seow, E. K. (2022). Detecting adulteration of stingless bee honey using untargeted 1H NMR metabolomics with chemometrics. *Food Chemistry, 368*, 130808.
37. Akhtar, M. T., Samar, M., Shami, A. A., Mumtaz, M. W., Mukhtar, H., Tahir, A., Shahzad-ul-Hussan, S., Chaudhary, S. U., & Kaka, U. (2021). 1H-NMR-based metabolomics:

An integrated approach for the detection of adulteration in chicken, chevon, beef, and donkey meat. *Molecules, 26*(15), 4643.

38. Sobolev, A. P., Thomas, F., Donarski, J., Ingallina, C., Circi, S., Marincola, F. C., & Mannina, L. (2019). Use of NMR applications to tackle future food fraud issues. *Trends in Food Science and Technology, 91*, 347–353.
39. Santos, M. de F. C., Rech, K. S., Dutra, L. M., Menezes, L. R., Santos, A. D. D. C., Nagata, N., Stefanello, M. E. A., & Barison, A. (2023). 1H HR-MAS NMR chemical profile and chemometric analysis as a tool for quality control of different cultivars of green tea (*Camellia sinensis*). *Food Chemistry, 408*, 135016.
40. Montoya-García, C. O., García-Mateos, R., Magdaleno-Villar, J. J., Volke-Haller, V. H., Villa-Ruano, N., Zepeda-Vallejo, L. G., & Becerra-Martínez, E. (2023). NMR-based metabolomics to determine the fluctuation of metabolites in hydroponic purslane crops at different harvesting times. Food Research International, *166*, 112489.
41. Lanzotti, V., Anzano, A., Grauso, L., Zotti, M., Sacco, A., Senatore, M., & Mazzoleni, S. (2022). NMR metabolomics and chemometrics of lettuce, *Lactuca sativa* L., under different foliar organic fertilization treatments. *Plants, 11*(16), 2164.
42. Murti, R. H., Afifah, E. N., & Nuringtyas, T. R. (2021). Metabolomic response of tomatoes (*Solanum lycopersicum* L.) against bacterial wilt (*Ralstonia solanacearum*) using 1H-NMR spectroscopy. *Plants, 10*(6), 1143.
43. Mandrone, M., Chiocchio, I., Barbanti, L., Tomasi, P., Tacchini, M., & Poli, F. (2021). Metabolomic study of sorghum (*Sorghum bicolor*) to interpret plant behavior under variable field conditions in view of smart agriculture applications. *Journal of Agricultural and Food Chemistry, 69*(3), 1132–1145.
44. Giampaoli, O., Sciubba, F., Conta, G., Capuani, G., Tomassini, A., Giorgi, G., & Miccheli, A. (2021). Red beetroot's NMR-based metabolomics: Phytochemical profile related to development time and production year. *Foods, 10*(8), 1887.
45. Ingallina, C., Di Matteo, G., Spano, M., Acciaro, E., Campiglia, E., Mannina, L., & Sobolev, A. P. (2023). Byproducts of globe artichoke and cauliflower production as a new source of bioactive compounds in the green economy perspective: An NMR study. *Molecules, 28*(3), 1363.
46. Montgomery, K. H., Brar, G., & Krishnan, V. V. (2021). Metabolomics study at the postharvest conditions of cold storage and fungicide (Imazalil Sulfate) treatment in navel oranges and clementine mandarins. *ACS Agricultural Science and Technology, 2*(1), 79–89.
47. Date, Y., Ishikawa, C., Umeda, M., Tarumoto, Y., Okubo, M., Tamura, Y., & Ono, H. (2022). Sugarcane metabolome compositional stability in pretreatment processes for NMR measurements. *Metabolites, 12*(9), 862.
48. Shakiba, N., Lösel, H., Wenck, S., Kumpmann, L., Bachmann, R., Creydt, M., Seifert, S., Fischer, M., & Hackl, T. (2023). Analysis of hazelnuts (*Corylus avellana* L.) stored for extended periods by 1H NMR spectroscopy monitoring storage-induced changes in the polar and nonpolar metabolome. *Journal of Agricultural and Food Chemistry, 71*(6), 3093–3101.
49. Migues, I., Hodos, N., Moltini, A. I., Gambaro, A., Rivas, F., Moyna, G., & Heinzen, H. (2021). 1H NMR metabolic profiles as selection tools of new mandarin cultivars based on fruit acceptability. *Scientia Horticulturae, 287*, 110262.

50. Song, J., Jeong, J., Kim, E. H., & Hong, Y. S. (2023). A strategy for healthy eating habits of daily fruits revisited: A metabolomics study. Current Research in Food Science *6*, 100440.
51. Zainal, B., Ding, P., Ismail, I. S., & Saari, N. (2019). 1H NMR metabolomics profiling unveils the compositional changes of hydro-cooled rockmelon (*Cucumis melo* L. reticulatus cv glamour) during storage related to in vitro antioxidant activity. *Scientia Horticulturae, 246*, 618–633.
52. FAO. (2022). Agricultural production statistics. 2000–2020. FAOSTAT Analytical Brief. Series No. 41. Rome. Retrieved from https://openknowledge.fao.org/server/api/core/bitstreams/b75223dd-4e30-43aa-85a9-4c587753b027/content.

CHAPTER 8

General Perspective of Iron Biofortification: A Promising Tool Against Mineral Disorder Nutrition

ALEJANDRO PALACIO-MÁRQUEZ,[1] RAMIRO BAEZA JIMÉNEZ,[2] JOSÉ JUAN BUENROSTRO-FIGUEROA,[2] and EDWIN ROJO-GUTIÉRREZ[2*]

[1]Facultad de Ciencias Agrícolas y Forestales, Universidad Autónoma de Chihuahua, Chihuahua, Mexico

[2]Centro de Investigación en Alimentación y Desarrollo, Unidad Delicias, Chihuahua, México

[*]Corresponding author.

ABSTRACT

Iron is an essential element involved in several metabolic processes of vital importance in humans, such as oxygen transport, DNA synthesis and repair, cell cycle regulation, and electron transport. It is a substantial mineral for the healthy growth and development of the body. To meet iron requirements, a daily intake amount is needed that varies depending on the age, sex, and diet (from 0.27 to 27 mg/day). Despite iron being mostly recycled and preserved by the body (losing only ≈1–2 mg/day), it generally presents low absorption concentrations (5–35%). Moreover, iron requirements increase during growth spurts or menstruation periods. Approximately 33% of the world's population is at risk of iron deficiency, and an estimated 42% of children (<5 years old) and 40% of pregnant women are anemic according to the World Health Organization, representing a serious global public health disorder. The main causes of iron deficiency are attributed to a low intake, excess blood loss due to different diseases (pathologic infections), and absorption issues of the mineral. Supplementation and food-based intervention strategies are

Food Security, Safety, and Sustainability: Validated Innovative Approaches and Technological Advancements.
José Juan Buenrostro-Figueroa and Julio César Tafolla-Arellano (Eds.)

promising approaches to preventing and overcoming iron deficiency. An innovative strategy for the problem of iron deficiency in the diet is biofortification, defined as the process of increasing the bioavailable concentrations of essential elements in edible parts of cultivated plants through agronomic management. Previous works have demonstrated that biofortified staple crops with iron are effective in increasing their concentration levels in the human diet, placing these biofortified crops as one of the key factors in achieving sustainable development goals. In the present chapter, the current status of iron biofortification will be explored, discussing biofortification's strategy potential in comparison to supplementation or functional foods in terms of effectiveness and sustainability to prevent or diminish iron deficiency.

8.1 INTRODUCTION

Iron is an essential micronutrient for the correct functioning of the human body due to its substantial physiological functions. Iron is involved in the formation of iron- and heme-containing proteins, essential for oxygen transport (respiration) and energy metabolism (electron transfer), and the synthesis of neurotransmitters, collagen, DNA, and steroid hormones. Other functions also include the detoxification of xenobiotics, and the provision of nonspecific resistance of the body for proper immune function, among others [1–3]. Since the body cannot synthesize iron, food is the main source to obtain it. However, it is often a growth-limiting mineral in the human diet, even when it is one of the most abundant elements on Earth [4]. The low bioavailability (5–35%) of dietary iron (heme and nonheme) makes it difficult to maintain the recommended values in the human body. The daily intake of iron varies depending on age, sex, diet, and physical activities [5]. According to the Institute of Medicine [6], recommended dietary allowances are 8 and 18 mg/day for males and females, respectively (19–50 years). Factors such as growth spurts, menstruation periods, pregnancy, or intensive physical exercise increase iron requirements. Blood loss (pathologic infections, abnormally heavy menstruation), inadequate iron absorption, and low iron intake are the main causes of iron deficiency (ID) [7–9]. The highest risk of ID is found in populations that have insufficient access to foods rich in absorbable iron during periods of high iron demand. These categories include children, adolescents, and women of reproductive age, particularly during pregnancy [6]. Approximately 33% of the world's population is at risk of ID, and an estimated 42% of children (<5 years old) and 40% of pregnant women are anemic according to the World Health Organization [10], representing a serious global public health disorder. It is known that food production is currently insufficient to feed the world's growing population, and what is available is deficient in several essential elements. In order to mitigate ID, a considerable amount of effort has been devoted to developing strategies to improve iron intake and its bioavailability, as well as food security. Such strategies include mainly food diversification (FD), food fortification, micronutrient supplementation, and biofortification, applied either alone or in combination. Biofortification is the process of increasing the amount and bioavailability of vitamins and minerals during the plant growth cycle through agronomic management [11]. Is one of the most practical and sustainable methods to address nutritional issues. Moreover, biofortified crops are expected to reach rural residents who have limited access to a variety of dietary options or other micronutrient therapies [12]. In the present chapter, the current state of iron biofortification will be reviewed, discussing its potential as a sustainable ID solution, its advantages, and challenges, in contrast to the rest of the aforementioned strategies.

8.2 MICRONUTRIENT DEFICIENCY PROBLEM

Micronutrients are the vitamins and minerals that we consume in trace amounts but are essential for the physiological development of the human body. They are essential for the correct use of energy biomolecules such as amino acids, lipids, and glucose [13, 14]. Despite the minimum amounts of micronutrients required, it is estimated that approximately 3 billion people have deficiencies, representing a serious global problem with a strong social impact. This problem mainly affects children, women, and the elderly, especially those in low-income countries or with unbalanced diets [15–17].

Micronutrient deficiencies generally have inconspicuous implications for human health, which is why it is called "hidden hunger," since their development is progressive over time and their effects are not noticeable until they have caused significant damage. In addition, biomarkers for micronutrient status are not available for the vast majority of the population, leaving a statistical gap for micronutrient deficiencies globally [16, 18]. It is believed that this deficiency is related to malnutrition; however, a metabolism altered by obesity causes changes in the absorption, distribution, or excretion of nutrients. This problem occurs mainly in high-income countries where people get a third of their daily calories from ultra-processed foods that are generally poor in micronutrients [19].

Among the main conditions due to a lack of micronutrients are birth defects, blindness, growth issues, emaciation, as well as cognitive deterioration, which decreases school performance and work productivity [18]. Gorji & Ghadiri [20] reported that micronutrient deficiency decreases the ability to resist infections and is a common cause of immunodeficiency, mainly in developing countries since it decreases the production of antibodies by altering the function and number of various immune cells. In addition, various studies correlate micronutrient deficiency with a higher rate of suffering from depression, finding that a deficient micronutrient content influences the modulation of the hypothalamic-pituitary-adrenal axis and alters the function of the nervous system through the activation of stress circuits [21–23].

Deficiencies in vitamin A can become the main cause of blindness in children. In addition, it has been found that this deficiency affects growth, and organ development, and generates thyroid dysfunction [13, 24]. On the other hand, iodine deficiency is related to thyroid problems and effects on IQ, causing a loss of about 13 points [25, 26]. Meanwhile, zinc deficiency is associated with impaired immune response and an increase in respiratory viral infections, particularly in the elderly [20]. There are also other essential nutrients that cause various problems in human health, such as iron [13].

8.2.1 IRON DEFICIENCY

Iron is one of the essential nutrients for the development of the human body. Within the body, iron can be a structural part of various proteins such as myoglobin, and enzymes of the oxidative system, and mainly, it is the central component of hemoglobin [27–29]. In addition, it participates in multiple enzymatic functions involved in oxygen transport, energy metabolism, cell integrity, and DNA synthesis; Fe deposits can also be found inside the cells in the form of ferritin and hemosiderin [28, 30]. Recommended dietary allowances for iron are 8 and 18 mg/day for males and females, respectively. Daily requirements for pregnant women increase to 27 mg. Iron is present in the human body as the most abundant micronutrient (≈4 g), where 75% constitutes hemoglobin [29, 31]. Anemia is a major global public health issue and one of the main consequences of ID, which affects approximately 35% of the world's population. Surprisingly, statistics have proven that most of the cases occur in developing countries (>89%), with the possibility of still increasing significantly. On the other hand, in low-income countries such as those located in the African continent, 48% of the population is affected by IDA [32–34]. According to the Global Burden of Disease Study carried out in 2016, iron deficiency anemia (IDA) is one of the 5 causes of instability in people's daily lives, especially in children under 5 years of age, adolescents, women, pregnant women, and lactating mothers [34, 35]. IDA is the cause of brain deterioration in children and the elderly, birth problems for both newborns and mothers and decreased physical abilities in adults. Among the symptoms presented by patients with IDA are paleness, fatigue, breathing difficulties, frequent headaches, diffuse and moderate alopecia, swollen tongue, dry skin, and hair. Sometimes cardiovascular problems may occur, such as heart murmur or tachycardia. ADI is frequently associated with chronic problems such as intestinal inflammation, heart failure, obesity, kidney disease, rheumatoid arthritis, and cancer [35, 36]. ID is also related to a deficient immune system, especially in the face of viral infections, since iron is a key precursor for the synthesis of reactive oxygen species (ROS) and the myeloperoxidase activity of neutrophils in defense against viruses. Adequate iron supplementation could inhibit viral infections caused by the influenza A virus or HIV [20]. Also, ID has been linked to increased absorption of toxic heavy metals such as cadmium [27].

8.3 PREVENTION STRATEGIES FOR IRON DEFICIENCY

As it was previously stated in the present chapter, ID is a serious global health problem that requires immediate and sustained intervention. Current intervention strategies include FD, food fortification, micronutrient supplementation, and biofortification. ID occurs when iron absorption cannot keep up with the metabolic demands for iron to sustain growth and replenish iron loss over an extended period, due to the depletion of iron stores [37]. In that matter, all strategies focus on enhancing the intake and bioavailability of iron. Each of them presents advantages and challenges (Table 8.1) that need to be considered.

8.3.1 FOOD DIVERSIFICATION

It is known that a healthy diet cannot be constituted by a single food to satisfy the human body's dietary requirements. Therefore, a balanced, varied, and sufficient diet (for all age and gender groups) is needed. Balanced nutrition refers to a variety of natural foods (red and white meat, fruits, vegetables, legumes, and cereals, among others) that provide essential nutrients for the correct functioning of the human body. Long-term diet diversification ensures healthy nutrition with

TABLE 8.1 Main Challenges of the Iron Strategy Interventions Used to Prevent and Treat Iron Deficiency

Strategy	Main Challenges	References
Food diversification	• Crop diversification. • Processing, marketing, and distribution of diverse food. • High prices of healthy food. • Food inhibitors. • Unhealthy eating habits. • Need robust integrated programs.	[43–45]
Iron food fortification	• Monitoring and evaluation of the status data from food fortification over targeted groups is complicated. • Gastrointestinal issues (adults and children). • Degradation of nutrients and undesirable sensorial changes. • Toxic effects. • Increase of production and final prices of food. • Massive fortification still presents ethical conflicts since the freedom of choice is compromised. • Collective efforts from the government, food industry, and scientists to develop efficient programs and regulations.	[104]
Supplementation	• Several gastrointestinal side effects. • Unpropped supplementation can lead to toxicity or ineffective effects.	[74, 75]
Biofortification	• Building consumer demand. • Integration of biofortification programs in public and private policies. • Develop crop varieties with high production and capacity to hyperaccumulate micronutrients. • Limitations in gene editing methodologies for most crops. • Increasing the bioavailability and content of micronutrients in crops simultaneously.	[87, 104]

a feasible combination of macronutrients and micronutrients [38]. Food diversification not only focuses on addressing the production and consumption of food of interest, but also its processing, marketing, and distribution potential as well as its limitations. The goal of food diversification strategies is to increase the variety of foods that are accessible and acceptable to groups at risk of malnutrition [39], or already suffering nutritional disorders such as iron deficiency. Therefore, food diversification intends to increase the intake of essential elements through diet. In that matter, iron intake could be increased by employing different food diversification strategies to overcome iron deficiency. These strategies include increasing the intake of iron-rich foods (seafood, meat, beans, dark green leafy vegetables, among others), consuming foods with components that enhance iron bioavailability, and avoiding the intake of antinutrients that could affect the iron absorption supplied by the daily diet (*see* Table 8.2). Iron derived from the food matrix can be found in two forms: Heme (animal source) and non-heme iron (plant source). Heme iron presents a higher bioavailability than non-heme iron [40, 41]. Food-based interventions aim to be part of sustainable food systems; however, in recent decades, there has been a transition from diversified cropping systems to cereal-based systems to satisfy the increasing demand for staple crops, affecting global food diversity, and thus compromising accessibility and inducing price fluctuations of food [42]. Consequently, households are unable to purchase healthy food items since their purchasing power is below the high prices of food. This limits the dietary diversification of the people, leading to an unbalanced intake of macro- and micronutrients that compromises their nutritional security [43, 44]. However, despite high income, people can easily access a more diverse diet. Unfavorable food choices lead to poor eating habits, abusing processed foods, alcohol, carbonated beverages, snacks, and junk food rich in sugars, fats, as well as red meat [19, 45]. Food diversification could be the most desirable and sustainable method to encourage healthy eating habits and prevent nutrition disorders. Nevertheless, there is still a long way ahead, and a lot of effort needs to be fully implemented around the world, since there are many factors (crop diversification, processing, distribution) involved and require robust integrated programs [43, 44].

TABLE 8.2 Food Substances That Affect Iron Absorption

Enhancers	**Iron Food Type Influenced**
Vitamin C (ascorbic acid)	Non-heme
Folic acid	
Citric acid	
Amino acids rich in cysteine	
Vitamin A	
Organic acids (acetic acid, lactic acids, tartaric acid, malic acid)	
Animal tissue (meat, fish, poultry)	Heme and non-heme
Inhibitors	
Phytates	
Tannins	
Polyphenols	
Calcium salts	
Oxalic acid	
Competitors	
Heavy metals (lead, cobalt, strontium, manganese, zinc)	

Source: Adapted from Ref. [5 & 53].

8.3.2 FORTIFICATION OF FOOD

Fortified food has been cataloged as one of the safest and most cost-effective strategies [46, 47]. It consists of adding essential micronutrients (vitamins, minerals, amino acids) to food during processing to enhance its nutrient quality intake for the benefit of public health with minimal risk [48, 49]. Three major types of food fortification (mass, targeted, and market-driven fortification) can be implemented depending on how they are designed, controlled, and delivered. Mass or universal fortification involves adding one or more micronutrients to staple foods or condiments typically consumed by the general public who have a low or deficient intake of the selected micronutrients. This kind of fortification is commonly mandatory but can also be conducted voluntarily. Either way, both are commonly regulated by governments. Targeted fortification is applied to a specific population group such as infants, children, and pregnant women, among others. Unlike mass-fortified foods, targeted fortification is rarely fortified with a single or even a small number of micronutrients. They are typically fortified with a variety of essential micronutrients that the targeted population lacks and is at risk of deficiency. Lastly, market-driven fortification focuses on the consumer's nutritional benefit from the manufactured fortified food as a marketing advantage of the company. Consumers can address nutritional issues through market-driven fortified foods, which contain essential micronutrients that lower the risk of micronutrient insufficiency [49–53]. Proper identification and suitability of a food vehicle is one of the most important requirements of general food fortification. Understanding the possible chemical interactions of the food matrix selected as a vehicle with the iron compounds intended to be added is a key factor to achieve high bioavailability of micronutrients. Moreover, the food vehicle should ideally be a popular and highly consumed food by the target population. Dietary patterns, food habits, geographical regions, and socioeconomic status are all factors to consider when selecting foods for fortification [54]. The main objective of a food vehicle used for fortification is to focus on the target population without causing significant changes in dietary patterns but being consumed on a regular basis, allowing an increase in body iron levels [55]. Among the most common iron-fortified foods are salt, sugar, cereal-based products, milk, and dairy products. However, the potential effects of iron enhancers and inhibitors must be considered to improve the efficacy of fortification [54].

Developing effective iron-fortified foods has proven to be a challenge since traditional salt-based ($FeSO_4$, $FeCl_3$) fortification methods have gastro-intestinal stability and bioavailability issues, and iron typically produces undesirable sensory effects like color, texture, and flavor changes due to the peroxidation of lipids and degradation of bioactive compounds (anthocyanins, flavonoids, tannins, and phenolic compounds), caused by the chemical interaction of iron compounds. These effects are more pronounced when iron soluble compounds (iron powder) are employed (more bioavailability) in comparison to less soluble iron compounds (lower bioavailability) [56, 57]. Additionally, some studies even have reported toxicity risks when high concentrations of heme iron are consumed [58]. The use of iron nanoparticles could also lead to health hazards since it is possible, due to their small size, to cross physiological barriers, potentially damaging tissues, or cells [59]. Another concern is that massive fortification of iron may negatively affect healthy people who already meet daily iron recommendation amounts, producing an excessive intake of iron that can lead to life-threatening conditions related to liver and heart conditions, as well as diabetes [6]. To overcome these issues, iron fortification technologies have been developed, such as encapsulation, chelation, emulsification, food-to-food, and nanotechnology. Baldelli et al. [60] encapsulated iron gluconate via spray drying. They could increase iron bioavailability by 60% and iron absorption by 38% in comparison to control (non-encapsulated powder). Darwish et al. [61] fortified yogurt employing nanocapsules (bovine serum albumin nanoparticles) containing iron. Their results showed that hemoglobin and iron plasma values of albino rats with IDA were successfully and safely restored without altering the physical or sensorial properties of the fortified yogurt. Garcés et al. [62] produced functional bacteria (*Streptococcus thermophilus* and *Lactobacillus acidophilus*) incorporating iron oxide nanoparticles into their exopolysaccharides to obtain fortified yogurt. The presence of nanoparticles had no effect on the bacteria's ability to ferment milk. Magnetic yogurts (yogurt with iron oxide nanoparticles) demonstrate a new concept in functional foods and a potential vehicle for the treatment of IDA. Athira et al. [63] developed iron-chelating peptides from whey protein as a novel alternative to traditional chemical iron fortificants for food products. These peptides could not only improve iron intake, absorption, and bioavailability but also confer biofunctional benefits including antioxidant and antidiabetic activities. Mandeka et al. [64] produced a fortified yogurt by adding casein-iron chelates, without statistically affecting its sensory properties. The fortified yogurt exhibited an iron content value of 12.3 mg/100 g, meeting the daily recommendation of a male adult (10 mg). Barbosa & García-Rojas [65] analyzed microencapsulated double emulsions (water-in-oil-in-water) as delivery systems for iron food fortification. As emulsifying agents, they used whey protein isolate and polyglycerol polyricinoleate. The emulsion presented high stability for 7 days, and high *in vitro* bioaccessibility for adults (49.54%) and infants (39.71%). Kubuga et al. [66] proposed the indigenous leaf meal *Hibiscus sabdariffa* as an alternative food for iron fortification. Their results suggested that the leaf meal enhanced women's iron status over time and may promote the healthy growth of toddlers during food shortage seasons. Agrahar-Murugkar [67] developed food-to-food fortified bread biscuits by substituting refined wheat flour with millet, oilseeds, spices, and herbs. Two treatments showed noticeable results in their study. The consumption of 50 g of treatment 1 (sesame, cumin, and moringa bread) or 100 g of treatment 4 (salty bread with soy butter) could provide up to 26–28% of the recommended daily iron amounts for children (7–9 y).

Currently, only the iron fortification of wheat flour is mandatory in 81 countries, whereas in low- and middle-income countries the results have not been as expected, resulting in only minor reductions in IDA prevalence [68]. There is no universal vehicle that alone could meet the daily recommended iron amounts for all countries and people worldwide. However, the combination of multiple fortified foods (mandatory and voluntary), where each one provides a portion of the recommended iron values, maybe a much safer and better strategy, avoiding the excessive consumption of a single food [51].

8.3.3 IRON SUPPLEMENTATION

Oral iron supplementation has settled as the gold standard intervention approach for women, infants, and children (mainly from low socio-economic status) to combat micronutrient shortages and maintain their adequate iron status [69, 70] as a safe, cost-effective, and long-term strategy [53]. Different presentations of oral iron supplements can be found in the market. Doses and formulations are varied, with the most common being ferrous iron salts like ferrous sulfate and ferrous gluconate due to their low cost and high bioavailability. Similarly to targeted food fortification, supplementation of iron is also focused on public groups at risk of iron issues such as children, adolescents (especially girls), and pregnant women. Iron supplementation has already shown successful interventions in public health, implemented by numerous nations as a strategy to treat or decrease anemia occurrence [71–73]. Despite supplemented iron having high absorption, it presents a high incidence of gastrointestinal side effects, including nausea, constipation, epigastric pain, dysbiosis, and diarrhea [5, 74, 75]. Choosing the best oral iron formulations, dosing regimens, and monitoring procedures remains a challenge. The use of unnecessarily high doses of oral iron is likely to result in toxicity with no added benefit; therefore, effective and proper use of iron supplementation is critical [76]. Stoffel et al. [77] carried out an in-depth analysis of the recommended dose and dose frequency of oral iron supplementation. They concluded that single morning doses (60–120 mg ferrous salt) at alternate-day schedules increase iron absorption while reducing the incidence of side effects. Given the evident limitations of gastrointestinal adverse effects on oral iron supplements, there is a need for novel optimized products that present high bioavailability and are suitable for all ages or genders. Pergola et al. [78] conducted a review of alternative iron supplements to conventional ferrous salts. In their study, they featured ferric citrate, ferric maltol, and sucrosomial iron as novel oral iron agents as promising therapies for patients with IDA in chronic kidney diseases. Nano-supplements are another emerging oral iron supplementation system that has shown good stability through the gastrointestinal tract and effective bioavailability in *in-vitro* trials [79]. Also, novel iron polysaccharide complexes have been explored as an iron supplement. Wang et al. [80] developed and characterized an AGSP-FE(III) complex. The complex exhibited good iron supplementary efficiency during simulated gastrointestinal digestion, as well as good stability.

8.4 IRON BIOFORTIFICATION

Biofortification has shown a great impact on developing nutritionally enhanced food crops to benefit human health. It is defined as the process of increasing bioavailable concentrations of essential elements in edible portions of plants, through traditional genetic improvement, molecular biology tools, or proper fertilization management [11–81].

Biofortification seeks to increase the nutrient content of easily accessible crops for low-income populations. For every dollar invested in biofortification, 17 dollars can be obtained as revenue, considered the most cost-effective and sustainable strategy to solve micronutrient deficiency [82, 83].

However, it is necessary to thoroughly evaluate the crop to be biofortified, considering environmental and human factors to be successfully implemented. Although mineral biofortification does not affect the sensory properties of food, highly consumed crop varieties are preferred, since they present a higher availability in the market and are commonly more accessible to the general population, such as beans, corn, wheat, potatoes, sorghum, rice, among others [82, 84]. In addition, selected crops must present overall high yields, stability, quality, and resistance to pathogens, to encourage farmers to use biofortified crops or implement biofortification methods [84, 85].

In 2016, Howarth Bouis from the HarvestPlus organization, and María Andrade, Robert Mwanga, and Jan Low from the IPC (International Potato Center), were awarded the World Food Prize for their work related to food biofortification. Howarth Bouis, along with the HarvestPlus organization, successfully developed a wide variety of biofortified crops with outstanding results. Among the biofortified crops were corn, potatoes, and cassava containing vitamin A values that could provide over 50% of the recommended daily intake, beans and millets containing 60% of the daily recommended amounts of iron, and wheat and rice varieties with zinc levels so high that they could provide up to 80% of the daily recommendation [83–86]. Biofortification, due to its advantages and benefits, has become a relevant and sustainable tool in the last decade to overcome nutrient deficiencies such as zinc, selenium, iodine, vitamin A, or iron [87, 88].

8.4.1 IRON BIOFORTIFICATION STRATEGIES OF CROPS

In the case of iron, staple crops usually have low levels of this micronutrient (Table 8.3). The diets of most of the world's population are based on these starch-rich crops but poor in nutrients. Additionally, foods from plants present significantly lower bioavailability of micronutrients than meat due to the presence of antinutrients such as phytic acid and some polyphenols. Another factor is that the parts where the micronutrient is concentrated are not consumed. Such is the case of cereals where iron is concentrated in the aleurone and the seed embryo, which are discarded during processing to make flour [88, 89].

TABLE 8.3 Iron Concentration in the Different Staple Crops

Crop	Iron Content	References
Bean	55 mg. kg^{-1}	[106]
Wheat	25 to 56 μg. g^{-1}	[107]
Maize	4.2 mg. kg^{-1} (raw)	[108]
Rice	38 mg. kg^{-1} in rough rice and 4.1 mg. kg^{-1} in milled rice	[91]
Millet	<3 to 8.2 mg/100 g	[89]
Potato	54.95 mg.kg^{-1}	[109]
Cassava	2.7 mg.kg^{-1} (raw)	[108]
Sorghum	33.6 mg.kg^{-1}	[108]
Beef steak	35 μg. g^{-1}	[88]

The most common strategy to deliver biofortified crops is conventional breeding. This is because, in most crops, the varieties that accumulate iron are not adapted for modern agriculture, especially staple crops [84, 90]. For example, in countries like Rwanda, plant breeders belonging to the Harvest Plus group have managed to develop bean varieties with iron levels above 94 μg. g^{-1}, while crosses of iron-accumulating rice with high-yielding have developed varieties with competitive yields for current agriculture and with iron values of 1.2 to 1.5 times above the average [84, 91].

On the other hand, the application of iron fertilizers is presented as an economical and effective alternative to increase the iron content in edible portions of staple crops [85]. Ali et al. [92] reported that the application of iron sulfate increased the iron content in mung bean grains by 81% in relation to the control. Sida-Arreola et al. [93] could increase iron levels of common bean plants, by up to 29% by applying 25 μmol. L^{-1} of iron chelate. Whereas Aciksoz et al. [94] achieved a 19% increase in iron content in wheat grains when a mix of iron sulfate plus urea was applied edaphically. Agronomic biofortification with iron is more complicated due to the difficulties it presents for its absorption. In that matter, the foliar application has emerged as a promising solution according to the results reported by the literature [11].

Another strategy with promising results to raise iron levels in basic crops is the use of genetic engineering or molecular biology, especially in situations where traditional plant breeding does not achieve the proposed objectives like rice. In a study, it was intended to obtain 14 μg. g^{-1} of iron content in processed rice grains through traditional techniques, but it was not possible to achieve. However, by inserting the *AtIRT* gene, it was possible to increase the iron content in grains up to 2.2 times more than the varieties without genetic modification [90, 95]. Through this technology, various processes such as the absorption of iron by the roots, its transport, remobilization, storage, and higher bioavailability can be adapted, and with these, produce plant foods with a significantly higher iron content [90].

8.4.2 IRON PHYSIOLOGY IN THE PLANT

Iron is one of the 17 essential elements for the growth and development of plants and is involved in key metabolic processes including photosynthesis, respiration, nitrogen fixation, chlorophyll synthesis, electron transfer, and reversible redox reactions, being the mineral most abundant in the earth [96, 97]. Various metabolic pathways are dependent on iron, and it forms a structural part of a large number of enzymes, being key in DNA synthesis and in maintaining the structure of chloroplasts; its deficiency in the plant is one of the main causes of chlorosis, generating low production and poor quality, being the most limiting micronutrient for agriculture [98].

Plants can take iron in the form of Fe^{+2} and Fe^{+3}. The latter is the most present in the soil, but also the least available due to the high pH that most croplands present. In that matter, plants have adapted and developed strategies to convert Fe^{+3} to Fe^{+2} or increase the solubility and absorption of Fe^{+3}. These strategies are described as follows: (i) rhizodermal cells release protons through the ATPase enzyme into the rhizosphere to acidify the medium and allow the conversion of Fe^{+3} to Fe^{+2}, by means of iron reduction oxidase enzymes (non-grass plants); and (ii) the plant release of Phyto siderophores through the roots, which are compounds that cover Fe^{+3} and allow its absorption (grass plants) [99, 100].

Once inside the plant, iron is transported by the symplastic route and taken up by different transporters depending on the type of plant, the main ones being iron-regulated transporter (IRT) and yellow stripe-like (YSL) transporters. When the iron in its different forms has reached the xylem, it binds to chelators that allow its translocation within the plant, since iron in its free state can be toxic to the plant. Various studies have found that the main molecules to which iron binds are nicotianamine, citric acid, mugineic acid, and some phenolic compounds [97]. Inside the cells, the main sources of storage are the vacuoles and a molecule called ferritin, which stores about 5% of the total iron and is key in the precise release of this nutrient for metabolic purposes [100].

Nicotianamine and 2-deoxymugineic acid participate in the displacement of iron within the phloem, which allows its translocation to different parts of the plant such as shoots, fruits, and seeds in the formation stage. This process has captured a lot of attention since it's critical to fully exploit biofortification programs, so a lot of efforts have focused on studies that aim to elucidate thoroughly the mechanism by which iron moves from old tissues to fruits [99].

8.4.3 STATUS AND PERSPECTIVES

Since 2004, there have been various programs that try to improve the nutrient content of basic crops. The clearest example is the HarvestPlus project led by the International Food Policy Research Institute (IFPRI). However, recent studies have shown that the objectives set out in these programs have not been fully achieved, especially in studies related to iron content. A survey carried out in East Africa revealed that bean varieties offered as biofortified to consumers did not present any differences in iron content compared to the common varieties that were also distributed by the market in that area [101–103].

Iron biofortification of crops is influenced by several aspects, such as the genotype, the medium in which it is grown, the employed biofortification methodology, and biotic and abiotic factors [81]. Based on this, recent works of iron biofortification are based on high-tech strategies (Table 8.4) such as the use of nano-fertilizers, microorganisms that promote the solubilization and growth of the plants, and molecular techniques that facilitate the absorption, translocation, and accumulation of iron within the plant [81, 87].

TABLE 8.4 Biofortification Works Using High Technology Methods in Basic Crops

Staple Crop	Methodology	Results	References
Wheat	Priming with Fe_2O_3 nanoparticles at a dose of 25 to 600 ppm to two wheat varieties.	Increases of 26 and 46% in the Fe content in grains with the dose of 25 ppm.	[110]
	Inoculation with *Providencia* sp. *Anabaena* sp. and *Calothrix* sp.	Increase in protein, Fe, Mn, and Cu content in grain.	[111]
	Expression of the rice nicotianamine synthase 2 (OsNAS2) gene in two wheat varieties.	Significant differences in iron content compared to unmodified varieties reaching levels above 93 mg. kg^{-1}.	[112, 113]
	Application of Fe nanoparticles in combination with *Bacillus aryabhattai*.	Increases from 33 to 35% in the Fe content.	[114]
Rice	Genetic modification to reduce phytic acid levels.	The Fe concentration of the modified seeds increased up to 1.3 times in polished rice.	[115]
	Release of iron bound to phytic acid by overexpression of the appA gene (phytase).	Increased Fe accumulation by 100% in polished rice grains and up to 4 times higher inorganic P content.	[116]
	Multiple combinations of genes related to transport (AtIRT1, AtNAS1) and storage (Pvferritin).	Increased 4.7 times the Fe content in polished rice grains.	[117]
Bean	Inoculation with *Pantoea dispersa* MPJ9 and *Pseudomonas putida* MPJ6 to increase the Fe content.	Increases in the iron content in fruits of up to 3.4 times with MPJ9 and 2.8 times with MPJ6.	[118]
	Foliar and edaphic application of iron chelate at different doses.	Increases of up to 41.3% in grain iron content.	[119]
Maize	Use of improved varieties of maize in combination with various microorganisms that promote plant growth.	The application of *Trichoderma viride* increased 2.1 times the content of Fe in the grain in relation to the control without application.	[120]
	Edaphic application of N, Fe, Zn, and Se.	Increase in the content of the four applied elements without finding significant effects on biomass production.	[121]

However, studies indicate that to have a real impact on the enhancement of the daily dietary iron intake of the population, it is necessary to focus on biofortification techniques to obtain more iron available at the time of consumption. Efforts should be directed towards genetic modifications of crops to lower the content of compounds that limit the iron content, such as phytic acid or various polyphenols; the improvement of postharvest handling techniques that influence factors such as the rupture of the cotyledon cell wall to release intracellular iron, and the reduction of phytic acid through the activation of endogenous phytase. The method of food preparation and the influence of other components in the diet should also be considered [102–104].

8.5 CONCLUSION

Regarding the intervention strategy to address iron deficiency, all pursuits aim to eradicate iron malnutrition with minimal risk, efficiently addressing social, economic, and environmental perspectives. While it is unlikely that a single strategy could fully address iron deficiency, the development of a more efficient intervention may provide benefits to a larger number of people, regardless of their geographical locations, ages, incomes, and genders. Increasing the real impact to treat or preventing iron deficiency, as well as its cost-benefit, is crucial. The progress obtained as a result of scientific studies focused on enhancing the iron content of plants has been exponential in recent years. In general, biofortification is presented as an efficient and sustainable alternative to overcome iron deficiency, capable of increasing both the content and availability of iron in a wide variety of staple crops. However, despite these promising results, there are still variations in results within studies from biofortification programs, leading to a stagnation point. Therefore, more studies are needed to successfully improve the nutritional quality of staple crops and ensure the adequate iron status of the world population in a sustainable way. From a general perspective, biofortification may combine the best of food diversification and fortification in the coming years.

KEYWORDS

- **anemia**
- **biofortification**
- **cell cycle regulation**
- **electron transport**
- **iron deficiency**
- **metabolic processes**
- **microorganisms**

REFERENCES

1. Milto, I. V., Suhodolo, I. V., Prokopieva, V. D., & Klimenteva, T. K. (2016). Molecular and cellular bases of iron metabolism in humans. *Biochemistry (Moscow), 81*(6), 549–564.
2. Gafter-Gvili, A., Schechter, A., & Rozen-Zvi, B. (2019). Iron deficiency anemia in chronic kidney disease. *Acta Haematologica, 142*(1), 44–50.
3. UCSF Health. University of California San Francisco. Patient education hemoglobin and functions of iron. https://www.ucsfhealth.org/education/hemoglobin-andfunctions-of-iron (accessed on 28 May 2024)..
4. Quintero-Gutiérrez, A. G., González-Rosendo, G., Sánchez-Muñoz, J., Polo-Pozo, J., & Rodríguez-Jerez, J. J. (2008). Bioavailability of heme iron in biscuit filling using piglets as an animal model for humans. *International Journal of Biological Sciences, 4*, 58–62.
5. Abbaspour, N., Hurrel, R., & Kelishadi, R. (2014). Review on iron and its importance for human health. *Journal of Research in Medical Sciences, 19*(2), 164–174.
6. National Institutes of Health. Iron fact sheet for health professionals. https://ods.od.nih.gov/factsheets/Iron-HealthProfessional/#en2 (accessed on 28 May 2024).
7. Aggett, P. J. (2012). Iron. In J. W. Erdman, I. A. Macdonald, & S. H. Zeisel (Eds.), *Present Knowledge in Nutrition* (10th ed., pp. 506–520). Wiley-Blackwell.
8. Dainty, J. R., Berry, R., Lynch, S. R., Harvey, L. J., & Fairweather-Tait, S. J. (2014). Estimation of dietary iron bioavailability from food iron intake and iron status. *PLoS One, 9*(10), e111824.
9. National Library of Medicine. Biochemistry, iron absorption. https://www.ncbi.nlm.nih.gov/books/NBK448204/ (accessed May 27, 2024).
10. World Health Organization. Anemia. https://www.who.int/healthtopics/anaemia#tab=tab_1 (accessed May 27, 2024)
11. de Valença, A. W., Bake, A., Brouwer, I. D., & Giller, K. E. (2017). Agronomic biofortification of crops to fight hidden hunger in sub-Saharan Africa. *Global Food Security, 12*, 8–14.
12. Srivastav, P., Vutukuru, M., Ravindran, G., Awad, M. M. (2022). Biofortification-present scenario, possibilities, and challenges: A scient metric approach. *Sustainability, 14*(18), 11632.
13. Ritchie, H., & Roser, M. (n.d.). Our World in Data. Micronutrient deficiency. https://ourworldindata.org/micronutrient-deficiency?utm_medium=syndication (accessed May 27, 2024).
14. Feng, H., Zhang, T., Yan, W., Lu, L., Tao, Y., Cai, W., & Wang, Y. (2020). Micronutrient deficiencies in paediatric short bowel syndrome: A 10-year review from an intestinal rehabilitation centre in China. *Paediatric Surgery International, 36*, 1481–1487. https://doi.org/10.1007/s00383-020-04764-3.
15. Peña-Rosas, J. P., Mithra, P., Unnikrishnan, B., Kumar, N., De-Regil, L. M., Nair, N. S., & Solon, J. A. (2019). Fortification of rice with vitamins and minerals for addressing micronutrient malnutrition. *Cochrane Database of Systematic Reviews, 10*, CD009902.
16. Praharaj, S., Skalicky, M., Maitra, S., Bhadra, P., Shankar, T., Brestic, M., & Hossain, A. (2021). Zinc biofortification in food crops could alleviate the zinc malnutrition in human health. *Molecules, 26*(12), 3509.
17. Kiran, A., Wakeel, A., Mahmood, K., Mubarak, R., & Haefele, S. M. (2022). Biofortification of staple crops to alleviate human malnutrition: Contributions and potential in developing countries. *Agronomy, 12*(2), 452.

18. Stevens, G. A., Beal, T., Mbuya, M. N., Luo, H., Neufeld, L. M., Addo, O. Y., & Young, M. F. (2022). Micronutrient deficiencies among preschool-aged children and women of reproductive age worldwide: A pooled analysis of individual-level data from population-representative surveys. *The Lancet Global Health, 10*(11), e1590–e1599.
19. Kobylińska, M., Antosik, K., Decyk, A., & Kurowska, K. (2022). Malnutrition in obesity: Is it possible? *Obesity Facts, 15*(1), 19–25.
20. Gorji, A., & Ghadiri, M. K. (2021). Potential roles of micronutrient deficiency and immune system dysfunction in the coronavirus disease 2019 (COVID-19) pandemic. *Nutrition, 82*, 111047.
21. Song, E. K., Moser, D. K., Kang, S. M., & Lennie, T. A. (2015). Association of depressive symptoms and micronutrient deficiency with cardiac event–free survival in patients with heart failure. *Journal of Cardiac Failure, 21*(12), 945–951.
22. Blampied, M., Bell, C., Gilbert, C., & Rucklidge, J. J. (2020). Broad spectrum micronutrient formulas for the treatment of symptoms of depression, stress, and/or anxiety: A systematic review. *Expert Review of Neurotherapeutics, 20*(4), 351–371.
23. Campisi, S. C., Zasowski, C., Shah, S., Shah, A., Bradley-Ridout, G., Korczak, D. J., & Szatmari, P. (2020). Assessing the evidence of micronutrients on depression among children and adolescents: An evidence gap map. *Advances in Nutrition, 11*(4), 908–927.
24. Timoneda, J., Rodríguez-Fernández, L., Zaragozá, R., Marín, M. P., Cabezuelo, M. T., Torres, L., & Barber, T. (2018). Vitamin A deficiency and the lung. *Nutrients, 10*(9), 1132.
25. Yadav, K., & Pandav, C. S. (2018). National iodine deficiency disorders control programme: Current status & future strategy. *Indian Journal of Medical Research, 148*(5), 503.
26. Martínez, S. G., Morillo, E. M., Llano, N. A., Gomis, A. I. E., Álvarez, E. D., & Torre, E. M. (2023). Iodine status in pregnant women in northern Spain: Effect of iodised salt and iodine supplements on thyroid function. *Endocrinología, Diabetes y Nutrición, 70*(3), 162–170.
27. Bailey, R. L., West, K. P., Jr., & Black, R. E. (2015). The epidemiology of global micronutrient deficiencies. *Annals of Nutrition and Metabolism, 66*(Suppl. 2), 22–33.
28. Tostado-Madrid, T., Benítez-Ruiz, I., Pinzón-Navarro, A., Bautista-Silva, M., & Ramírez-Mayans, J. A. (2015). Actualidades de las características del hierro y su uso en pediatría. *Acta Pediátrica de México, 36*(3), 189–200.
29. Sermini, C. G., Acevedo, M. J., & Arredondo, M. (2017). Biomarcadores del metabolismo y nutrición de hierro. *Revista Peruana de Medicina Experimental y Salud Pública, 34*, 690–698.
30. Forrellat-Barrios, M. (2016). Regulación del metabolismo del hierro: dos sistemas, un mismo objetivo. *Revista Cubana de Hematología, Inmunología y Hemoterapia, 32*(1), 4–14.
31. U.S. Department of Health and Human Services. (2023). National Institute of Health. Iron. https://ods.od.nih.gov/factsheets/Iron-HealthProfessional/ (accessed May 26, 2024).
32. Wegmüller, R., Bentil, H., Wirth, J. P., Petry, N., Tanumihardjo, S. A., Allen, L., & Rohner, F. (2020). Anemia, micronutrient deficiencies, malaria, hemoglobinopathies and malnutrition in young children and non-pregnant women in Ghana: Findings from a national survey. *PloS One, 15*(1), e0228258.

33. Pasricha, S. R., Tye-Din, J., Muckenthaler, M. U., & Swinkels, D. W. (2021). Iron deficiency. *The Lancet, 397*(10270), 233–248.
34. Wiafe, M. A., Ayenu, J., & Eli-Cophie, D. (2023). A review of the risk factors for iron deficiency anaemia among adolescents in developing countries. *Anemia, 2023*, 6406286.
35. Camaschella, C. (2019). Iron deficiency. *Blood, 133*(1), 30–39.
36. Lopez, A., Cacoub, P., Macdougall, I. C., & Peyrin-Biroulet, L. (2016). Iron deficiency anaemia. *The Lancet, 387*(10021), 907–916.
37. Wood, R. J., & Ronnenberg A. (2005). Iron. In *Modern Nutrition in Health and Disease* (10th ed., pp. 248–270). Lippincott Williams & Wilkins.
38. Zulfiqar, U., Maqsood, M., & Hussain, S. (2020). Rice research for quality improvement: genomics and genetic engineering. In A. Roychoudhury (Ed.), *Biofortification of Rice with Iron and Zinc: Progress and Prospects* (pp. 605–627). Springer Nature Singapore Pte Ltd.
39. Hélène, D. (2003, 23–28). Food diversification strategies are neglected in spite of their potential effectiveness: why is it so and what can be done? 2nd International Workshop, Food-based approaches for a healthy nutrition. Ouagadou, Kulikoro, Mali. https://www.researchgate.net/publication/228813695_Food_diversification_strategies_are_neglected_in_spite_of_their_potential_effectiveness_why_is_it_so_and_what_can_be_done.
40. Evlash, V., Aksonova, O., & Gubsky, S. (2022). Food-based intervention strategies for iron deficiency prevention. *Biology and Life Sciences Forum, 18*, 52.
41. Arshad, M. K., Fatima, I., Ahmad, W., Ellahi, S., Mumtaz, M., Akhtar, M. U., & Ali, W. S. (2023). Mint (*Mentha*): A herb and used as a functional ingredient. *Scholarly International Journal of Traditional and Complementary Medicine, 6*(3), 38–52.
42. Farsi, Z., Sajadi, S. A., Afaghi, E., Fournier, A., Aliyari, S., Ahmadi, Y., & Hazrati, E. (2021). Explaining the experiences of nursing administrators, educators, and students about education process in the COVID-19 pandemic: a qualitative study. *BMC Nursing, 20*, 151.
43. Gashu, D., Demment, M. W., & Stoecker, B. J. (2019). Challenges and opportunities to the African agriculture and food systems. *African Journal of Food, Agriculture, Nutrition and Development, 19*(1), 14190–14217.
44. Zhao, A., Li, Z., Ke, Y., Huo, S., Ma, Y., Zhang, Y., & Ren, Z. (2020). Dietary diversity among chinese residents during the covid-19 outbreak and its associated factors. *Nutrients, 12*(6), 1699.
45. Singh, A., & Vellapandian, C. (2023). High risk of metabolic complications due to high consumption of processed foods. *Current Nutrition and Food Science, 19*(3), 198–208.
46. Hoddinott, J.; Rosegrant, M.; Torero, M. (2012). Investments to reduce hunger and undernutrition. Copenhagen Consensus Center [Online] April 9, 2012. https://copenhagenconsensus.com/sites/default/files/hungerandmalnutrition.pdf (accesed May 28, 2024).
47. Sánchez, Esteban, Monica garcia-bañuelos, and Juan Sida Arreola. "Biofortification - Promising Approach to Increasing the Content of Iron and Zinc in Staple Food Crops." *Journal of Elemntology* 19 (09/24 2012). https://doi.org/10.5601/jelem.2014.19.3.708.
48. WHO. The global prevalence of anaemia in 2011. (2015). Geneva: World Health Organization [Online]. https://apps.who.int/iris/bitstream/handle/10665/177094/9789241564960_eng.pdf;jsessionid= 096DEEBEC3083DF7B9E3E0A3162EED10?sequence=1 (accessed May 29, 2024).

49. Olson, R., Gavin-Smith, B., Ferraboschi, C., & Kraemer, K. (2021). Food fortification: The advantages, disadvantages and lessons from sight and life programs. *Nutrients, 13*(4), 1118.
50. Klassen-Wigger, P., & Barclay, D. V. (2018). Market-driven fortification. In M. G. M. Venkatesh & R. F. Hurrel (Eds.), *Food Fortification in a Globalized World* (pp. 63–67). Academic Press.
51. Mannar, M. G. V., & Hurrell, R. F. (2018). Food fortification. In M. G. M. Venkatesh & R. F. Hurrel (Eds.), *Food Fortification in a Globalized World* (pp. 3–11). Academic Press.
52. Jan, Y., Malik, M., Yaseen, M., Ahmad, S., Imran, M., Rasool, S., & Haq, A. (2019). Vitamin D fortification of foods in India: present and past scenario. *Journal of Steroid Biochemistry and Molecular Biology, 193*, 105417.
53. Bathla, S., & Arora, S. (2022) "Prevalence and Approaches to Manage Iron Deficiency Anemia (Ida)." [In eng]. *Crit Rev Food Sci Nutr 62*(32), 8815–8828. https://doi.org/10.1080/10408398.2021.1935442.
54. Kumar, S., Anukiruthika, T., Dutta, S., Kashyap, A. V., Moses, J. A., & Anandharamakrishnan, C. (2020). Iron deficiency anemia: A comprehensive review on iron absorption, bioavailability and emerging food fortification approaches. *Trends in Food Science & Technology, 98*, 58–75.
55. Detzel, P., & Wieser, S. (2015). Food fortification for addressing iron deficiency in Filipino children: Benefits and cost-effectiveness. *Annals of Nutrition & Metabolism, 66*(Suppl. 2), 35–42.
56. World Health Organization. Guidelines on food fortification with micronutrients. Retrieved from: https://www.who.int/publications/i/item/9241594012 (accessed May 24, 2024).
57. Singh, A., Pratap Singh, A., & Ramaswamy, H. S. (2016). A controlled agitation process for improving quality of canned green beans during agitation thermal processing. *Journal of Food Science, 81*(6), E1399–E1411.
58. Blanco-Rojo, R., & Vaquero, M. P. (2019). Iron bioavailability from food fortification to precision nutrition: A review. *Innovative Food Science & Emerging Technologies, 51*, 126–138.
59. Kumari, A., & Chauhan, A. K. (2022). Iron nanoparticles as a promising compound for food fortification in iron deficiency anemia: A review. *Journal of Food Science and Technology, 59*(9), 3319–3335.
60. Baldelli, A., Liang, D. Y., Guo, Y., & Pratap-Singh, A. (2023). Effect of the formulation on mucoadhesive spray-dried microparticles containing iron for food fortification. *Food Hydrocoll., 134*, 107906.
61. Darwish, A. M., Soliman, T. N., Elhendy, H. A., & El-Kholy, Wedad. (2021). Nano-encapsulated iron and folic acid fortified functional yogurt enhance anemia in albino rats. *Frontiers in Nutrition, 8*, 1–23.
62. Garcés, V., González, A., Sabio, L., Sánchez-Arévalo, C. M., Gálvez, N., & Dominguez-Vera, J. M. (2020). Magnetic and golden yogurts. Food as a potential nanomedicine carrier. *Materials, 13*(2), 481.
63. Athira, S., Mann, B., Sharma, R., Pothuraju, R., & Bajaj, R. K. (2021). Preparation and characterization of iron-chelating peptides from whey protein: An alternative approach for chemical iron fortification. *Food Research International, 141*, 110133.

64. Mandecka, A., Dąbrowska, A., Bobak, L., & Szoltysik, M. (2022). Casein hydrolysate and casein-iron chelate as natural bioactive compounds for yogurt fortification. *Appl. Sci., 12*(24), 12903.
65. Barbosa, B. S. T., & García-Rojas, E. E. (2022). Double emulsions as delivery systems for iron: Stability kinetics and improved bio accessibility in infants and adults. *Curr. Res. Nutr. Food Sci., 5*, 718–725.
66. Kubuga, C. K., Hong, H. G., & Song, W. O. (2019). Hibiscus sabdariffa meal improves iron status of childbearing age women and prevents stunting in their toddlers in Northern Ghana. *Nutrients, 11*, 198.
67. Agrahar-Murugkar, D. (2020). Food to food fortification of breads and biscuits with herbs, spices, millets and oilseeds on bio-accessibility of calcium, iron and zinc and impact of proteins, fat and phenolics. *LWT, 130*, 109703.
68. Barkley, J. S., Wheeler, K. S., & Pachón, H. (2015). Anaemia prevalence may be reduced among countries that fortify flour. *Br. J. Nutr., 114*, 265–273.
69. Brittenham, G. (2018). Disorders of iron homeostasis: iron deficiency and overload. In R. Hoffman, L. E. Silberstein, J. I. Weitz, M. E. Salama, E. J. Benz, H. E. Heslop, J. Anastasi, & S. A. Abutalib (Eds.), *Hematology: Basic Principles and Practice* (pp. 478–490). Elsevier.
70. Srivastava, S., & Kumar, S. (2021). Does socio-economic inequality exist in micronutrients supplementation among children aged 6–59 months in India? Evidence from National Family Health Survey 2005–06 and 2015–16. *BMC Public Health, 21*, 545.
71. Pasricha, S. R., Hayes, E., Kalumba, K., & Biggs, B. A. (2013). Effect of daily iron supplementation on health in children aged 4-23 months: A systematic review and meta-analysis of randomized controlled trials. *Lancet Glob. Health, 1*, e77–86.
72. Kassebaum, N. J., Jasrasaria, R., Naghavi, M., Wulf, S. K., Jhons, N., & Lozano, R. (2014). A systematic analysis of global anemia burden from 1990 to 2010. *Blood, 123*, 615–624.
73. Paulino, C. T. D. S., Nishijima, M., & Sarti, F. M. (2021). Association of iron supplementation programs with iron-deficiency anemia outcomes among children in Brazil. *Nutrients, 13*(5), 1524.
74. Tolkien, Z., Stecher, L., Mander, A. P., Pereira, D. I., & Powell, J. J. (2015). Ferrous sulfate supplementation causes significant gastrointestinal side-effects in adults: a systematic review and meta-analysis. *PLoS One, 10*, e0117383.
75. Paganini, D., & Zimmermann, M. B. (2017). The effects of iron fortification and supplementation on the gut microbiome and diarrhea in infants and children: A review. *American Journal of Clinical Nutrition, 106*(6), 1688S–1693S.
76. Lo, J. O., Benson, A. E., Martens, K. L., Hedges, M. A., McCurry, H. S., DeLoughery, T., Aslan, J. E., & Shatzel, J. J. (2022). The role of oral iron in the treatment of adults with iron deficiency. *European Journal of Haematology, 110*(2), 123–130.
77. Stoffel, N. U., Siebenthal, H. K., Moretti, D., & Zimmerman, M. B. (2020). Oral iron supplementation in iron-deficient women: How much and how often? *Molecular Aspects of Medicine, 75*, 100865.
78. Pergola, P. E., Fishbane, S., & Ganz, T. (2019). Novel oral iron therapies for iron deficiency anemia in chronic kidney disease. *Advances in Chronic Kidney Disease, 26*(4), 272–291.
79. Singh, K., Chopra, D. S., Singh, D., & Singh, N. (2022). Nano-formulations in treatment of iron deficiency anemia: An overview. *Clinical Nutrition ESPEN, 52*, 12–19.

80. Wang, L., Wang, L., Su, C., Wen, C., Gong, Y., You, Y., Zhao, J., Han, Y., Song, S., & Xiao, H. (2020). Characterization and digestion features of a novel polysaccharide-Fe(III) complex as an iron supplement. *Carbohydrate Polymers, 249*, 116812.
81. Szerement, J., Szatanik-Kloc, A., Mokrzycki, J., & Mierzwa-Hersztek, M. (2022). Agronomic biofortification with Se, Zn, and Fe: An effective strategy to enhance crop nutritional quality and stress defense—A review. *Journal of Soil Science and Plant Nutrition, 22*(1), 1129–1159.
82. Khush, G. S., Lee, S., Cho, J. I., & Jeon, J. S. (2012). Biofortification of crops for reducing malnutrition. *Plant Biotechnology Reports, 6*, 195–202.
83. Bouis, H. (2018). Reducing mineral and vitamin deficiencies through biofortification: Progress under HarvestPlus. In H. K. Biesalski (Ed.), *Hidden Hunger: Strategies to Improve Nutrition Quality* (Vol. 118, pp. 112–122). Stuttgart: Karger Publishers.
84. Petry, N., Boy, E., Wirth, J. P., & Hurrell, R. F. (2015). The potential of the common bean (*Phaseolus vulgaris*) as a vehicle for iron biofortification. *Nutrients, 7*(2), 1144–1173.
85. Ghosh, A., Hasim-Reja, M., Nalia, A., Kanthal, S., Maji, S., Venugopalan, V., & Nath, R. (2019). Micronutrient biofortification in pulses: An agricultural approach. *Current Journal of Applied Science and Technology, 35*(4), 1–12.
86. Andersson, M. S., Saltzman, A., Virk, P. S., & Pfeiffer, W. H. (2017). Progress update: Crop development of biofortified staple food crops under HarvestPlus. *African Journal of Food, Agriculture, Nutrition and Development, 17*(2), 11905–11935.
87. Jha, A. B., & Warkentin, T. D. (2020). Biofortification of pulse crops: Status and future perspectives. *Plants, 9*(1), 73.
88. Connorton, J. M., & Balk, J. (2019). Iron biofortification of staple crops: Lessons and challenges in plant genetics. *Plant and Cell Physiology, 60*(7), 1447–1456.
89. Anitha, S., Kane-Potaka, J., Botha, R., Givens, D. I., Sulaiman, N. L. B., Upadhyay, S., & Bhandari, R. K. (2021). Millets can have a major impact on improving iron status, haemoglobin level, and in reducing iron deficiency anaemia–A systematic review and meta-analysis. *Frontiers in Nutrition, 8*, 725529.
90. Vasconcelos, M. W., Gruissem, W., & Bhullar, N. K. (2017). Iron biofortification in the 21st century: Setting realistic targets, overcoming obstacles, and new strategies for healthy nutrition. *Current Opinion in Biotechnology, 44*, 8–15.
91. Majumder, S., Datta, K., & Datta, S. K. (2019). Rice biofortification: High iron, zinc, and vitamin-A to fight against "hidden hunger." *Agronomy, 9*(12), 803.
92. Ali, B., Ali, A., Tahir, M., & Ali, S. (2014). Growth, seed yield and quality of mungbean as influenced by foliar application of iron sulphate. *Pakistan Journal of Life and Social Sciences, 12*(1), 20–25.
93. Sida-Arreola, J. P., Sánchez-Chávez, E., Ávila-Quezada, G. D., Zamudio-Flores, P. B., & Muñíz, C. A. (2015). Iron biofortification and its impact on antioxidant system, yield and biomass in common bean. *Plant, Soil and Environment, 61*(12), 573–576.
94. Aciksoz, S. B., Yazici, A., Ozturk, L., & Cakmak, I. (2011). Biofortification of wheat with iron through soil and foliar application of nitrogen and iron fertilizers. *Plant and Soil, 349*, 215–225.
95. Boonyaves, K., Gruissem, W., & Bhullar, N. K. (2016). NOD promoter-controlled AtIRT1 expression functions synergistically with NAS and FERRITIN genes to increase iron in rice grains. *Plant Molecular Biology, 90*, 207–215.
96. Kobayashi, T., & Nishizawa, N. K. (2012). Iron uptake, translocation, and regulation in higher plants. *Annual Review of Plant Biology, 63*, 131–152.

97. Nikolic, M., & Pavlovic, J. (2018). Plant responses to iron deficiency and toxicity and iron use efficiency in plants. In M. A. Hossain, T. Kamiya, D. J. Burritt, L. S. Phan-Tran, & T. Fujiwara (Eds.), *Plant Micronutrient Use Efficiency* (pp. 55–69). Academic Press.
98. Rout, G. R., & Sahoo, S. (2015). Role of iron in plant growth and metabolism. *Reviews in Agricultural Science, 3*, 1–24.
99. Kobayashi, T., Nozoye, T., & Nishizawa, N. K. (2019). Iron transport and its regulation in plants. *Free Radical Biology and Medicine, 133*, 11–20.
100. Zhang, X., Zhang, D., Sun, W., & Wang, T. (2019). The adaptive mechanism of plants to iron deficiency via iron uptake, transport, and homeostasis. *International Journal of Molecular Sciences, 20*(10), 2424.
101. International Food Policy Research Institute. Biofortification: A food-systems solution to help end hidden hunger. https://www.fao.org/publications/card/es/c/CA8711EN/ (accessed May 27, 2024).
102. Glahn, R. P., & Noh, H. (2021). Redefining bean iron biofortification: A review of the evidence for moving to a high Fe bioavailability approach. *Frontiers in Sustainable Food Systems, 5*, 682130.
103. Glahn, R. P., Wiesinger, J. A., & Lung'aho, M. G. (2020). Iron concentrations in biofortified beans and nonbiofortified marketplace varieties in East Africa are similar. *Journal of Nutrition, 150*(11), 3013–3023.
104. Huertas, R., Karpinska, B., Ngala, S., Mkandawire, B., Maling'a, J., Wajenkeche, E., & Foyer, C. H. (2023). Biofortification of common bean (*Phaseolus vulgaris* L.) with iron and zinc: Achievements and challenges. *Food and Energy Security, 12*(2), e406.
105. Ohanenye, I. C., Emenike, C. U., Mensi, A., Medina-Godoy, S., Jin, J., Ahmed, T., & Udenigwe, C. C. (2021). Food fortification technologies: Influence on iron, zinc and vitamin A bioavailability and potential implications on micronutrient deficiency in sub-Saharan Africa. *Science in Africa, 11*, e00667.
106. Beebe, S. (2020). Biofortification of common bean for higher iron concentration. *Frontiers in Sustainable Food Systems, 4*, 573449.
107. Harrington, S. A., Connorton, J. M., Nyangoma, N. I., McNelly, R., Morgan, Y. M., Aslam, M. F., & Balk, J. (2023). A two-gene strategy increases iron and zinc concentrations in wheat flour, improving mineral bioaccessibility. *Plant Physiology, 191*(1), 528–541.
108. Okwuonu, I. C., Narayanan, N. N., Egesi, C. N., & Taylor, N. J. (2021). Opportunities and challenges for biofortification of cassava to address iron and zinc deficiency in Nigeria. *Global Food Security, 28*, 100478.
109. Singh, B., Goutam, U., Kukreja, S., Siddappa, S., Sood, S., Sharma, J., & Bhardwaj, V. (2022). Biofortification strategies to improve iron concentrations in potato tubers: lessons and future opportunities. *Potato Research*, *65*(1), 51–64.
110. Sundaria, N., Singh, M., Upreti, P., Chauhan, R. P., Jaiswal, J. P., & Kumar, A. (2019). Seed priming with iron oxide nanoparticles triggers iron acquisition and biofortification in wheat (*Triticum aestivum* L.) grains. *Journal of Plant Growth Regulation*, *38*, 122–131.
111. Rana, A., Joshi, M., Prasanna, R., Shivay, Y. S., & Nain, L. (2012). Biofortification of wheat through inoculation of plant growth promoting rhizobacteria and cyanobacteria. *European Journal of Soil Biology*, *50*, 118–126.
112. Beasley, J. T., Bonneau, J. P., Sánchez-Palacios, J. T., Moreno-Moyano, L. T., Callahan, D. L., Tako, E., & Johnson, A. A. (2019). Metabolic engineering of bread wheat

improves grain iron concentration and bioavailability. *Plant Biotechnology Journal*, *17*(8), 1514–1526.

113. Singh, S. P., Keller, B., Gruissem, W., & Bhullar, N. K. (2017). Rice Nicotianamine Synthase 2 expression improves dietary iron and zinc levels in wheat. *Theoretical and Applied Genetics*, *130*, 283–292.
114. Merinero, M., Alcudia, A., Begines, B., Martínez, G., Martín-Valero, M. J., Pérez-Romero, J. A., & Pajuelo, E. (2022). Assessing the biofortification of wheat plants by combining a plant growth-promoting rhizobacterium (PGPR) and Polymeric Fe-Nanoparticles: Allies or Enemies? *Agronomy*, *12*(1), 228.
115. Karmakar, A., Bhattacharya, S., Sengupta, S., Ali, N., Sarkar, S. N., Datta, K., & Datta, S. K. (2020). RNAi-mediated silencing of ITPK gene reduces phytic acid content, alters transcripts of phytic acid biosynthetic genes, and modulates mineral distribution in rice seeds. *Rice Science*, *27*(4), 315–328.
116. Bhattacharya, S., Sengupta, S., Karmakar, A., Sarkar, S. N., Gangopadhyay, G., Datta, K., & Datta, S. K. (2019). Genetically engineered rice with appA gene enhanced phosphorus and minerals. *Journal of Plant Biochemistry and Biotechnology*, *28*, 470–482.
117. Boonyaves, K., Wu, T. Y., Gruissem, W., & Bhullar, N. K. (2017). Enhanced grain iron levels in rice expressing an iron-regulated metal transporter, nicotiana mine synthase, and ferritin gene cassette. *Frontiers in Plant Science*, *8*, 130.
118. Patel, P., Trivedi, G., & Saraf, M. (2018). Iron biofortification in mungbean using siderophore producing plant growth promoting bacteria. *Environmental Sustainability*, *1*, 357–365.
119. Felix, J. W., Sanchez-Chavez, E., de-la-Cruz-Lazaro, E., & Marquez-Quiroz, C. (2021). Edaphic and foliar biofortification of common black bean (*Phaseolus vulgaris* L.) with iron. *Legume Research*, *44*(2), 192–196.
120. Nishanth, S., Prasanna, R., Hossain, F., Muthusamy, V., Shivay, Y. S., & Nain, L. (2021). Interactions of microbial inoculants with soil and plant attributes for enhancing Fe and Zn biofortification in maize genotypes. *Rhizosphere*, *19*, 100421.
121. Grujcic, D., Yazici, A. M., Tutus, Y., Cakmak, I., & Singh, B. R. (2021). Biofortification of silage maize with zinc, iron and selenium as affected by nitrogen fertilization. *Plants*, *10*(2), 391.

CHAPTER 9

Bacterial Nanocellulose: Production and Perspectives for Future Applications in Food Industry

JUAN DANIEL RIVALDI,[1,2,3*] PRISCILLA VERGARA,[2] NURIA GÓMEZ,[2] JUAN CARLOS VILLAR,[2] and FÉLIX GARCIA-OCHOA[3]

[1]*Industrial Applications Department, Faculty of Chemical Sciences, Universidad Nacional de Asunción, San Lorenzo, Paraguay*

[2]*Cellulose and Paper Group, Forest Product Department, Forest Research Center (ICIFOR-CSIC), Madrid, España*

[3]*Chemical and Materials Engineering Department, Faculty of Chemical Sciences, Universidad Complutense de Madrid, Madrid, España*

*Corresponding author.

ABSTRACT

Recently, the food industry has shown high interest in nanocellulose as a sustainable feedstock for additives and the development of intelligent food packaging. Bacterial nanocellulose (BNC) is a natural exopolysaccharide synthesized by a few aerobic bacteria (mainly *Komagataeibacter* species) with a high degree of polymerization, high crystallinity, and hydrophilicity. Currently, extensive studies are focused on BNC production and its application as reinforcement in composite materials. However, there are few reports on production and the implications of its use in food production, and it is still under study. BNC can be applied in the food industry as a stabilizer of emulsions, rheology modifier, immobilizer (of probiotic bacteria, enzymes, microbicides), and biofilms for sustainable food packaging. Culturing using static and stirred bioreactors produces BNC with different physical properties that affect its applicability. Nonetheless, laboratory-scale processes have

Food Security, Safety, and Sustainability: Validated Innovative Approaches and Technological Advancements.
José Juan Buenrostro-Figueroa and Julio César Tafolla-Arellano (Eds.)

shown low BNC yield and long cultivation time that have limited its scale-up. Thus, the current major challenge is to establish optimal parameters for the scale-up process aiming to increase productivity with the desirable characteristics of BNC at a low cost. Furthermore, BNC usually presents poor mechanical properties and a lack of antioxidant and antimicrobial activity. However, these naturally occurring polymers can be easily functionalized or create a BNC matrix composite to obtain new biomaterials for food and innovative packaging. This work presents an integrative review of the bioprocess conditions for BNC production and its application for food and packaging.

➢ **Highlights:**

- Bacterial cellulose is an eco-friendly nano-biomaterial which has unique properties for many applications.
- Static or agitated fermentation affects the productivity, structural, and mechanical properties of BNC.
- BNC is a sustainable biomaterial with great potential as an ingredient, additive, and packaging for the food industry.
- Knowledge and technology have to be developed to understand the BNC production process.

9.1 INTRODUCTION

In the last decades, the development of new, renewable, and sustainable materials has been, and still is, one of the major present challenges. The markets require new biocompatible and sustainable polymers with chemical and mechanical properties for applications in the medical, environmental, and food sectors. Cellulose (poly-β-1,4-glucose) is the most abundant natural and renewable homopolymer on the planet. This macromolecule constitutes one of the main structural polymers of plants, along with lignin and hemicellulose. Cellulose is used for several industrial applications and is usually obtained from wood pulp, which requires numerous unit operations and intensive use of energy. In this sense, nanocellulose extraction from wood or other lignocellulosic materials (sugarcane bagasse, soybean husk, etc.) usually requires alkaline hydrolysis for lignin removal followed by acid hydrolysis steps to transform the pure cellulose into nanocellulose [1, 106]. Nowadays, the production of bacterial nanocellulose (BNC) is receiving the most attention as a feasible alternative to cellulose production from vegetable materials. BNC is an extracellular biodegradable, biocompatible, and nanostructured polymer of glucose generated by specific, aerobic, Gram-negative bacteria [2, 3]. This macromolecule, which demonstrates important mechanical and chemical properties, is recognized as a generally safe (GRAS) polymer and is appropriate for applications in the food industry [4]. Bacterial cellulose has competitive advantages over cellulose from plants since it is produced in a crystalline and pure form by the cell (without lignin, pectin, or other polymers) and under moderate process conditions. Despite the advances in its characterization and the use for obtaining new composites, usually, the studies are focused on the production and functionalization of BNC for medical applications [5]. In the food area, BNC could be useful for modifying the rheological properties of food products, as an immobilizer of bacteria and/or enzymes, and for the production of BNC food packaging films. One of the most important current issues in bioprocessing is the scale-up of production at low cost. Biotechnological processes for BNC production usually require a medium with expensive components, several steps for inoculum preparation, and later nanocellulose purification. Furthermore, kinetics studies of cell growth and BNC production are still under exploration, and many research papers do not address in-depth studies of important biochemical parameters and process modeling. The following chapter aims to combine the most current research on bacterial nanocellulose production and applications in the food industry.

9.2 BNC PRODUCERS

Several bacterial species secrete extracellular polysaccharides, such as xanthan [6]; however, only a few have shown the capability to produce cellulose. Bacterial nanocellulose is a homopolymer of glucopyranose units linked by β-1,4-glycosidic bonds, forming a highly crystalline macrostructure, which is produced by aerobic Gram-negative bacteria of various genera, including *Aerobacter, Azotobacter, Agrobacterium, Achromobacter, Alcaligenes, Gluconacetobacter, Komagataeibacter, Pseudomonas, Rhizobium*, and *Sarcina* [107]. Additionally, bacterial cellulose production was described in the genus *Rhodococcus* [2].

Gluconoacetobacter xylinus (formerly *Acetobacter xylinum*), a rod-shaped, aerobic, Gram-negative bacterium, is the most extensively studied organism as a model system for the synthesis of nanocellulose [7]. Figure 9.1 shows the schematic representation of the possible multi-step biosynthetic pathways of cellulose in *Gluconoacetobacter* sp. The assimilation of glucose and its conversion into cellulose is promoted by a group of four enzymes, including glucokinase (GK) (EC 2.7.1.2) that catalyzes its phosphorylation into glucose-6-phosphate, the isomerization of glucose-6-phosphate into glucose-1-phosphate by the phosphoglucomutase (PGM) (EC 5.4.2.2), and its conversion into uridine diphosphate glucose in the presence of uridine diphosphate-glucose pyrophosphorylase (UGP) (EC 2.7.7.9); this is followed by the synthesis of cellulose on the bacterium surface through the linearly ordered assembly of glucose units by the cellulose synthase (CS) (EC 2.4.1.12) attached to the cytoplasmic membrane. The linear cellulose is extruded through the pores of the membrane and assembled into other linear fibrils of cellulose by intermolecular and intramolecular hydrogen bonding to form a complex and crystalline network of nanocellulose [108, 109]. The BNC fibril remains attached to the membrane cell during the cell division, and the bacteria are entrapped into the network surface that facilitates close contact with the oxygen [110].

9.3 BNC PRODUCTION BIOPROCESSES

9.3.1 BIOCHEMICAL PARAMETERS FOR STATIC AND DYNAMIC PROCESSES

Currently, BNC can be produced under static, agitated, and aerated/agitated culture conditions, which have a significant influence on the shape and mechanical properties of the final bioproduct. The productivity and physical characteristics of BNC can be affected by several factors, such as the strain of bacteria, cultivation methods, carbon and nitrogen sources, temperature, and pH. The Hestrin-Schramm (HS) medium is commonly used for BNC production, which is expensive and contains glucose as a carbon and energy source, yeast extract, peptone, citric acid, and disodium phosphate. Besides glucose utilization, other single carbon sources or complex raw materials have been reported for BNC synthesis, including fructose, mannitol, glycerol, xylose, maltose, lactose, sorbitol, sucrose, methanol, ethanol, and butanol [8, 9]. Also, low-cost media such as agricultural waste, fruit juices or pulps (coconut, pineapple, banana), molasses from beets or sugarcane, starch from potatoes, corn steep liquor, and other inedible carbon sources (lignocellulosic hydrolysates) can be utilized [111, 112]. Table 9.1 shows some recent studies carried out for BNC production.

In the static method, production results in pellicles or a network of cellulose fibrils [10, 11]. Under these conditions, cellulose is formed by the bacteria near the medium/pellicle interface. The static method is usually

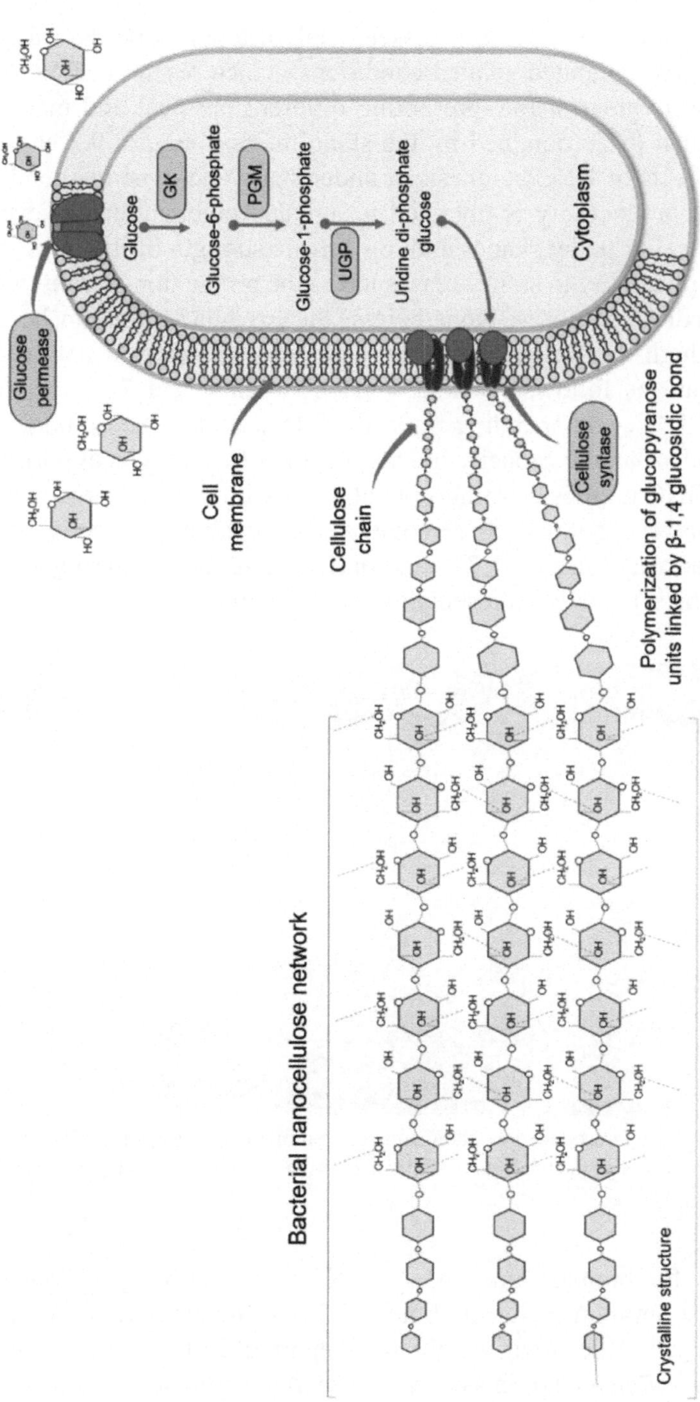

FIGURE 9.1 Schematic representation of bacterial nanocellulose (BNC) synthesis. [Abbreviations: GK: Glucokinase; PGM: Phosphoglucomutase; and UGP: Uridine diphosphate-glucose pyrophosphorylase].

conducted in trays or large-area vessels. Only a few reports describe BNC production under agitated/aerated conditions, which results in BNC with a spherical or irregular shape, presenting different physical and mechanical properties than those obtained by the static method. Figure 9.2 shows the different aspects of BNC synthesized under static and dynamic conditions. The highest productivity is obtained in agitated cultivation; however, the small spherical or pellet shapes and low tensile strength of the BNC reduce its direct applicability in some areas, such as the production of wound dressings for medical purposes. Nonetheless, the irregular form can be useful to produce high-quality paper or food additives. The more usual operating conditions are as follows: an initial pH between 4 and 7.5, temperature from 28 to 37°C, and spending time for 1–15 days are the optimal conditions dependent on the capacity of each strain [12]. The typical purification process of BNC implies washing several times with distilled water, washing with 1 N NaOH (25–80°C) to remove residual medium and cells, washing with distilled water until pH 7.0, and drying at 40–60°C overnight. Figure 9.3 shows the schematic representation of BNC production.

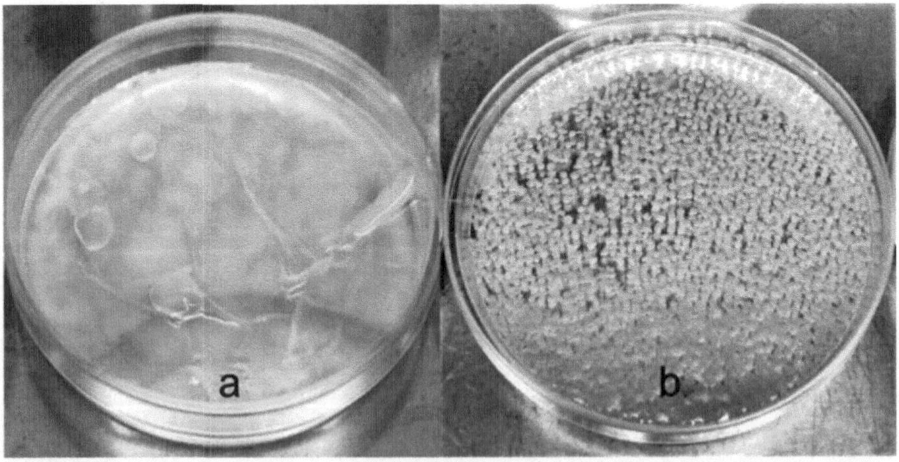

FIGURE 9.2 BNC is produced by static and agitated methods. Pellicles of BNC formed in the air-liquid interface of the medium by static method (a); and irregular shapes of BNC under agitation (b).

The pH for biomass growth and BNC synthesis ranges from 4.0 to 7.0, with the optimum value at about pH 5.0. In this regard, an increase of 60% in the cellulose dried weight was reported during the cultivation of *Acetobacter xylinum* 0416 in a controlled medium compared to uncontrolled

TABLE 9.1 Recent Studies of Bacterial Cellulose Production

Genus, Specie, and/or Strain	Method	Medium	Carbon Source	Maximum Production (g/L)	References
Kosakonia oryzendophytica FY-07 (former *Enterobacter* sp. FY-07)	Static	HS-modified/corn stover total hydrolysate (CSTH)	HS medium supplemented with xanthan gum and KNO_3/(CSTH: glucose, xylose, arabinose).	17.13 (HS+ xanthan gum + KNO_3) 13.96 (CSTH)	[13]
K. xylinus ATCC 53582	Static	HS-modified/carbohydrates.	Lactose, oligomers, starch.	5.15	[3]
K. xylinus IITR DKH20	Static	HS-modified/fruit extract	Extract of peel waste of pineapple, orange, sweet lime, and banana.	11.4 (extract of pineapple peel).	[14]
Lactiplantibacillus plantarum AS.6	Static	HS-modified	Hydrolyzed prickly pear peels.	6.01	[15]
Lactiplantibacillus plantarum AS.6	Static	HS/HS-modified/Yamanaka medium-modified/GEM (glucose-ethanol medium).	Glucose, fructose, mannitol, xylose, galactose, sucrose, and starch.	4.51 (modified Yamanaka medium)	[16]
Komagataeibacter sp. NFXK3	Static	HS-modified	Glucose, glycerol, ethanol.	9.61 (Glucose + 3% ethanol, *strain NFXK3*). 11.46 (glycerol +1.5% ethanol, strain FXK3).	[17]
G. xylinus LYP25 (mutant)	Static	BCEL-GC01	Glucose	14.0	[18]
G. xylinus NRRL B-759	Static	Sugar beet molasses	Sucrose	10.25	[19]
K. europaeus BCRC 14148	Static	HS-modified	Papaya and orange extract.	3.48 (Papaya extract). 3.47 (Orange extract).	[20]
G. xylinus ATCC 53524	Stirred	BCEL-GC01	Hydrolysates (*Miscanthus*, barley straw, and pine tree).	16.70 (*Miscanthus*) 13.09 (Barley straw) 12.54 (Pine tree)	[21]
K. rhaeticus MSCL 1463	Stirred	Whey	Galactose	1.95	[22]

Note: BCEL-GC01: 50 g/L glucose; 20 g/L corn steep liquor (CSL); 4 g/L ammonium sulfate; 2 g/L dibasic potassium phosphate; and 0.4 g/L magnesium sulfate $7H_2O$.

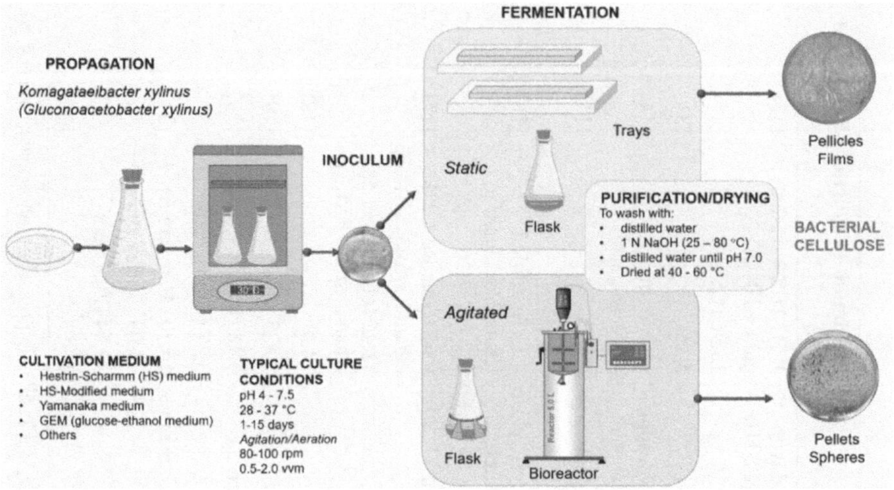

FIGURE 9.3 Schematic representation of bacterial nanocellulose (BNC) production.

pH. Nevertheless, cellulose production decreases by 30% when the pH drops from 5 to 3.5 [23].

In general, a decrease in pH is observed after 3–4 days of cultivation in non-buffered media, due to the accumulation of secondary metabolites, such as gluconic, lactic, and/or acetic acids, resulting from the consumption of carbohydrates or nitrogen sources [10, 24]. At a pH lower than 3.5, the viability of cells, and consequently the productivity of bacterial cellulose, is markedly affected.

Oxygen is a key substrate for growth and the activation of metabolic pathways, and an optimal oxygen supply is fundamental for the synthesis of bioproducts in overall aerobic microorganisms. The oxygen uptake is from the suspended gas bubble to the liquid phase, then transferred to the cell cytoplasm and organelles. Oxygen is typically supplied into the culture medium by bubbling air (enriched or not in pure oxygen) through a sparge, and then usually distributed in the liquid by agitation using a mechanical device. To improve oxygen transfer, two alternatives may be applied: one is to supply a higher concentration of oxygen to the bioreactor, and the second is to increase agitation and/or gas flow rate, increasing the interfacial area (liquid/gas) [25]. As mentioned above, bacterial cellulose producers are strictly aerobic organisms; therefore, the dissolved oxygen and mainly the oxygen uptake are crucial for the process rate of synthesis of this biopolymer [26].

In static culture, usually, the dissolved oxygen is limited, and mass transfer is restricted to the gas-liquid (cell culture medium) interface, affecting the metabolic process, cell viability, and productivity of BNC [113]. Dynamic culture supplies oxygen to the culture medium, increasing the cell density and promoting high productivity of BNC, although resulting in irregular shape and without the formation of a gel-like membrane [27, 28]. It was found that in flask culture, a shaking speed of 80 to 100 rpm promotes a high amount of bacterial cellulose production (about 5.0 g/L) after 6 days, probably associated with low-shear stress. However, at agitation equal to or higher than 150 rpm, the dissolved oxygen in the medium increases but does not necessarily increase the BNC production. In general, after 4 days of cultivation in a flask, the viscosity of the culture medium increases due to BNC accumulation, making oxygen transfer difficult [29].

Submerged and agitated cultivation is applied in several industrial bioprocesses because of the higher oxygen transfer rate that produces higher biomass and productivity of bioproducts when using aerobic microorganisms [30]. However, bacterial cells can be disrupted or at least affected by hydrodynamic stress (shear stress) generated by vigorous aeration, turbulence, and shear rate. Oxygen transfer and uptake rates, together with shear effects, are crucial in bioreactor design, performance, and scale-up [25, 30]. Due to these reasons, different types of bioreactors have been considered to produce BNC [11, 31].

Reiniati et al. [24] reported BNC production by *Komagataeibacter xylinus* in a stirred-tank bioreactor (STBR), achieving yields of 1.13 and 0.54 g/L at agitation rates of 700 rpm and 500 rpm after 72 h, respectively. The authors demonstrated that agitation has a fundamental effect on the productivity and shapes of the bacterial cellulose produced. In this respect, Chen et al. [26] studied the effect of 13 different impeller configurations on BNC production by *Komagataeibacter xylinus* ATCC 23770 grown in STBR. The authors demonstrated an increase in cellulose production using single-pitched-blade and twin-pitched-blade (large) impeller configurations at an agitation of 300 rpm and maintaining pH at 5.0. The authors also found that high cellulose productivity (1.25 g/L.h) was reached under the optimized cultivation conditions using a pitched-blade (large) impeller with a high volumetric oxygen-transfer coefficient ($k_La = 23.53$/h).

Cheng et al. [114] reported the production of elliptical cellulose pellets with a final concentration of 7.72 g/L and a productivity of 0.107 g/L·h after 72 hours of cultivation using a modified airlift reactor equipped with three wire-mesh draft tubes. The authors mentioned that the bacterial nanocellulose (BNC) production was three times higher than in a conventional column bubble reactor (cellulose 2.82 g/L and productivity of 0.039 g/L·h).

Additionally, the k_La values were highest in the modified airlift reactor and increased with the superficial air velocity (Vs). For example, with a Vs of 2.5 cm/s, the k_La was 160/h in the airlift reactor and only 100/h in a conventional bubble column reactor. However, k_La decreased with increasing viscosity of the medium due to the accumulation of bacterial cellulose. The k_La was higher in a medium containing cellulose as elliptical pellets compared to a fibrous shape, for the same airflow rate.

In another study, fermentation was conducted with a modified 5.0 L airlift bioreactor containing six rectangular net draft plates, resulting in a production of 2.6 g/L of membrane-type bacterial cellulose after 4 days of culturing, with productivity of 0.027 g/L·h [32]. The reactor configuration is crucial for process control because the accumulation of bacterial cellulose on the oxygen and pH electrodes is more noticeable in airlift reactors, as the agitation in these reactors is weaker compared to stirred-tank reactors.

A key factor to be considered in BNC production is that bacteria are sensitive to shear stress; consequently, a low shear rate can improve the productivity and characteristics of the cellulose produced. To reduce the shear stress, some reports show the effectiveness of the addition of polysaccharides, such as agar, xanthan, and acetan [33, 34]. In a 50-L airlift reactor, BNC production by *Acetobacter xylinumin* reached 8.7 g/L in 40 h of cultivation in a medium containing 0.1% (w/w) agar in the culture medium, compared with 6.3 g/L in the control run without agar, achieved after 65 h of cultivation. However, when a 10-L STBR was used for the same culture medium, the addition of 0.1% (w/w) agar yields the same cellulose (6.9 g/L) as the control. The low agar concentration used in the STBR was not effective to reduce the shear stress and increasing productivity. However, the productivity in the airlift reactor was 2 times (0.20 g/L h) that obtained in the stirred tank [33]. In another study using STBR, by increasing the agar concentration in the culture medium to 0.4% (w/w), the maximum cellulose concentration was 12.8 g/L compared with 8 g/L without agar [35]. Aswini et al. [36] evaluated 12 different additives to reduce the shear stress and increase the productivity of cellulose by *Acetobacter senegalensis*. The authors claimed that the addition of polysaccharides increases the relative viscosity and promotes the formation of large BNC agglomerations, facilitating oxygen diffusion and increasing k_La and productivity [36]. Nevertheless, it is more realistic to assume that the cells are subjected to less dynamic stress, increasing the number of viable cells, or at least not negatively affected.

Despite the diverse production methods cited, which are useful for obtaining BNC for different uses, the biggest challenge is to find alternative methods to increase the efficiency and productivity of the process and/or improve the properties of the BNC, considering the final application.

9.3.2 CHALLENGES FOR BIOPROCESS DESIGN AND SCALE-UP

The development of the static BNC production process has been studied but not extensively. Different carbon sources (mainly sugars) and the composition of the culture medium in terms of the main nutrients (basically nitrogen source) have been tested [37, 115]. The operating conditions have also been studied: mainly temperature and initial pH, as previously mentioned. However, the kinetics of the process are not well known; long reaction times (usually weeks) are used without distinguishing between biomass growth, substrate consumption, and biopolymer production, basically due to the lack of safe analytical methods for the compounds cited. It is not frequent to even have data at different times, but only at the end of the process. To our knowledge, there is not a rigorous kinetic model able to simulate the bioprocess; only a very recent publication [11] proposes a simple model describing the variation over time of the key compounds: biomass, carbon substrate, and cellulose. A rigorous kinetic model is essential for the design of the reactor, although, in the case of static BNC production, this concept of reactor design is not usually referred to; the formation of a film is usually sought using different film growth systems.

Regarding the dynamic BNC production process, the medium composition employed is similar to that used in the static process. Various carbon sources and essential nutrients have also been tested, as described above, considering that good static performance can be extrapolated to dynamic production. Something similar happens with the operating conditions, temperature, and pH, although in dynamic production there are studies that affect pH programming, normally not allowing it to drop below certain values, as previously mentioned.

In dynamic production, there is some data on the evolution of some of the compounds in the system over time, although not much. The main difficulty is found in the measurement of biomass and polysaccharide concentrations separately. Therefore, there is also no kinetic model describing the evolution of the aforementioned key compounds over time.

The type of reactor that is usually used in dynamic production is the STBR, although there are some works in airlift. Reactor design, without a kinetic model, can only be very elementary. In addition, there is a new phenomenon, concerning static production, which is agitation and the phenomena associated with it. Agitation can be carried out in agitated flasks or stirred tanks. In both cases, the agitation will influence the oxygen transport rate (OTR), and this can usually mark the oxygen uptake rate (OUR), as corresponds to two phenomena in the series [25].

In STBR, OTR (and surely OUR for the mentioned reason) will depend on [30]: (i) the geometry of the system: height-diameter reactor ratio, size and number of agitators, number and width of baffles, design of the air distributor, etc.; (ii) the operating conditions: agitation and supply airflow; and (iii) the properties of the culture, mainly its viscosity, which will increase over time due to the production of BNC.

The availability of oxygen (OTR-OUR balance) will have a decisive influence on the evolution of the process, and therefore on its performance. The fluid dynamic conditions (mainly given by the stirred speed) influence OTR-OUR values, but they can also generate hydrodynamic stress, acting negatively on cells [115]. By analogy with other bioprocesses, such as xanthan production [6, 38], it is very likely that cells producing a biopolymer, such as BNC, are sensitive to shear stress when not protected by the polymer. Therefore, it is likely that in a dynamic process, it is necessary to start with low agitation, to maintain cell viability, and then agitation must be increased as the polysaccharide is produced because the increase of the viscosity provokes the decrease of the OTR, which can affect the performance of the bioprocess, because oxygen is an essential nutrient for this strictly aerobic bioprocess.

These phenomena are very little studied, including the level of agitation, the value of the oxygen transport coefficient (k_La), the OTR and OUR, the possibility of occurrence of hydrodynamic stress, etc., and their influence on both the biomass growth and the BNC production rates require a deeper study than that conducted till now. Of course, due to the lack of knowledge about the mentioned aspects, there are no reliable kinetic models to produce BNC in dynamic operation.

The scale-up of bioprocesses must be done following some basic rules [115]: (i) geometrical, following well-established criteria, maintaining the geometric analogy between the different equipment sizes: height-diameter ratio, relative size of agitator and baffles, number and type of agitators, etc.; (ii) it is necessary to ensure the composition of the medium and the value of the previously determined operating conditions: temperature and pH or their programming; and (iii) the fluid dynamic conditions must be reproduced on different scales, this being the most difficult rule. Furthermore, to reproduce the conditions of shear stress and oxygen availability (OTR-OUR) there are two criteria: (a) to keep the power dissipated in the medium constant (P/V constant); or (b) to keep the oxygen supply constant ($k_L a$ constant, or what is equivalent, constant maximum OTR). This last criterion has been proposed as safer [115].

9.4 CHARACTERISTICS AND PROPERTIES OF BNC

The BNC is a three-dimensional highly pure reticular biodegradable and biocompatible polymer with carbon and hydrogen content of about 44 and 6%, respectively. It contains a large number of hydroxyl groups easily linkable with other natural polymers to create composites [9, 39]. This biopolymer presents two cellulosic subunits (Cellulose I and II) with a degree of polymerization ranging from 500 to 18,000 that form a crystalline structure (gel-like membrane) that protects the cell from adverse environments [40, 41]. The diameter of the nanofibers of bacterial cellulose is smaller than that of plant cellulose, ranging from 10 to 90 nm, and 100 to 1,000 nm in length, forming an ultrafine reticular structure. In the case of dynamic cultures, the BNC presents special shapes, like spheres, ellipsoids, and/or pellets with different sizes ranging from 10 μm to 10 mm. After purification, both the film and pellet are uncolored and translucent but not optically transparent material.

BNC shows tensile strength (TS) from 33 to 422 MPa, elongation at break ranging from 0.5 to 5.0%, and Young's modulus of 5 to 17 GPa [42–44]. These properties are affected by different factors, including bacterial strain, culture conditions, and drying conditions. In this regard, water activity (a_w) and drying temperatures strongly affect the mechanical properties, as mentioned by Cazón et al. [45] who reported that a_w higher than 0.32 continuously decreased the TS of the bacterial cellulose films. Studies on dehydration/rehydration of BNC by convection drying showed that low drying temperature (25°C) results in higher TS, thermal stability, and dry mass content than cellulose dried at high temperature (105°C) [46].

Besides, the selection of a cultivation method and purification affects the commercial applications of bacterial cellulose, depending on the morphology and physical properties. Crystallographic studies have shown that native cellulose is composed of two crystalline phases known as I_α (with a triclinic unit cell) and I_β (with a monoclinic unit cell) [47]. In bacterial cellulose, phase I_α is predominant; however, the crystallinity index (Cr) is substantially affected by the culture conditions and varies from 0.30 to 0.89 [10, 24]. In general, cellulose obtained in stationary cultures shows higher crystallinity than that produced in agitated cultures; this is due to the fact that the cellulose microfibrils are parallelly oriented in the former method [10, 24, 27].

Another important property, the Young's modulus that denotes the flexibility and tensile strength of a material, can be manipulated as a function of the culture conditions. Thus, BNC produced in stationary culture exhibits a higher Young's modulus (about 18 GPa) than in agitated conditions (2.1 GPa) [32, 48, 49]. Furthermore, the water-holding capacity (WHC) is a crucial parameter to be considered when applying BNC in some fields. Values of WHC of 185 to 300 g of water per gram of cellulose were reported for BNC synthesized under agitated conditions [32]. The viscosity of the gel-membrane (pellicle) obtained under static conditions has been reported to be about 10 to 129 cP [50, 51]. A higher viscosity (around 520 cP) of BNC in the homogenized (disintegrated) form was described for cellulose produced in an agitated flask [52]. Therefore, BNC produced under agitated conditions could be added to improve the ability of food to hold or add water during processing or storage, and to control the viscosity in fluid foods [49, 51, 52].

A fundamental property to be considered in the BNC is the water vapor permeability (WVP), which corresponds to the amount of water that permeates through a film per unit of area and time. This parameter is dependent on a_w, pressure, and thickness and is useful to determine the moisture management capability for application as packaging material. In general, water vapor permeation through polymer films follows the solution-diffusion mechanism, and the permeability coefficient increases with the temperature and relative humidity [116]. The permeability of dry cellulose towards dry gases is relatively low (about 10^{-3} g/m.s.Pa) and increases in contact with moisture. WVP of bacterial cellulose was reported ranging from $1.35 \cdot 10^{-12}$ to $3.13 \cdot 10^{-11}$ g/m s Pa for a_w in the range of 0 to 0.9 [53]. These values are relatively high for direct use in packaging; however, the combination with other polymers and solids can result in various composites with adequate WVP values and higher mechanical properties for food packaging [39, 54].

9.5 APPLICATIONS IN FOOD INDUSTRY

Bacterial cellulose can be used as part of a food ingredient or additive because it is a GRAS substance. It is an indigestible dietary polysaccharide that may be used for food applications in the form of films/membranes, pellets, or fibers, homogenized dispersion (disintegrated fibers in liquid), or pure crystals of BNC (obtained by chemical and physical treatment of BC for removal of the amorphous fraction). In most cases, it should be combined with other polymers (carbohydrates, proteins, pectin, PVA, and others) to be used as an emulsifier/stabilizer of emulsion, rheological modifier, as an antimicrobial carrier, and for composites for the food-packaging sector [14, 53, 55, 56].

9.5.1 EMULSIFICATION

The emulsification consists of a mechanical dispersion of two or more immiscible liquids in a semi-stable mixture. In food applications, the emulsion is a homogeneous mixture composed generally of two phases, an organic (oil) phase and an aqueous (water) phase, stabilized by an emulsifier. Amphiphilic molecules and particles have been used as emulsifiers in the food industry, including soy lecithin, mono- and di-glycerides, polysorbates, and vegetable oils. Most of the cited compounds are good emulsifiers; however, they may not provide long-term stability during food storage [57]. The emulsions can also be stabilized by adding insoluble particles, microgels, and Janus particles (nanoparticles with two or more chemical or physical properties on the surface) called Pickering stabilizers [58]. These particles absorb at the oil-water interphase forming interfacial layers that prevent the coalescence of drops and inhibit lipid oxidation in food products [55, 56]. Some polymers and particles such as starch, pectin, Arabic gum, modified starch, proteins, and galactomannans are good emulsifiers and stabilizers [55, 59]. Among the Pickering stabilizers, polymers, cellulose shows high potential because of its amphiphilic nature and sustainability; however, this material has low dispersion stability and hydrophilicity with a limited capacity for emulsification [56, 60].

Nanocellulose generated by *K. sucrofermentans* DSM 15973, when it is used in a disintegrated and then homogenized form, showed high effectiveness in stabilizing olive oil-in-water emulsions. This was due to the promotion of a stronger gel-like viscoelastic network of fibrils on the surface of the oil droplet, acting as a barrier to avoid coalescence. Compared with commercial cellulose commonly used in the food industry, BNC promoted more emulsion stability against changes in pH, temperature, and ionic strength than carboxymethyl cellulose (CMC) and hydroxypropyl methylcellulose (HMC) [61]. Recently, cellulose produced by *Komagataeibacter hansenii* CGMCC 3917 has been combined with soybean protein isolate through hydrogen bonds to form colloidal particles and used as a stabilizer of emulsions containing 75% oil in water. In general, the addition of cellulose improves the emulsification capacity of the soy protein. Moreover, increased stability of the emulsion was observed when increasing the cellulose to soy protein isolate ratio, achieving the highest emulsion stability with a 7:25 ratio during two months of storage [62].

Bacterial cellulose may also be oxidized by 2,2,6,6- tetramethylpiper-idine-1-oxyl (TEMPO), a method that selectively oxidizes the primary hydroxyl group and introduces sodium carboxyl groups in the cellulose to promote the dispersion of the nanofibrils in water [63]. Soybean protein isolate and TEMPO-oxidized bacterial cellulose (TOBC) complex showed high capability as a stabilizer of emulsions (oil/water) due to the strong electrostatic attraction of protein on the anionic TOBC surface at pH 3.0. Therefore, emulsions containing TOBC-soybean protein resulted in a low creaming index (<5%) observed with the increase of solid content in the emulsion at 3% after 14 days of storage (values of 0 to 20% indicate low serum separation and higher emulsion stability). Furthermore, at pH 3.0, the stabilized emulsions showed a higher droplet size than emulsions at pH 7.0 [58]. It is well-known that pH is an important parameter that regulates the structure and properties of the emulsions and affects interactions between charged particles [117]. In general, emulsions containing BNC showed resistance to variations in ionic strength, pH, and temperature, contrary to the traditionally used emulsifiers such as surfactants [56, 58].

Consumer demand for safe and nutritious food has led industries to explore natural and immune-boosting novel additives. In this context, BNC emerges as an alternative to reduce high-calorie components such as fats (reduction of 5–15 wt.%) and provide stability to meat emulsions (an oil-in-water mixture containing a phase of solid or liquid fat particles) and a continuous phase of water with dissolved salts and/or proteins [64, 65]. Cellulose synthesized by *Gluconacetobacter xylinus* NRRL B-42, added in small amounts (0.267 g/100 g product), showed high potential to produce low-lipid and low-sodium meat sausages emulsified with high-oleic sunflower oil, enhancing water holding capacity, hardness of the product, and cohesiveness. Additionally, the cellulose improved the textural parameters, and the product remained stable for 45 days of cold vacuum storage [66].

Moreover, bacterial cellulose was reported as a replacer of fat in many prepared foods, such as Sukuk sausage [64], Nile tilapia (*Oreochromis niloticus*) mechanically separated meat sausage [67], chicken Chinese sausage [68], increasing the quality of the product by improving water content, color, firmness, and overall acceptability of the products.

9.5.2 IMMOBILIZATION OF MICROORGANISMS AND MOLECULES

Probiotic bacteria (microorganisms conferring health benefits to the host) are usually entrapped or encapsulated in natural homopolymers to protect them from adverse environments, enhancing cell viability and preserving physiological properties when delivered in the intestinal tract of humans and animals. It is recommended that probiotic products contain at least 10^7 CFU per gram or milliliter to guarantee the passage and survival of cells in the digestive system [69, 70].

To increase the survival of probiotics, strategies of co-encapsulation of both probiotics with prebiotics (synbiotic products) have been extensively reported [71–74]. Recently, the International Scientific Association for Probiotics and Prebiotics (ISAPP) updated the definition of synbiotic (sometimes incorrectly referred to as symbiotic) as a "mixture comprising live microorganisms and substrate(s) utilized by host microorganisms that confers a health benefit on the host" [75]. As a biocompatible polymer, BNC shows potential as a co-polymer not only for the production of synbiotic products but also for encapsulating active pharmaceutical and nutraceutical molecules.

In this sense, films of cellulose produced by *Komagataeibacter xylinus* C3, previously dissolved in an aqueous solution of NaOH-urea-H_2O at low temperature, were successfully used as a coating for encapsulation of *Lactobacillus rhamnosus* GG. The probiotic bacteria were previously entrapped in alginate and alginate/pullulan (exopolysaccharide produced by *Aureobasidium pullulans*) matrix and covered with a layer of solution of bacterial cellulose (0.5–2%, m/v). This capsule provided protection to probiotics against simulated gastrointestinal conditions, achieving 74% and 94% of survival rate after 2 hours of contact at pH 2.0, respectively. Apparently, the highest protective effect offered by the matrix pullulan was due to the ability of this prebiotic to clog the pores of the cellulose network, reducing the acid diffusion into the capsules. Furthermore, gut enzymes were able to digest the capsule coating, releasing cells and prebiotics in the simulated gut fluid. That fact is important since the effect of probiotics on the colon depends on the number of viable cells and the amount of prebiotic delivery for the growth of beneficial bacteria [76].

Non-dairy lactobacilli (*Lactiplantibacillus pentosus* B329 and *Lactiplantibacillus plantarum* 820) were immobilized in bacterial cellulose and used as novel starter cultures for sour milk production, showing high fermentative activity after freeze-drying and five-month storage. Viability of the bacteria (>10^7 log cfu/g) in the functional sour milk was enhanced by immobilization in BNC during storage at 4°C for 28 days [77].

Nisin, a bacteriocin produced by *Lactococcus lactis subsp. lactis*, was successfully encapsulated by gelation with BNC (produced by *Komagataeibacter xylinus* ATCC 53524). BNC at a concentration of 5 mg/ml achieved high nisin encapsulation efficiency (80–93%) and showed considerable antimicrobial activity against *Lactobacillus rhamnosus* and *Leuconostoc mesenteroides*, which are responsible for spoilage in meat products [78]. BNC has also been used to immobilize glucoamylase [79], β-galactosidase [80], papain [118], and lipase [81] with potential applications in the food industry.

9.5.3 PACKAGING

The hydroxyl groups and the hydrogen bonding of the cellulose give special properties that create a three-dimensional semi-crystalline network insoluble in water. The large number of hydroxyl groups in the entire molecules may enable the combination with various macromolecules to form semi-crystalline structures and composites [39]. Furthermore, to satisfy the requirements of BNC for different applications, the films produced by different microorganisms are frequently modified by *in situ* or *ex situ* processes to enhance their properties and provide new functionality [54].

The ideal packaging material must act as a barrier towards gases and vapor and be resistant enough to mechanical forces to estimate the packed product's shelf life. Water vapor permeability (WVP) is an important parameter for packaging applications; it evaluates the water vapor mass transference through the polymer package used as a barrier [45, 53]. BNC membrane is naturally hydrophilic and a poor barrier for water vapor when compared to petroleum-based polymers such as polyethylene and polypropylene. The esterification of the bacterial cellulose with molecules containing a fatty aliphatic chain decreases the WVP, which could be an interesting solution. According to Tomé et al. [119], cellulose membrane produced by *Gluconacetobacter xylinus* shows a WVP of about 1.10^{-13} and $1.1.10^{-12}$ g/m.s.Pa for relative humidity (RH) of 31% and 85%, respectively. The insertion of hexanoyl chloride on the bacterial cellulose surface decreased the WVP and permeability of humidified carbon dioxide, oxygen, and nitrogen by 50%.

Bacterial cellulose can be functionalized with antimicrobial and antioxidant molecules to create active packaging to prevent microbial propagation and preserve the organoleptic characteristics of packed food. Composite films of bacterial cellulose with chitosan were successfully impregnated with 0.5% of curcumin, increasing the hydrophobicity, oxygen permeability, and mechanical properties of the film [82]. The composite was used as a sealer and achieved better preservation effectiveness of fresh strawberries than polyethylene film and low oxidation of edible oil. In a recent study, Gao et al. [54] reported *in situ* functionalization of cellulose produced by *Kosakonia oryzendophytica* FY-07 (*Enterobacteriaceae*) with magnolol (Mag, lignan with antibacterial activity extracted from *Magnolia officinalis*) previously dispersed in xanthan gum (XG), as a supplement of the fermentation medium. The biopolymer produced can effectively inhibit fungi and inactivate Gram-positive bacteria, although with a lower effect on Gram-negative bacteria. The cellulose film presented a crystallinity index of 60% and a homogeneous structure, evidenced by the near-linear thermal degradation curve, showing potential for applications in the packaging industry.

Films of bacterial cellulose combined with curdlan, a polysaccharide produced by *Acaligenes fecalis* (a repeat unit of three β (1→3)-linked D-glucose residues), and cinnamon essential oil showed high tensile strength, crystallinity, and thermal stability, with appreciable reduction of WVP and moisture content. The incorporation of cinnamon essential oil was effective against bacterial growth and lipid oxidation of chilled chicken during the shelf life. Therefore, the blending of bacterial cellulose/curdlan/cinnamon oil showed effectiveness for food packaging applications [83].

Within the considerations to validate food packaging materials, transparency is a crucial technical parameter that affects the product's attractiveness. In general, pure BNC is non-transparent and needs to be modified to meet food packaging specifications [53, 84]. Films of cellulose (generated by *Komagataeibacter xylinus*) were immersed in a 5% (w/w) solution of glycerol and polyvinyl alcohol and dried to obtain combined films with high transparency and low opacity. Transparency was measured in the UV-VIS spectra range by a spectrophotometer and calculated from the logarithm of the percent of transmittance (600 nm) per unit of thickness, while opacity was determined by the absorbance (500 nm) per unit of thickness. To measure color and calculate color differences within films or other materials, usually the three values of the CIE color space are considered, where L* represents lightness from black to white (scale of 0 to 100), while a* and b* indicate the chromaticity (without numerical limits) [85]. Regarding the colors of the modified bacterial cellulose, the L*, a*, and b* parameters (set at 28 standard/light source of D65) showed values dependent on the relative humidity, with L* and transparency ranging from 64.4 to 72.6 and 14.4 to 40.4, respectively [53].

Unfortunately, there is not extensive literature that analyzes the transparency and colors of BNC and its composites under standard conditions; therefore, there is insufficient information to compare these parameters. Shelf life and quality of foods may also be extended by edible film active packaging based on bacterial cellulose. Yanti et al. [120] successfully modified films of cellulose produced by *Acetobacter xylinum* LKN6 with carboxymethyl cellulose (CMC) (as stabilizer) and glycerol (as plasticizer) by casting method. By addition of 1% of the CMC and 1% glycerol, a film was created with considerable tensile strength (17.47 MPa), elongation at a break (25.60%), and Young's modulus (6.54 GPa). Furthermore, the film preserved the quality of meat sausage, evidenced by the stability of the protein content comparable to those achieved with low-density polyethylene (LDPE) film. However, the evaporation in the form of water vapor through the modified film was higher than with LDPE, but lower than with the unmodified cellulose. Regarding the protection of the meat sausage from microbial contamination, the modified BNC showed superior protection than LDPE, maintaining the quality of the packed product. Table 9.2 shows some recent applications of BNC in food technology.

9.5.4 OTHER USES IN FOOD INDUSTRY

BNC can also be added as a rheological modifier of food, promoting characteristics and properties such as gelling, water binding, and thickening. For example, in the ice cream industry, fluctuations in temperature during fabrication and storage result in recrystallization and partial coalescence of fat globules, which affect the texture, stability, and shelf life of the product [93]. Recently, cellulose obtained from *Gluconacetobacter xylinus* and inulin were combined to formulate eight dietary fiber-enriched ice creams, varying in inulin concentration (0.7 and 1.4 wt.%) and cellulose concentration (17 and 30 wt.%). The addition of homogenized cellulose in 150 g of ice cream showed a significant reduction in the melting rate (about 52–57 g/60 min) compared to the control (125 g/60 min), improving the viscosity and stability of the product. The hydrocolloid nature of cellulose allowed for the reduction of water mobility (water holding capacity) of the product, enhancing gelation, viscosity, and controlling crystallization in ice creams [121]. In another study, when a 20% BNC/protein isolate complex (1:20) gel was added to ice

TABLE 9.2 Recent Applications of Bacterial Nanocellulose (BNC) in Food Technology

Product/ Application	BC Concentration	BC Origin/ Producer	Supplementary Molecule/Compound	Conditions	Properties/Food Attributes	References
Ice cream	17–30% (ww)	*G. xylinus*	Inulin (0.7–1.4%, dw)	BC homogenized and blended with fruit concentrate/fresh fruits pulp.	Reduction of melting rate, increase fiber content, overall acceptability.	[86]
Sour milk starter	2.0% (m/v) in medium.	*A. xylinum*	–	BC incorporated in fermentation medium (immobilization of probiotics).	High efficacy of cell immobilization.	[77]
Encapsulation/ immobilization	5% (m/v) in trifluoroacetic acid (TFA).	*A. senegalensis*	Polyvinyl alcohol	BC-PVA fiber formed by electrospinning.	71.1% of the immobilized *L. acidophilus* survived to SGF.	[87]
	10% (w/v)	NS (strain from Kombucha).	Inulin 1%	Probiotic immobilized by adsorption-incubation method in inulin (1%) and Kombucha-BC (10%).	64.6% of the immobilized. *L. plantarum* TISTR 541 survived to SGF.	[88]
	0.5–2% (w/v)	*K. xylinus* C3	Alginate, pullulan	Alginate/Pullulan beads containing probiotic coated with BC dissolved in NaOH–urea–H_2O at a low temperature.	94% of the immobilized *L. rhamnosus* GG in alginate/BC/ pullulan survived to SGF.	[76]
Food packaging	5% (w/w)	Commercial (NS)	Chitosan/curcumin	Dispersion and homogenization of components. Chitosan with different molecular weight.	Chitosan/BC/curcumin composite presented high contact angles and mechanical properties, a low moisture content, WVP, and oxygen transmission rate.	[89]
	2% (w/w)	*Acetobacter* sp.	Curdlan/bacterial cellulose/cinnamon.	BC dissolved and mixture with CD.	Increase TS, EAB, and gas barrier. Conservation of chilled chicken packed with film.	[83]
	0.1% (w/v)	*A. xylinum*	Hydrolyzed protein obtained from pine tree fruit.	BNC obtained by TEMPO. 0.1% TOBC suspension with hydrolyzed protein (0.5, 1, and 1.5%).	Prevent spoilage, inhibit oxidation, and improve physical characteristics of pacific white shrimp.	[90]

TABLE 9.2 *(Continued)*

Product/ Application	BC Concentration	BC Origin/ Producer	Supplementary Molecule/Compound	Conditions	Properties/Food Attributes	References
Moisture barrier	NS	*K. rhaeticus*	Tomato puree, BC:CMC (weight ratio 5:1) and palm olein at a weight ratio of 5:3:2.	BC/CMC, palm olein and tomato. Puree in glycerol were homogenized. Dispersions were then vacuum degassed, cast onto glass plates dried at 50 ◦C for 24 h.	Film with low water vapor permeability, reduced the moisture migration from the guacamole to the nachos.	[91]
Sensor for pathogens.	NS	*G. xylinum*	Carboxylated multiwalled carbon nanotubes (c-MWCNTs), polyethyleneimine (PEI). *S. aureus*-specific lytic phage.	Adsorption of nanotubes on BC films and polymerization with PEI.	Biosensor effectively detected 3 and 5 CFU/mL of *S. aureus* in PBS and milk samples.	[92]

Note: ww: Fresh/wet weight; dw: dry weight; NS: not specified; SGF: simulated gastrointestinal fluid; TOBC: TEMPO-oxidized bacterial cellulose; TS: tensile strength; EAB: elongation at break; and CMC: carboxymethyl cellulose.

cream, textural and rheological improvements in the product were observed, with sensory attributes similar to the control. BNC was chosen as the fat substitute due to its highest overall similarity to cream [94].

BNC has been suggested as low-calorie ingredients and additives in food technology due to its potential in reducing serum lipids and cholesterol in *in vitro* and *in vivo* studies [56, 95]. It has also been reported that BNC may inhibit obesity and the inflammatory effect of a high-fat diet in mice [60]. Hydrogels of bacterial cellulose have been traditionally commercialized in the Philippines as *Nata de coco,* a food-grade cellulose gel produced by the fermentation of coconut water (supplemented with amino acids and other nutrients) with *Komagataeibacter xylinus*. This dessert is extensively consumed around the world [96, 97]. BNC has also been used to improve the hardness and cohesiveness of Kamaboko (processed Japanese seafood), to conserve the viscosity of the cocoa powder dispersion (chocolate drink) after sterilization, to increase the gel strength of *Tofu* (soybean curd), to preserve palatability when sugar is substituted by Aspartame in the traditional Japanese sweets *An* [98], as lard replacement, and to improve water-holding capacity in dolphin-fish (mahi-mahi) surimi [121].

9.6 FUTURE TRENDS

BNC is a promising biopolymer with multiple and interesting applications, which should be expanded. Until now, the BNC produced in static has been used more than that obtained under dynamic conditions, so a deep study of dynamic production seems necessary, especially in STBR. As already indicated, a description of the kinetics of the bioprocess is lacking, which requires the development of analysis techniques for the key compounds and the formulation of increasingly complex kinetic models, as proposed for the production of other biopolymers [99–101]. In addition, it is necessary to clarify the role of oxygen, how OTR-OUR influences the rate and evolution of the process, as well as the possible existence of hydrodynamic stress, which would make us reconsider the operating conditions (agitation and air flow) and perhaps even the type of reactor (STBR or airlift).

The properties of high crystallinity, high mechanical strength, thermal stability, water retention, numerous hydroxyl groups suitable to be combined with other molecules, reduced dry gas permeation, make BNC a promising additive or ingredient for food and packaging technology. Alternative low-cost medium, structural modification of surface and combination with other natural polymers to create new composites are currently under investigation.

It is clearly interesting to delve into the use of BNC as an emulsifier and emulsion stabilizer, as well as its possible use in other fields, not enough studied, such as the immobilization of microorganisms and molecules. Finally, it is necessary to comment on the great possibilities that this polymer presents in packaging, the general lines of application having already been described above, but which require determined development. Therefore, further research should be carried out to establish optimized strategies for efficient production of BNC at lower cost and should be promoted structural modification of this biopolymer to find new functional capabilities for application in the food industry.

ACKNOWLEDGMENT

Juan D. Rivaldi gratefully acknowledges the postdoctoral fellowship support from Programa Nacional de Becas Don Carlos Antonio López – BECAL-Paraguay (Contract N° 278/2022).

KEYWORDS

- **bacterial nanocellulose**
- **bioproduction**
- **biopolymer**
- **emulsification**
- **food applications**
- **functionalization**
- **immobilization**
- **microorganisms**

REFERENCES

1. Cheng, M., Qin, Z., Hu, J., Liu, Q., Wei, T., Li, W., & Liu, B. (2020). Facile and rapid one–step extraction of carboxylated cellulose nanocrystals by H2SO4/HNO3 mixed acid hydrolysis. *Carbohydrate Polymers, 231*, 115701. doi: 10.1016/j.carbpol.2019.115701.
2. Tanskul, S., Amornthatree, K., & Jaturonlak, N. (2013). A New Cellulose-producing bacterium, *Rhodococcus* Sp. MI 2: Screening and optimization of culture conditions. *Carbohydrate Polymers, 92*(1), 421–428. doi: 10.1016/j.carbpol.2012.09.017.

3. Przygrodzka, K., Szymańska, M., Banaszek, A., Zielińska, B., Ekiert, E., & Drozd, R. (2022). Bacterial cellulose production by *Komagateibacter xylinus* with the use of enzyme-degraded oligo- and polysaccharides as the substrates. *Applied Sciences, 12*(24), 12673–12673. doi: 10.3390/app122412673.
4. Shi, Z., Zhang, Y., Phillips, G. O., & Yang, G. (2014). Utilization of bacterial cellulose in food. *Food Hydrocolloids, 35*, 539–545. doi: 10.1016/j.foodhyd.2013.07.012.
5. Volova, T. G., Prudnikova, S. V., Sukovatyi, A. G., & Shishatskaya, E. I. (2018). Production and properties of bacterial cellulose by the strain *Komagataeibacter xylinus* B-12068. *Applied Microbiology and Biotechnology, 102*(17), 7417–7428. doi: 10.1007/s00253-018-9198-8.
6. Garcia-Ochoa, F., Castro, E. Gómez., & Santos, V. E. (2000). Oxygen transfer and uptake rates during xanthan gum production. *Enzyme and Microbial Technology, 27*(9), 680–690. doi: 10.1016/s0141-0229(00)00272-6.
7. Zhong, C. (2020). Industrial-scale production and applications of bacterial cellulose. *Frontiers in Bioengineering and Biotechnology, 8,* 1–19. doi: 10.3389/fbioe.2020.605374.
8. Lu, Z., Zhang, Y., Chi, Y., Xu, N., Yao, W., & Sun, B. (2011). Effects of alcohols on bacterial cellulose production by acetobacter xylinum 186. *World Journal of Microbiology and Biotechnology, 27*(10), 2281–2285. doi: 10.1007/s11274-011-0692-8.
9. Suryanto, H., Muhajir, M., Sutrisno, T. A., Mudjiono, Zakia, N., & Yanuhar, U. (2019). The mechanical strength and morphology of bacterial cellulose films: The effect of NaOH concentration. *IOP Conference Series: Materials Science and Engineering, 515*, 012053. doi: 10.1088/1757-899x/515/1/012053.
10. Castro, C., Zuluaga, R., Álvarez, C., Putaux, J. -L., Caro, G., Rojas, O. J., Gañán, P. (2012). Bacterial cellulose produced by a new acid-resistant strain of gluconacetobacter genus. *Carbohydrate Polymers, 89*(4), 1033–1037. doi: 10.1016/j.carbpol.2012.03.045.
11. Konopacki, M., Grygorcewicz, B., Kordas, M., Ossowicz-Rupniewska, P., Nowak, A., Perużyńska, M., & Rakoczy, R. (2022). Intensification of bacterial cellulose production process with sequential electromagnetic field exposure aided by dynamic modelling. *Biochemical Engineering Journal, 182*, 108432. doi: 10.1016/j.bej.2022.108432.
12. Çoban, E. P., & Biyik, H. (2011). Evaluation of different ph and temperatures for bacterial cellulose production in Hs (Hestrin-Scharmm) medium and beet molasses medium. *African Journal of Microbiology Research*, *5*(9), 1037–1045. doi: 10.5897/ajmr11.008.
13. Gao, G., Liao, Z., Cao, Y., Zhang, Y., Zhang, Y., Wu, M., & Ma, T. (2021). Highly Efficient Production of Bacterial Cellulose from Corn Stover Total Hydrolysate by Enterobacter Sp. FY-07. *Bioresource Technology, 341*, 125781. doi: 10.1016/j.biortech.2021.125781.
14. Khan, H., Saroha, V., Raghuvanshi, S., Bharti, A. K., & Dutt, D. (2021). Valorisation of fruit processing waste to produce high value-added bacterial nanocellulose by a novel strain Komagataeibacter xylinus IITR DKH20. *Carbohydrate Polymers, 260*, 117807. doi: 10.1016/j.carbpol.2021.117807.
15. El-Gendi, H., Salama, A., El-Fakharany, E. M., & Saleh, A. K. (2023). Optimization of bacterial cellulose production from prickly pear peels and its *ex situ* impregnation with fruit byproducts for antimicrobial and strawberry packaging applications. *Carbohydrate Polymers, 302*, 120383. doi: 10.1016/j.carbpol.2022.120383.

16. Saleh, A. K., El-Gendi, H., Soliman, N. A., El-Zawawy, W. K., & Abdel-Fattah, Y. R. (2022). Bioprocess development for bacterial cellulose biosynthesis by novel *Lactiplantibacillus plantarum* Isolate along with Characterization and Antimicrobial Assessment of Fabricated Membrane. *Scientific Reports, 12*(1), 1–17. doi: 10.1038/s41598-022-06117-7.
17. Fatima, A., Paloma Ortiz-Albo, Neves, L. A., Nascimento, F. X., & Crespo, J. G. (2023). Biosynthesis and characterization of bacterial cellulose membranes presenting relevant characteristics for air/gas filtration. *Journal of Membrane Science, 674*, 121509–121509. doi: 10.1016/j.memsci.2023.121509.
18. Lee, J., Lee, K. H., Kim, S., Son, H., Chun, Y., Park, C., & Yoo, H. Y. (2023). Improved production of bacterial cellulose using gluconacetobacter sp. LYP25, a strain developed in UVC mutagenesis with limited viability conditions. *International Journal of Biological Macromolecules, 232*, 123230. doi: 10.1016/j.ijbiomac.2023.123230.
19. ÖZ, Y. E., & Kalender, M. (2022). A novel static cultivation of bacterial cellulose production from sugar beet molasses: Series static culture (SSC) System. *International Journal of Biological Macromolecules, 225*, 1306–1314. doi: 10.1016/j.ijbiomac.2022.11.190.
20. Tseng, Y. S., Patel, A. K., Chen, C. -W., Dong, C. -D., & Singhania, R. R. (2022). Improved production of bacterial cellulose by *Komagataeibacter europaeus* employing fruit extract as carbon source. *Journal of Food Science and Technology. 60*, 1054–1064. doi: 10.1007/s13197-022-05451-y.
21. Kim, H., Son, J., Lee, J., Yoo, H. Y., Lee, T., Jang, M., & Park, C. (2021). Improved production of bacterial cellulose through investigation of effects of inhibitory compounds from lignocellulosic hydrolysates. *GCB Bioenergy, 13*(3), 436–444. doi: 10.1111/gcbb.12800.
22. Kolesovs, S., Ruklisha, M., & Semjonovs, P. (2023). Synthesis of bacterial cellulose by Komagataeibacter Rhaeticus MSCL 1463 on Whey. *3 biotech, 13*(3). doi: 10.1007/s13205-023-03528-9.
23. Zahan, K. A., Pa'e, N., & Muhamad, I. I. (2015). Monitoring the effect of pH on bacterial cellulose production and *Acetobacter xylinum* 0416 growth in a Rotary Discs Reactor. *Arabian Journal for Science and Engineering, 40*(7), 1881–1885. doi: 10.1007/s13369-015-1712-z.
24. Reiniati, I., Hrymak, A. N., & Margaritis, A. (2017). Kinetics of cell growth and crystalline nanocellulose production by *Komagataeibacter xylinus. Biochemical Engineering Journal, 127*, 21–31. doi: 10.1016/j.bej.2017.07.007.
25. Garcia-Ochoa, F., Gomez, E., & Santos, V. E. (2010). Fluid dynamic conditions and oxygen availability effects on microbial cultures in STBR: An overview. *Biochemical Engineering Journal, 164*, 107803. doi: 10.1016/j.bej.2020.107803.
26. Chen, G., Chen, L., Wang, W., Chen, S., Wang, H., Wei, Y., & Hong, F. F. (2019b). Improved bacterial nanocellulose production from glucose without the loss of quality by evaluating thirteen agitator configurations at low speed. *Microbial Biotechnology, 12*(6), 1387–1402. doi: 10.1111/1751-7915.13477.
27. Czaja, W., Romanovicz, D., & Brown, R. M. (2004). Structural investigations of microbial cellulose produced in stationary and agitated culture. *Cellulose, 11*(3/4), 403–411. doi: 10.1023/b.0000046412.11983.61.

28. Chen, G., Wu, G., Chen, L., Wang, W., Hong, F. F., & Jönsson, L. J. (2019a). Performance of nanocellulose-producing bacterial strains in static and agitated cultures with different starting pH. *Carbohydrate Polymers, 215*, 280–288. doi: 10.1016/j.carbpol.2019.03.080.
29. Tantratian, S., Tammarate, P., Krusong, W., Bhattarakosol, P., & Phunsri, A. (2005). Effect of dissolved oxygen on cellulose production by *Acetobacter* sp. *Journal of Scientific Research, Chulalongkorn University, 30*(2), 179–186.
30. Garcia-Ochoa, F., Gomez, E., Santos, V. E., & Merchuk, J. C. (2009). Oxygen uptake rate in microbial processes: An overview. *Biochemical Engineering Journal, 49*(3), 289–307. doi: 10.1016/j.bej.2010.01.011.
31. Zywicka, A., Peitler, D., Rakoczy, R., Konopacki, M., Kordas, M., & Fijalkowski, K. (2015). The effect of different agitation modes on bacterial cellulose synthesis by Gluconacetobacter xylinus strains. *Acta Sci. Pol. Zootechnica, 14*(1), 137–150.
32. Wu, S. -C., & Li, M. -H. (2015). Production of bacterial cellulose membranes in a modified airlift bioreactor by *Gluconacetobacter xylinus*. *Journal of Bioscience and Bioengineering, 120*(4), 444–449. doi: 10.1016/j.jbiosc.2015.02.018.
33. Chao, Y., Mitarai, M., Sugano, Y., & Shoda, M. (2001). Effect of addition of water-soluble polysaccharides on bacterial cellulose production in a 50-l airlift reactor. *Biotechnology Progress, 17*(4), 781–785. doi: 10.1021/bp010046b.
34. Song, H. -J., Li, H., Seo, J. -H., Kim, M. -J., & Kim, S. -J. (2009). Pilot-scale production of bacterial cellulose by a spherical type bubble column bioreactor using saccharified food wastes. *Korean Journal of Chemical Engineering, 26*(1), 141–146. doi: 10.1007/s11814-009-0022-0.
35. Bae, S., Sugano, Y., & Shoda, M. (2004). Improvement of bacterial cellulose production by addition of agar in a jar fermentor. *Journal of Bioscience and Bioengineering, 97*(1), 33–38. doi: 10.1016/s1389-1723(04)70162-0.
36. Aswini, K., Gopal, N. O., & Uthandi, S. (2020). Optimized culture conditions for bacterial cellulose production by acetobacter senegalensis ma1. *BMC Biotechnology, 20*(1). doi: 10.1186/s12896-020-00639-6.
37. Santos, S. M., Carbajo, J. M., Gómez, N., Ladero, M., & Villar, J. C. (2017). Paper Reinforcing by in situ growth of bacterial cellulose. *Journal of Materials Science, 52*(10), 5882–5893. doi: 10.1007/s10853-017-0824-0.
38. Casas, J. A., Santos, V. E., & García-OchoaF. (2000). Xanthan gum production under several operational conditions: Molecular structure and rheological properties. *Enzyme and Microbial Technology, 26*(2-4), 282–291. doi: 10.1016/s0141-0229(99)00160-x.
39. Volova, T. G., Prudnikova, S. V., Kiselev, E. G., Nemtsev, I. V., Vasiliev, A. N., Kuzmin, A. N., & Shishatskaya, E. I. (2022). Bacterial Cellulose (BC) and BC composites: Production and properties. *Nanomaterials, 12*(2), 192–192. doi: 10.3390/nano12020192.
40. Tahara, N., Tabuchi, M., Watanabe, K., Yano, H., Morinaga, Y., & Yoshinaga, F. (1997). Degree of polymerization of cellulose from *Acetobacter xylinum* BPR2001 decreased by cellulase produced by the strain. *Bioscience, Biotechnology, and Biochemistry, 61*(11), 1862–1865. doi: 10.1271/bbb.61.1862.
41. Lahiri, D., Nag, M., Dutta, B., Dey, A., Sarkar, T., Pati, S., & Abdul Kari, Z. (2021). Bacterial cellulose: production, characterization, and application as antimicrobial agent. *International Journal of Molecular Sciences, 22*(23), 12984. doi: 10.3390/ijms222312984.
42. Suratago, T., Taokaew, S., Kanjanamosit, N., Kanjanaprapakul, K., Burapatana, V., & Phisalaphong, M. (2015). Development of bacterial cellulose/alginate nanocomposite

membrane for separation of ethanol–water mixtures. *Journal of Industrial and Engineering Chemistry, 32*, 305–312. doi: 10.1016/j.jiec.2015.09.004.

43. Potivara, K., & Phisalaphong, M. (2019). Development and characterization of bacterial cellulose reinforced with natural rubber. *Materials, 12*(14), 1–17. doi: 10.3390/ma12142323.
44. Abol-Fotouh, D., Hassan, M. A., Shokry, H., Roig, A., Azab, M. S., & Kashyout, A. E.-H. B. (2020). Bacterial nanocellulose from Agro-Industrial Wastes: Low-cost and enhanced production by *Komagataeibacter Saccharivorans* MD1. *Scientific Reports, 10*(1), 1–14. doi: 10.1038/s41598–020–60315–9.
45. Cazón, P., Velázquez, G., & Vázquez, M. (2020a). Bacterial cellulose films: Evaluation of the water interaction. *Food Packaging and Shelf Life, 25*, 100526. doi: 10.1016/j.fpsl.2020.100526.
46. Stanisławska, A., Staroszczyk, H., & Szkodo, M. (2020). The effect of dehydration/rehydration of bacterial nanocellulose on its tensile strength and physicochemical properties. *Carbohydrate Polymers, 236*, 116023. doi: 10.1016/j.carbpol.2020.116023.
47. Atalla, R. H., & Vanderhart, D. L. (1984). Native cellulose: A composite of two distinct crystalline forms. *Science, 223*(4633), 283–285. doi: 10.1126/science.223.4633.283.
48. Krystynowicz, A., Czaja, W., Wiktorowska-Jezierska, A., Gonçalves-Miśkiewicz, M., Turkiewicz, M., & Bielecki, S. (2002). Factors affecting the yield and properties of bacterial cellulose. *Journal of Industrial Microbiology and Biotechnology, 29*(4), 189–195. doi: 10.1038/sj.jim.7000303.
49. Shi, Q. S., Feng, J., Li, W. R., Zhou, G., Chen, A. M., Ouyang, Y. S., & Chen, Y. B. (2013). Effect of different conditions on the average degree of polymerization of bacterial cellulose produced by *Gluconacetobacter intermedius* BNC-41. *Cellulose Chemistry and Technology, 47*(7–8), 503–508.
50. Keshk, S., & Sameshima, K. (2006). The utilization of sugar cane molasses with/without the presence of lignosulfonate for the production of bacterial cellulose. *Applied Microbiology and Biotechnology, 72*(2), 291–296. doi: 10.1007/s00253-005-0265-6.
51. Gayathry, G., & Gopalaswamy, G. (2014). Production and characterisation of microbial cellulosic fibre from *Acetobacter xylinum. IJFTR, 39*(1), 93–96. http://nopr.niscpr.res.in/handle/123456789/27372 (accessed on 27 May 2024).
52. Watanabe, K., Tabuchi, M., Morinaga, Y., & Yoshinaga, F. (1998). Structural features and properties of bacterial cellulose produced in agitated culture. *Cellulose, 5*(3), 187–200. doi: 10.1023/a:1009272904582.
53. Cazón, P., Velazquez, G., & Vázquez, M. (2020b). UV-protecting films based on bacterial cellulose, glycerol and polyvinyl alcohol: Effect of water activity on barrier, mechanical and optical properties. *Cellulose, 27*(14), 8199–8213. doi: 10.1007/s10570-020-03346-9.
54. Gao, G., Niu, S., Liu, T., Zhang, Y., Zhao, X., Shi, Z., & Ma, T. (2023). Fabrication of bacterial cellulose composites with antimicrobial properties by in situ modification utilizing the specific function-suspension containing water-insoluble magnolol. *International Journal of Biological Macromolecules, 239*, 124329–124329. doi: 10.1016/j.ijbiomac.2023.124329.
55. Costa, C. R., Medronho, B., Lindman, B., Edlund, H., & Norgren, M. (2021). Cellulose as a natural emulsifier: From nanocelluloses to macromolecules. *IntechOpen eBooks*. p. 116. doi: 10.5772/intechopen.99139.

56. Zhang, W., Zhang, Q. Y., Wang, J. J., Zhang, L. L., & Dong, Z. Z. (2022). Efficiency assessment of bacterial cellulose on lowering lipid levels in vitro and improving lipid metabolism in Vivo. *Molecules, 27*(11), 3495. doi: 10.3390/molecules27113495.
57. McClements, D. J., & Jafari, S. M. (2018). Improving emulsion formation, stability and performance using mixed emulsifiers: A review. *Advances in Colloid and Interface Science, 251*, 55–79. doi: 10.1016/j.cis.2017.12.001.
58. Zhang, X., Luo, X., Wang, Y., Li, Y., Li, B., & Liu, S. (2020). Concentrated O/W pickering emulsions stabilized by soy protein/cellulose nanofibrils: Influence of pH on the emulsification performance. *Food Hydrocolloids, 108*, 106025. doi: 10.1016/j.foodhyd.2020.106025.
59. Niu, H., Chen, X., Luo, T., Chen, H., & Fu, X. (2022). The interfacial behavior and long-term stability of emulsions stabilized by gum arabic and sugar beet pectin. *Carbohydrate Polymers, 291*, 119623–119623. doi: 10.1016/j.carbpol.2022.119623.
60. Zhai, X., Lin, D., Liu, D., & Yang, X. (2018). Emulsions stabilized by nanofibers from bacterial cellulose: New potential food-grade pickering emulsions. *Food Research International, 103*, 12–20. doi: 10.1016/j.foodres.2017.10.030.
61. Paximada, P., Tsouko, E., Kopsahelis, N., Koutinas, A. A., & Mandala, I. (2016). Bacterial cellulose as stabilizer of o/w emulsions. *Food Hydrocolloids, 53*, 225–232. doi: 10.1016/j.foodhyd.2014.12.003.
62. Liu, Z., Lin, D., Shen, R., & Yang, X. (2021). Bacterial cellulose nanofibers improved the emulsifying capacity of soy protein isolate as a stabilizer for pickering high internal-phase emulsions. *Food Hydrocolloids, 112*, 106279. doi: 10.1016/j.foodhyd.2020.106279.
63. Isogai, A., Saito, T., & Fukuzumi, H. (2011). TEMPO-oxidized cellulose nanofibers. *Nanoscale, 3*(1), 71–85. doi: 10.1039/c0nr00583e.
64. Akoğlu, A., Çakır, İ., Akoğlu, İ. T., Karahan, A. G., & Çakmakçı, M. L. (2015). Effect of bacterial cellulose as a fat replacer on some quality characteristics of fat reduced sucuk. *GIDA, 40*, 133–139. doi: 10.15237/gida.gd15020.
65. Oliveira, A. A. N., Mesquita, E. F. M., & Furtado, A. A. L. (2021). Use of bacterial cellulose as a fat replacer in emulsified meat products: Review. Food Science and Technology. doi: 10.1590/fst.42621.p.27.
66. Marchetti, L., Muzzio, B., Cerrutti, P., Andrés, S. C., & Califano, A. N. (2017). Bacterial nanocellulose as novel additive in low-lipid low-sodium meat sausages. Effect on quality and stability. *Food Structure, 14*, 52–59. doi: 10.1016/j.foostr.2017.06.004.
67. Oliveira, A. A. N., Guimarães, J. T., & Furtado, A. A. L. (2023). Effect of the addition of bacterial cellulose on the texture and color properties of sausages obtained from mechanically separated meat from Nile Tilapia (*Oreochromis niloticus* L.). *Food Science and Technology, 43*. doi: 10.1590/fst.81522.
68. Jantathai, S., Thongbai, B., Sukboonyasatit, D., Tunakeaw, S., & Wannarot, N. (2022). Supplementation of combined tomato powder and bacterial cellulose in chicken chinese sausage. *Food Agricultural Sciences and Technology, 8*(1), 61–73. https://ph02.tci-thaijo.org/index.php/stej/article/view/245974 (accessed on 27 May 2024).
69. Auty, M. A. E., Gardiner, G. E., McBrearty, S. J., O'Sullivan, E. O., Mulvihill, D. M., Collins, J. K., & Ross, R. P. (2001). Direct in Situ viability assessment of bacteria in probiotic dairy products using viability staining in conjunction with confocal scanning laser microscopy. *Applied and Environmental Microbiology, 67*(1), 420–425. doi: 10.1128/aem.67.1.420-425.2001.

70. Poppi, L. B., Rivaldi, J. D., Coutinho, T. S., Astolfi-Ferreira, C. S., Ferreira, A. J. P., & Mancilha, I. M. (2015). Effect of *Lactobacillus* Sp. isolates supernatant on *Escherichia coli* O157. Enhances the role of organic acids production as a factor for pathogen control. *Pesquisa Veterinária Brasileira, 35*(4), 353–359. doi: 10.1590/s0100-736x2015000400007.
71. Homayouni, A., Azizi, A., Ehsani, M. R., Yarmand, M. S., & Razavi, S. H. (2008). Effect of microencapsulation and resistant starch on the probiotic survival and sensory properties of synbiotic ice cream. *Food Chemistry, 111*(1), 50–55. doi: 10.1016/j.foodchem.2008.03.036.
72. Yoha, K. S., Moses, J. A., & Anandharamakrishnan, C. (2020). Effect of encapsulation methods on the physicochemical properties and the stability of *Lactobacillus plantarum* (NCIM 2083) in synbiotic powders and In-Vitro digestion conditions. *Journal of Food Engineering, 283*, 110033. doi: 10.1016/j.jfoodeng.2020.110033.
73. Li, H., Zhang, T., Li, C., Zheng, S., Li, H., & Yu, J. (2020). Development of a microencapsulated synbiotic product and its application in yoghurt. *LWT, 122*, 109033. doi: 10.1016/j.lwt.2020.109033.
74. Moon, E. C., Kang, Y. R., & Chang, Y. H. (2023). Development of soy protein isolate/sodium carboxymethyl cellulose synbiotic microgels by double crosslinking with transglutaminase and aluminum chloride for delivery system of *Lactobacillus acidophilus*. *International Journal of Biological Macromolecules, 237*, 124122–124122. doi: 10.1016/j.ijbiomac.2023.124122.
75. Swanson, K. S., Gibson, G. R., Hutkins, R., Reimer, R. A., Reid, G., Verbeke, K., Scott, K. P., Holscher, H. D., Azad, M. B., Delzenne, N. M., & Sanders, M. E. (2020). The International Scientific Association for Probiotics and Prebiotics (ISAPP) consensus statement on the definition and scope of synbiotics. *Nature Reviews Gastroenterology & Hepatology, 17*(11), 687–701. doi: 10.1038/s41575-020-0344-2.
76. Kistaubayeva, A., Abdulzhanova, M., Sirina Zhantlessova, Savitskaya, I., Karpenyuk, T., Goncharova, A., & Sinyavskiy, Y. A. (2023). The effect of encapsulating a prebiotic-based biopolymer delivery system for enhanced probiotic survival. *Polymers, 15*(7), 1752–1752. doi: 10.3390/polym15071752.
77. Lappa, I. K., Kachrimanidou, V., Alexandri, M., Papadaki, A., & Kopsahelis, N. (2022). Novel probiotic/bacterial cellulose biocatalyst for the development of functional dairy Beverage. *Foods, 11*(17), 2586. doi: 10.3390/foods11172586.
78. Gedarawatte, S. T. G., Ravensdale, J. T., Al-Salami, H., Dykes, G. A., & Coorey, R. (2021). Antimicrobial efficacy of nisin-loaded bacterial cellulose nanocrystals against selected meat spoilage lactic acid bacteria. *Carbohydrate Polymers, 251*, 117096. doi: 10.1016/j.carbpol.2020.117096.
79. Wu, S. C., & Lia, Y. K. (2008). Application of bacterial cellulose pellets in enzyme immobilization. *Journal of Molecular Catalysis B: Enzymatic, 54*(3–4), 103–108. doi: 10.1016/j.molcatb.2007.12.021.
80. Estevinho, B. N., Samaniego, N., Talens-Perales, D., Fabra, M. J., López-Rubio, A., Polaina, J., & Marín-Navarro, J. (2018). Development of enzymatically-active bacterial cellulose membranes through stable immobilization of an engineered β-Galactosidase. *International Journal of Biological Macromolecules, 115*, 476–482. doi: 10.1016/j.ijbiomac.2018.04.081.
81. Brisola, J., Andrade, G., Dias, S., Viana, R., Cristina, P., & Tischer, C. A. (2022). Covalent immobilization of lipase on bacterial cellulose membrane and

nanocellulose. *Materials Research-ibero-american Journal of Materials, 25*. doi: 10.1590/1980-5373-mr-2021-0350.
82. Liu, X., Xu, Y., Liao, W., Guo, C., Gan, M., & Wang, Q. (2023). Preparation and characterization of chitosan/bacterial cellulose composite biodegradable films combined with curcumin and its application on preservation of strawberries. *Food Packaging and Shelf Life, 35*, 101006. doi: 10.1016/j.fpsl.2022.101006.
83. Zhou, L., Fu, J., Bian, L., Chang, T., & Zhang, C. (2022). Preparation of a novel curdlan/bacterial cellulose/cinnamon essential oil blending film for food packaging application. *International Journal of Biological Macromolecules, 212*, 211–219. doi: 10.1016/j.ijbiomac.2022.05.137.
84. Abral, H., Pratama, A. B., Handayani, D., Mahardika, M., Aminah, I., Sandrawati, N., & Ilyas, R. A. (2021). Antimicrobial edible film prepared from bacterial cellulose nanofibers/starch/chitosan for a food packaging alternative. *International Journal of Polymer Science, 2021*, e6641284. doi: 10.1155/2021/6641284.
85. Luo, M. R. (2011). The quality of light sources. *Coloration Technology, 127*(2), 75–87. doi: 10.1111/j.1478-4408.2011.00282.x.
86. Xavier, J. R., & Ramana, K. V. (2022). Development of slow melting dietary fiber-enriched ice cream formulation using bacterial cellulose and inulin. *Journal of Food Processing and Preservation. 46*(5), e15394. doi: 10.1111/jfpp.15394.
87. Jayani, T., Sanjeev, B., Marimuthu, S., & Uthandi, S. (2020). Bacterial Cellulose Nano Fiber (BCNF) as Carrier Support for the Immobilization of Probiotic, *Lactobacillus acidophilus* 016. *Carbohydrate Polymers, 250*, 116965. doi: 10.1016/j.carbpol.2020.116965.
88. Charoenrak, S., Charumanee, S., Sirisa-ard, P., Bovonsombut, S., Kumdhitiahutsawakul, L., Kiatkarun, S., & Bovonsombut, S. (2023). Nanobacterial cellulose from kombucha fermentation as a potential protective carrier of lactobacillus plantarum under simulated gastrointestinal tract conditions. *Polymers, 15*(6), 1356. doi: 10.3390/polym15061356.
89. Xu, Y., Liu, X., Jiang, Q., Yu, D., Xu, Y., Wang, B., & Xia, W. (2021). Development and properties of bacterial cellulose, curcumin, and chitosan composite biodegradable films for active packaging materials. *Carbohydrate Polymers, 260*, 117778. doi: 10.1016/j.carbpol.2021.117778.
90. Pirveisi, N., Ariaii, P., Esmaeili, M., & Ahmadi, M. (2023). Investigating active packaging based on cellulose nanofibers oxidized by tempo method containing hydrolyzed protein obtained from pine tree fruit on the quality of pacific white shrimp (*Litopenaeus Vannamei*) during the storage period. *Journal of Food Measurement and Characterization*. doi: 10.1007/s11694-023-01843-6.
91. Freitas, J. A. M., Mendonça, G. M. N., Santos, L. B., Alonso, J. D., Mendes, J. F., Barud, H. S., & Azeredo, H. M. C. (2022). Bacterial cellulose/tomato puree edible films as moisture barrier structures in multicomponent foods. *Foods, 11*(15), 2336. doi: 10.3390/foods11152336.
92. Farooq, U., Ullah, M. W., Yang, Q., Aziz, A., Xu, J., Zhou, L., & Wang, S. (2020). High-density phage particles immobilization in surface-modified bacterial cellulose for ultra-sensitive and selective electrochemical detection of *Staphylococcus Aureus. Biosensors and Bioelectronics, 157*, 112163. doi: 10.1016/j.bios.2020.112163.
93. Petrut, R. F., Danthine, S., & Blecker, C. (2016). Assessment of partial coalescence in whippable oil-in-water food emulsions. *Advances in Colloid and Interface Science, 229*, 25–33. doi: 10.1016/j.cis.2015.12.004.

94. Guo, Y., Zhang, X., Hao, W., Xie, Y., Chen, L., Li, Z., & Feng, X. (2018). Nano-bacterial cellulose/soy protein isolate complex gel as fat substitutes in ice cream model. *Carbohydrate Polymers, 198*, 620–630. https://doi.org/10.1016/j.carbpol.2018.06.078.
95. Chau, C. -F., Yang, P. At., Yu, C. -M., & Yen, G. -C. (2008). Investigation on the Lipid- and Cholesterol-Lowering abilities of biocellulose. *Journal of Agricultural and Food Chemistry, 56*(6), 2291–2295. doi: 10.1021/jf7035802.
96. Nugroho, D. A., & Aji, P. (2015). Characterization of nata de coco produced by fermentation of immobilized *Acetobacter xylinum*. *Agriculture and Agricultural Science Procedia, 3*, 278–282. doi: 10.1016/j.aaspro.2015.01.053.
97. Zhang, J., Yichong, Y., Deng, J., Wang, Y., Hu, Q., Li, C., & Liu, S. (2017). Dynamic profile of the microbiota during coconut water pre-fermentation for nata de coco production. *Lebensmittel-Wissenschaft & Technologie, 81*, 87–93. doi: 10.1016/j.lwt.2017.03.036.
98. Okiyama, A., Motoki, M., & Yamanaka, S. (1993). Bacterial cellulose IV. Application to processed foods. *Food Hydrocolloids, 6*(6), 503–511. doi: 10.1016/s0268-005x(09)80074-x.
99. Garcia-Ochoa, F., Santos, V. E., & Alcón, A. (1998). Metabolic structured kinetic model for xanthan production. *Enzyme and Microbial Technology, 23*(1–2), 75–82. doi: 10.1016/s0141-0229(98)00014-3.
100. Garcia-Ochoa, F., Santos, V. E., & Alcon, A. (2004a). Structured kinetic model for *Xanthomonas campestris* growth. *Enzyme and Microbial Technology, 34*(6), 583–594. doi: 10.1016/j.enzmictec.2004.01.005.
101. Garcia-Ochoa, F., Santos, V. E., & Alcon, A. (2004b). Chemical structured kinetic model for xanthan production. *Enzyme and Microbial Technology, 35*(4), 284–292. doi: 10.1016/j.enzmictec.2003.11.024.
102. Cazón, P., Velázquez, G., & Vázquez, M. (2019). Characterization of bacterial cellulose films combined with chitosan and polyvinyl alcohol: Evaluation of mechanical and barrier properties. *Carbohydrate Polymers, 216*, 72–85. doi: 10.1016/j.carbpol.2019.03.093.
103. Chinma, C. E., Ariahu, C. C., & Alakali, J. S. (2015). Effect of temperature and relative humidity on the water vapor permeability and mechanical properties of cassava starch and soy protein concentrate based edible films. *Journal of Food Science and Technology, 52*(4), 2380–2386. doi: 10.1007/s13197-013-1227-0.
104. Lin, S. B., Chen, L. C., & Chen, H. H. (2009). Physical characteristics of surimi and bacterial cellulose composite gel. *Journal of Food Process Engineering, 34*(4), 1363–1379. doi: 10.1111/j.1745-4530.2009.00533.x.
105. Zhai, X., Lin, D., Zhao, Y., Li, W., & Yang, X. (2018). Effects of dietary fiber supplementation on fatty acid metabolism and intestinal microbiota diversity in C57BL/6J mice fed with a high-fat diet. *Journal of Agricultural and Food Chemistry, 66*(48), 12706–12718. doi: 10.1021/acs.jafc.8b05036.
106. Oliveira, A. A. N., Guimarães, J. T., & Furtado, A. A. L. (2023). Effect of the addition of bacterial cellulose on the texture and color properties of sausages obtained from mechanically separated meat from Nile Tilapia (*Oreochromis niloticus* L.). *Food Science and Technology, 43*, 1–9. doi: 10.1590/fst.81522.
107. Pirveisi, N., Ariaii, P., Esmaeili, M., Ahmadi, M. (2023). Investigating active packaging based on cellulose nanofibers oxidized by tempo method containing hydrolyzed protein obtained from pine tree fruit on the quality of pacific white shrimp

(*Litopenaeus vannamei*) during the storage period. *Journal of Food Measurement and Characterization*. doi: 10.1007/s11694–023–01843–6.

108. Cannon R. E., Anderson S. M. (1991). Biogenesis of bacterial cellulose. *Critical Reviews in Microbiology*, *17*(6), 435–447. doi: 10.3109/10408419109115207.
109. Kato, N., Sato, T., Kato, C. , Yajima, M., Sugiyama, J., Kanda, T., Mizuno, M., Nozaki, M., Yamanaka, S., Amano, Y. (2007) Viability and cellulose synthesizing ability of *Gluconacetobacter xylinus* cells under high-hydrostatic pressure. *Extremophiles, 11*, 693–698. doi: 10.1007/s00792–007–0085-y.
110. Ross P, Mayer R, Benziman M. (1991). Cellulose biosynthesis and function in bacteria. *Microbiology Reviews, 55*(1), 35–58. doi: 10.1128/mr.55.1.35–58.1991.
111. Kongruang, S. (2008). Bacterial Cellulose Production by *Acetobacter xylinum* Strains from Agricultural Waste Products. *Applied Biochemistry and Biotechnology*, *148*, 245–256, Doi: 10.1007/s12010–007–8119–6.
112. Noro, N., Sugano, Y., Shoda, M. (2004). Utilization of the buffering capacity of corn steep liquor in bacterial cellulose production by *Acetobacter xylinum. Applied Microbiology and Biotechnology, 64*(2),199–205. doi: 10.1007/s00253–003–1457–6
113. Shavyrkina, N. A., Skiba, E. A., Kazantseva, A. E., Gladysheva, E. K., Budaeva, V. V., Bychin, N. V., Gismatulina, Y. A., Kashcheyeva, E. I., Mironova, G. F., Korchagina, A. A., Pavlov, I. N., Sakovich, G. V. (2021) Static Culture Combined with Aeration in Biosynthesis of Bacterial Cellulose. *Polymers* 2021, *13*, 4241, 1–12 . doi: 10.3390/polym13234241
114. Cheng, H. P., Wang, P. M., Chen, J. W., Wu, W. T. (2002). Cultivation of *Acetobacter xylinum* for bacterial cellulose production in a modified airlift reactor. *Biotechnology and Applied Biochemistry*, *35*(2), 125–32. doi: 10.1042/ba20010066.
115. Garcia-Ochoa, F., Gomez, E., & Santos, V. E. (2020). Fluid dynamic conditions and oxygen availability effects on microbial cultures in STBR: An overview. *Biochemical Engineering Journal*, *164*, 1–25. doi: 10.1016/j.bej.2020.107803.
116. Chinma, C. E., Ariahu, C. C., & Alakali, J. S. (2015). Effect of temperature and relative humidity on the water vapor permeability and mechanical properties of cassava starch and soy protein concentrate based edible films. *Journal of Food Science and Technology*, *52*(4), 2380–2386. doi: 10.1007/s13197–013–1227–0.
117. Juttulapa, M., Piriyaprasarth, S., & Sriamornsak, P., (2013). Effect of pH on Stability of Oil-in-Water Emulsions Stabilized by Pectin-Zein Complexes. *Advanced Materials Research, 747*, 127–130. doi: 10.4028/www.scientific.net/AMR.747.127.
118. Vasconcelos, N. F., Cunha, A. P., Ricardo, N. M. P. S., Freire, R. S., Vieira, L. D. A. P., Santa Brigida, A. I., & Andrade, F. K. (2020). Papain immobilization on heterofunctional membrane bacterial cellulose as a potential strategy for the debridement of skin wounds. *International Journal of Biological Macromolecules*, *165*, 3065–3077. doi: 10.1016/j.ijbiomac.2020.10.200.
119. Tomé, L. C., Brandão, L., Mendes, A. M. et al. (2010). Preparation and characterization of bacterial cellulose membranes with tailored surface and barrier properties. *Cellulose 17*, 1203–1211. doi: 10.1007/s10570–010–9457-z
120. Yanti, N. A., Ahmad, S. W., Ramadhan, L. O. A. N., Jamili, Muzuni, Walhidayah, T., & Mamangkey, J. (2021). Properties and application of edible modified bacterial cellulose film based sago liquid waste as food packaging. *Polymers*, *13*(20), 3570. doi: 10.3390/polym13203570.

121. Xavier, J. R., & Ramana, K. V. (2022). Development of slow melting dietary fiber-enriched ice cream formulation using bacterial cellulose and inulin. *Journal of Food Processing and Preservation*, *46*(5), e15394. doi: 10.1111/jfpp.15394.
121. Lin, S. B., Chen, L. C., & Chen, H. H. (2011). Physical characteristics of surimi and bacterial cellulose composite gel. *Journal of Food Process Engineering*, *34*(4), 1363–1379. doi:10.1111/j.1745–4530.2009.00533.x.
122. Zhong, C. (2020). Industrial-scale production and applications of bacterial cellulose. *Frontiers in Bioengineering and Biotechnology*, *8,* 1–19 . doi: 10.3389/fbioe.2020.605374.
123. Çoban, E. P., & Biyik, H.(2011). Evaluation of different pH and temperatures for bacterial cellulose production in Hs (Hestrin-Scharmm) medium and beet molasses medium. *African Journal of Microbiology Research*, *5*(9), 1037–1045. doi: 10.5897/ajmr11.008.
124. Saleh, A. K., El-Gendi, H., Soliman, N. A., El-Zawawy, W. K., & Abdel-Fattah, Y. R. (2022). Bioprocess development for bacterial cellulose biosynthesis by novel *Lactiplantibacillus plantarum* Isolate along with Characterization and Antimicrobial Assessment of Fabricated Membrane. *Scientific Reports, 12*(1), 1–17. doi: 10.1038/s41598–022- 06117–7.
125.Tseng, Y. S., Patel, A. K., Chen, C. -W., Dong, C. -D., & Singhania, R. R. (2022). Improved production of bacterial cellulose by *Komagataeibacter europaeus* employing fruit extract as carbon source. *Journal of Food Science and Technology*, *60*, 1054–1064. doi: 10.1007/s13197–022–05451-y.
126. Aswini, K., Gopal, N. O., & Uthandi, S. (2020). Optimized culture conditions for bacterial cellulose production by acetobacter senegalensis ma1. *BMC Biotechnology*, *20*(1), 1–16 . doi: 10.1186/s12896–020–00639–6.
127. Potivara, K., & Phisalaphong, M. (2019). Development and characterization of bacterial cellulose reinforced with natural rubber. *Materials*, *12*(14), 1–17 . doi: 10.3390/ma12142323
128. Abol-Fotouh, D., Hassan, M. A., Shokry, H., Roig, A., Azab, M. S., & Kashyout, A. E.-H. B. (2020). Bacterial nanocellulose from Agro-Industrial Wastes: Low-cost and enhanced production by *Komagataeibacter Saccharivorans* MD1. Scientific Reports, *10*(1), 1–14 . doi: 10.1038/s41598–020–60315–9
129. Costa, C. R., Medronho, B., Lindman, B., Edlund, H., & Norgren, M. (2021). Cellulose as a natural emulsifier: From nanocelluloses to macromolecules. p 1–16. IntechOpen eBooks. doi: 10.5772/intechopen.9913
130. Akoğlu, A., Çakır, İ., Akoğlu, İ. T., Karahan, A. G., & Çakmakçı, M. L. (2015). Effect of bacterial cellulose as a fat replacer on some quality characteristics of fat reduced sucuk. *GIDA*, *40*, 133–139. doi: 10.15237/gida.gd15020
131. Oliveira, A. A. N., Mesquita, E. F. M., & Furtado, A. A. L. (2021). Use of bacterial cellulose as a fat replacer in emulsified meat products: Review. Food Science and Technology. doi: 10.1590/fst.42621.p.27
132. Oliveira, A. A. N., Guimarães, J. T., & Furtado, A. A. L. (2023). Effect of the addition of bacterial cellulose on the texture and color properties of sausages obtained from mechanically separated meat from Nile Tilapia (*Oreochromis niloticus* L.). *Food Science and Technology, 43*, 1–9. doi: 10.1590/fst.81522

133. Brisola, J., Andrade, G., Dias, S., Viana, R., Cristina, P., & Tischer, C. A. (2022). Covalent immobilization of lipase on bacterial cellulose membrane and nanocellulose. *Materials Research*, *25*, 1–10. doi: 10.1590/1980–5373-mr-2021–0350
134. Abral, H., Pratama, A. B., Handayani, D., Mahardika, M., Aminah, I., Sandrawati, N., & Ilyas, R. A. (2021). Antimicrobial edible film prepared from bacterial cellulose nanofibers/starch/chitosan for a food packaging alternative. *International Journal of Polymer Science*, e6641284. doi: 10.1155/2021/6641284
135. Pirveisi, N., Ariaii, P., Esmaeili, M., Ahmadi, M. (2023). Investigating active packaging based on cellulose nanofibers oxidized by tempo method containing hydrolyzed protein obtained from pine tree fruit on the quality of pacific white shrimp (*Litopenaeus vannamei*) during the storage period. *Journal of Food Measurement and Characterization*. doi: 10.1007/s11694–023–01843–6.
136. Velázquez, M. E., Ferreiro, O. B., Menezes, D. B., Corrales-Ureña, Y., Vega-Baudrit, J. R., & Rivaldi, J. D. (2022). Nanocellulose Extracted from Paraguayan Residual Agro-Industrial Biomass: Extraction Process, Physicochemical and Morphological Characterization. Sustainability, *14*, 11386, 1–15. doi: 10.3390/su141811386.
137. Morgan, J. L., Strumillo, J., & Zimmer, J. (2013). Crystallographic snapshot of cellulose synthesis and membrane translocation. *Nature*, *493*(7431), 181–186. doi: 10.1038/nature11744.

CHAPTER 10

Functionalized Starch Biomaterials as a Possible Alternative in the Future of Food Packaging: Advantages, Disadvantages, and Perspectives

FRANCISCO HERNÁNDEZ-CENTENO,[1]
MARÍA HERNÁNDEZ-GONZÁLEZ,[1]
HAYDEÉ YAJAIRA LÓPEZ-DE LA PEÑA,[1]
CLAUDIA ALEJANDRA RODRÍGUEZ-GONZÁLEZ,[2]
CLAUDIO RIOS-VELASCO,[3] PAUL BARUK ZAMUDIO-FLORES,[3] and
JUAN MANUEL TIRADO-GALLEGOS[4*]

[1]*Universidad Autónoma Agraria Antonio Narro, Food Science and Technology Department, Calzada Antonio Narro, Buenavista, Saltillo, Coahuila, C.P., Mexico*

[2]*Universidad Autónoma de Ciudad Juárez, Engineering and Technology Institute, Av. del Charro 610 Norte, Ciudad Juárez, Chihuahua, C.P., Mexico*

[3]*Centro de Investigación en Alimentación y Desarrollo, A.C. – Unidad Cuauhtémoc, Food Physiology and Technology of the Temperate Zone, Av. Río Conchos s/n, Parque Industrial, C.P., Ciudad Cuauhtémoc, Chihuahua, Mexico*

[4]*Universidad Autónoma de Chihuahua, Zootechnics and Ecology Faculty, Periférico Francisco R. Almada, Chihuahua, Chihuahua, C.P., Mexico*

[*]Corresponding author.

ABSTRACT

The excessive consumption of single-use non-biodegradable packaging materials for packaging and transporting food products has increased in recent

Food Security, Safety, and Sustainability: Validated Innovative Approaches and Technological Advancements.
José Juan Buenrostro-Figueroa and Julio César Tafolla-Arellano (Eds.)

decades, causing negative impacts on the environment and ecosystems. Starch-based film packaging, because it is biodegradable, could help reduce or diminish this ecological problem since it is safe, cheap, and easy to apply, as well as having the possibility of being modified and/or functionalized to become active packaging with application in the handling, packaging, or transport of food.

Starch is one of the carbohydrates most widely used for this purpose, due to its ease of production and low cost, and unconventional sources of this biopolymer represent novel and interesting options to produce biodegradable films with unique properties. Through the inclusion of bioactive botanical compounds, it is possible to transmit antioxidant and/or antimicrobial properties to them. Therefore, conventional and unconventional sources of starches reported in the literature, which have been used for the production of biodegradable films (functionalized or not), were reviewed, discussing their main physical properties (mechanical, thermal, and permeability) and the effects that are caused in them by adding antimicrobial and antioxidant compounds of botanical origin.

The main limitations and possible applications of these films in food packaging and their current trends in certain food applications were analyzed, since their use could, in the future, help ensure food safety throughout the production chain of these.

10.1 INTRODUCTION

Starches are biopolymeric compounds produced by plants from photosynthesis, so they represent the main energy reservoir in the plant kingdom [1] and they are also the main source of energy for the other links in food chains [2]. Starch production constitutes a fundamental productive activity [3]. They are mainly obtained from sources such as corn (*Zea mays* L.), potato (*Solanum tuberosum* L.) and yucca (*Manihot esculenta*), and in lower proportions from other cereals such as wheat (*Triticum* spp.), rice (*Oryza sativa*) and other intensively cultivated grasses [4]. Despite this, these types of crops are also the traditional food base of the population, so allocating them to their processing to obtain raw materials that are not from the food industry, or their direct consumption, results in a debatable fact, given the high rate of famine in some regions of the planet, which seriously compromises the food security of many people [5, 6].

Regarding its usefulness as a material, its industrial importance lies in the diversity of properties that it possesses [7, 8], since, according to Ai & Jane [9] reports, it may act as a thickener, gelling agent, stabilizer, and adhesives, among others. For example, once chemically modified, it can act as encapsulants of other compounds of interest [10]. Likewise, it is a fundamental raw material to produce a great variety of consumer goods [11], such as different foods and beverages, and materials considered "sustainable," including biofuels [12] and bioplastics [13]. This is due to its transformation feasibility, wide natural availability, low cost, biocompatibility, biodegradability, and being a renewable resource [14].

Fortunately, there haven't been efforts to find other botanical sources, furthermore the most common ones, able to provide starches with attractive properties for their application in the industry in general [15, 16]. Nowadays, research has highlighted the development and elaboration of organic-based plastic materials [14, 17–19], which represent a palliative to the problems caused by the abuse of conventional plastic usage derived from petroleum, historically used in the management, packaging, and conservation of fresh and processed food [20, 21], and which represent a danger in ecosystem contamination due to indiscriminate and disproportionate usage, combined with the growing human population [22–24, 29].

According to the above, the development of starch films extracted from conventional and non-conventional sources has gained importance in the search for a solution to the problem of excessive conventional plastic usage. Additionally, efforts have been made to develop third-generation materials, completely biodegradable, by adding bioactive compounds of botanical origin, resulting in a bio-based material, the definition of which is explained in the scheme in Figure 10.1. In this way, a new range of options in active packaging has been generated, allowing for the extension of product shelf life without compromising food biosecurity, integrity, or consumer health.

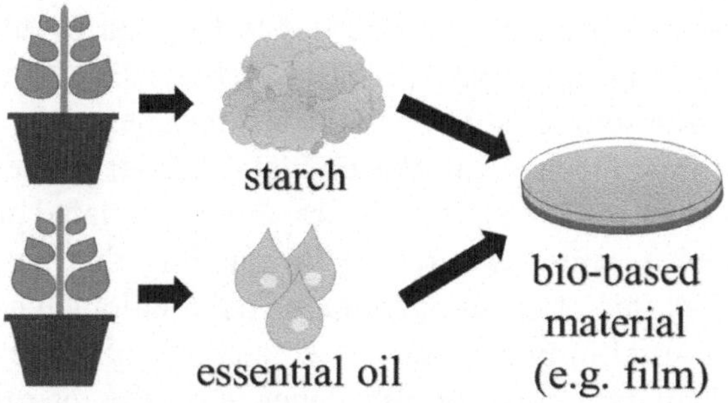

FIGURE 10.1 A bio-based material is that which is obtained from products derived from one or more living organisms, in this case, starches, and bioactive compounds, both of botanical sources. Its final properties will depend on those of the original materials.

Source: Adapted from Ref. [29]

In this review, some of the most current publications in this field have been considered, how and why it has been functionalized with natural bioactive compounds, as well as these types of bio-based materials' possible applications in the food industry, with the respective advantages and limitations that their natural origin represents, including in a special manner those materials developed from non-conventional source starches.

10.2 STARCH FILMS FROM CONVENTIONAL AND NON-CONVENTIONAL SOURCES: MECHANICAL, THERMAL, AND BARRIER PROPERTIES

In the last decades and nowadays, starch from non-conventional sources has been reported, which can be considered a novel raw material for this carbohydrate extraction. For instance, *Cucurbita fetidissima* Kunth roots [25]. Santana and Meireles exposed in a review a wide catalog of reported botanical species, which are considered susceptible to taking advantage of them in different industrial applications [16]. Starches vary in their physicochemical characteristics depending on the botanical sources, and these properties, at the same time, indicate the possible industrial applications, according to their functionality, as reported by Ogunsona et al. [15]; and Pfister & Zeeman [26].

Several starches from different botanical sources have been used for materials elaboration to formulate or make films and covers, due to the advantages of using this polymer [14, 15, 27, 28], including those obtained from non-conventional sources, such as mango (*M. indica*), okenia (*Okenia hypogaea*), sugar palm (*Arenga pinnata*), among others [16, 29]. Most of them have presented promising properties to be used in different industry fields, including food packaging [30, 31]. Table 10.1 shows several non-conventional starch sources studied in the last decade.

As for the films, mechanical, thermal, and barrier properties take technological importance [7, 18], which are directly related to the starches' molecular structure (amylose/amylopectin relation) that has been used for its elaboration [9]. Such structural components' relation, at a time, results in some starches' crystallinity percentage, which directly influences their functionality [32]. This functionality also depends on the interactions between these components with the used plasticizer materials [33] and the process temperature [34], which could define the viability of using these materials in a massive way.

10.2.1 MECHANICAL PROPERTIES

It is necessary to determine these properties in a material since the manufacture of the product will depend on them (as, for example, the packaging), the final use, the type of products it may contain, and the handling it may receive. According to Atarés & Chiralt [35], these properties include stress fracture (SF), elongation percentage (%E), and the elasticity module (ME), which tend to vary depending on the relationship between the base polymeric matrix and the additives. Among these additives are the type of plasticizer and the reinforcement material used. Additionally, the concentrations of additives also affect other characteristics, such as hydrophobicity or transparency [36], modifying the potential utility of the final films. According to Oropeza-González et al. films based solely on native starch have poor mechanical properties [22], so the use of plasticizers is necessary. This has been sought to improve the flexibility and reduce the brittleness of bioplastics made with pure starch [28, 37, 38].

Reinforcement components include polysaccharides obtained from the same plants, such as lignin, which can improve the mechanical, thermal, and barrier properties of native starch films. Espinoza-Acosta et al. reported in their study on composite films based on wheat starch and alcohol-soluble lignin that this reinforcing element decreased the values of SF and EM but increased the E% [39]. Other polysaccharides, such as chitosan, have also been widely used as mechanical reinforcement in starch films, as well as to decrease its solubility. For example, in a study conducted by Zhong et al. films of kudzu starch (*Pueraria lobata*) and chitosan were elaborated for this purpose, among others [40]. Tables 10.1–10.3 present studies in which the addition of biological reinforcing materials, such as cellulose and chitosan, has been added to bio-based films elaborated with starches from different sources.

Maniglia et al. evaluated films made of starch from babassu mesocarp (*Attalea speciosa*) having glycerol as a plasticizer (20% w/w dry starch) and found an improvement in the values of SF and EM, but a decrement in the E% compared to films elaborated with the same source of flour, attributing this to the loss of crystallinity of the isolated starch [41]. Hornung et al. evaluated Amazon turmeric (*Curcuma longa* L.) starch films plasticized with glycerol and found that the values of ST increased compared to the corn starch films, without significant changes in the EM and E% [42].

On the other hand, starch mixtures have also been explored to evaluate their effect on mechanical properties. Hornung et al. used mixtures of native starches from three varieties of yam tubers (*Dioscorea sp., Dioscorea piperifolia* Humb. var. Wild, and *Dioscorea trifida*) to make films, using glycerol as a plasticizer. Their results revealed that the mechanical properties strongly depend on the starch origin source, even among species of the same genus. In some of the films made with starch mixtures, the ST values increased up to 200% compared to those made with native starch, while the E% values exhibited the opposite effect, decreasing in some cases up to 232% compared to native starch films [43]. Table 10.1 presents the mechanical properties measurements obtained from several bio-based films elaborated with starches coming from different non-conventional sources.

10.2.2 THERMAL PROPERTIES

In this regard, the material's thermal properties identification such as the melting or glass transition temperature is important from the technological point of view since their possible practical applications will depend on it. Zhang et al. indicated that the starch bio-based materials (known as thermoplastic starch or TPS) [38] thermal properties are also affected according to the botanical source they come from, their molecular structure, as well as the plasticizer's usage and concentration, or by the chemical modifications they undergo [44]. These factors may increase or decrease the glass transition temperature (T_g) of the formulated materials. The inclusion of reinforcement materials also affects this type of properties. For instance, Espinoza-Acosta et al. used alcohol-soluble lignin to reinforce wheat starch films, resulting in an increase in the thermal degradation of the products compared to the non-reinforced ones [39].

Zhong & Li studied the glycerol and relative humidity storage effect on kudzu (*Pueraria lobata*) starch films and found that the film's thermal stability decreased as the plasticizer and relative humidity increased [37]. Luchese et al. reported in their study on pinhão (*Araucaria angustifolia*) native and hydrolyzed films having glycerol and xanthan gum as a plasticizer, that this chemical modification does not have a significant effect on the resulting film's T_g [32]. Maniglia et al. determined that babassu mesocarp (*A. speciosa*) starch films using glycerol as a plasticizer resulted in a lower WVP than those obtained with the starches of the same source because of the other components present in them, such as protein residues and lipids, among others, which are related to the material's base crystallinity percentage [41]. On his own, Gutiérrez observed that poli-ε-caprolactone addition on sago rhizomes (*Canna edulis* Kerr) thermoplastic starch films decreased T_g values up to around 54°C, in comparison with these additive-free films, which was attributed to a modification in the starch crystalline phase, which increased up to 20.5% with this additive presence [45]. Recently, it was found that Amazonian turmeric (*Curcuma longa* L.) starch films, having glycerol as a plasticizer, exhibited a T_g of around 48°C, similar to that determined for corn starch films, as reported by Hornung et al. [42], who commented that the differences in Tg, for films elaborated with starches from different sources, probably obey the source it comes from, as well as the plant's growing conditions and the analysis method.

In another study, Hornung et al. reported that the glass transition temperature (T_g) of native starch films depends on the botanical source of the starch used. In their study on films made from mixtures of yam (*Dioscoreaceae*) native starch, they found that those made with native starches had higher T_g values compared to those made with mixtures [43]. The authors related this to the percentage of crystallinity of the materials. Tables 10.1–10.3 present a series of studies that report the measurement and thermal properties of a wide variety of bio-based films made with starches from both conventional and non-conventional sources.

10.2.3 BARRIER PROPERTIES

The importance of knowing these properties, as previously described, lies in being aware of the packaging type that can be manufactured and the foods or products it may contain according to gas isolation needs such as oxygen and water vapor. As reported by Avérous & Fringant [46], the barrier properties of native starch-based films are deficient compared to those obtained with other polymers. This is due to one of the main technological disadvantages, which is the high hydrophilic character of starches. To minimize this, plasticizers and reinforcement materials have historically been included to some extent to reduce this disadvantage, as well as native starch modification [44]. Abdull-Khalil et al. indicated that the use of plasticizers improves this property in starch bio-based materials, reducing water vapor permeability (WVP) [30]. Zhong & Li found in a study on the effect of glycerol concentration in kudzu (*P. lobata*) films that the increase in plasticizer was directly proportional to the WVP, and this was attributed to the intermolecular hydrogen bonds between water and glycerol, which increased as the storage relative humidity increased [37]. Another important factor for this property is the material's thickness, so it is important to consider the density of the mixtures when producing the films [47].

On the other hand, researchers have been seeking a solution to the poor mechanical and barrier properties of native starch films through physical and chemical modification. Luchese et al. elaborated starch films with native and hydrolyzed pinhão (*A. angustifolia*) starch and found that the degree of hydrolysis is inversely proportional to the WVP, attributing this to the loss of starch crystallinity after chemical modification [32]. Gutiérrez et al. reported that cross-linked starches of yucca (*Manihot esculenta*) produced films with adequate WVP properties compared to those made with native starch of this species (using glycerol as a plasticizer). For the cross-linked starch films of black sweet potato or yam (*Dioscorea trifida*), they obtained a similar behavior, although with lower WVP values than those of cassava starch films [33]. Hornung et al. did not find significant differences between the WVP values of the different native starch films of different yam (*Dioscoreaceae*) species and those obtained using mixtures of them, so this variable probably does not depend on the botanical species, but rather on the film's production methodology [43]. Table 10.1 reports the barrier properties measured for non-conventional native starch films in investigations carried out in the last decade.

10.3 BOTANICAL ORIGIN ANTIMICROBIALS AND ANTIOXIDANTS USED TO FUNCTIONALIZE STARCH BIO-BASED PACKING PRODUCTS: ORIGIN AND FUNCTIONALITY

Several reviews have been carried out in this regard in the recent past, in which the maturity, structure, and function of these compounds are abundant. As published by Valdés et al. many spices and herbs have been used throughout history in the elaboration of food due to the sensory characteristics they impart, their antimicrobial and/or antioxidant capacities, and since they also act as agents for improving human health; their compounds of interest can be obtained from one or more plant organs (leaves, flowers, fruit, seeds, or roots) [48]. In accordance with Sánchez-González et al. the chemical composition of extracts and essential oils (which are the forms in which they are generally included in the formulations of biomaterials) varies according to many factors, among which are the plant's growing conditions, the harvest time, postharvest treatment, and even the extraction method of the compounds of interest [49].

TABLE 10.1 Mechanical, Thermal, and Barrier Properties of Non-Conventional Native Starch Films

Species	Film Composition	Material Conditioned Before and During Test	1Elastic Modulus (EM; MPa)	1Tensile Strength (TS; MPa)	1Elongation (%E)	1Tg (°C)	1WVP (g·m–1·s–1·Pa–1)·10n	References
Canna indica L.	Starch + glycerol	48 ± 2% RH, 22 ± 1°C, 5 h	–	6.8 ± 0.2	34.8 ± 9.5	–	1.40 ± 0.10 (109)	[65]
Curcuma longa	Starch	57% RH, 25°C, 48 h	27.8 ± 4.1	85.8 ± 9.6	22.6 ± 0.5	48.5 ± 1.1	–	[42]
Mangifera indica L. var. "Tommy Atkins"	Starch + glycerol	ASTM D882-12	839.2	14.46	15.93	–	3.012 (g mm $kPa^{-1}h^{-1}m^{-2}$)	[116]
Malus domestica	Starch + glycerol	ASTM-882-95a	3.10 ± 0.27	8.12 ± 0.36	52.12 ± 4.30	–	6.77 ± 0.85 (10^{11})	[111]
Araucaria angustifolia	Starch	50% RH, 25°C, 48 h	2,586 ± 175	46.54 ± 2.78	3.1 ± 0.1	77 ± 3.0	1.05 ± 0.07 (g mm/kPa h m^2)	[117]
Dioscorea sp.	Starch + glycerol	60% RH, 25C, 72 h	–	10.28 ± 0.26	43.24 ± 0.12	52.06	1.22 ± 0.01 (g mm m^2/h kPa)	[43]
Dioscorea piperifolia Humb. var. Wild	Starch + glycerol	60% RH, 25°C, 72 h	–	16.45 ± 0.19	24.50 ± 0.02	51.59	1.17 ± 0.01 (g mm m^2/h kPa)	[43]
Dioscorea trifida	Starch + glycerol	60% RH, 25°C, 72 h	–	18.21 ± 0.02	17.07 ± 0.03	48.18	1.15 ± 0.01 (g mm m^2/h kPa)	[43]
Attalea speciosa (mesocarp)	Starch + glycerol	58% RH, 25 ± 2°C, 48 h	4285.9 ± 208.76	63.00 ± 4.09	0.68 ± 0.18	–	3.52 ± 0.20 (10–10)	[41]
Sechium edule Sw.	Starch + glycerol + 0.8% w/w cellulose	57% RH, 25°C, 72 h	69.3 ± 4.24	8.2 ± 0.73	30.1 + 3.58	–	–	[118]
Sechium edule Sw.	Starch + glycerol + 0.12% w/w cellulose nanoparticle	57% RH, 25°C, 72 h	96.8 ± 9.32	9.2 ± 0.60	27.2 ± 1.25	–	–	[118]
Sechium edule Sw.	Starch + glycerol	57% RH, 25°C, 72 h	23	3	70	–	10.41 ± 0.06 (10^{-11})	[119]
Maranta arundinacea	Starch + glycerol	75 ± 3% RH, 25°C, 7 d	–	≈23	≈3	–	≈3 (g mm/m^2 day kPa)	[68]

TABLE 10.1 *(Continued)*

Species	Film Composition	Material Conditioned Before and During Test	1Elastic Modulus (EM; MPa)	1Tensile Strength (TS; MPa)	1Elongation (%E)	1Tg (°C)	1WVP (g·m–1·s–1·Pa–1)·10n	References
Curcuma longa L.	Starch + glycerol	58% RH, 25°C, 48 h	126	4	7	–	5.88 ± 0.04 (g $day^{-1}m^{-2}atm^{-1}$)	[98]
Avena sativa L. var. Bachíniva	Oxidized starch + glycerol	50 ± 5% RH, 25°C, 48 h	0.22	1.56	127.18	–	21.30 ± 0.57 (10^{11})	[69]
Musa paradisiaca L.	Oxidized starch + glycerol	50 ± 5% RH, 25°C, 48 h	7.18	5.66	25.15	–	14.50 ± 0.58 (10^{11})	[69]
Sechium edule Sw.	Starch + glycerol	57% HR, 25°C, 14 d	0.92 ± 0.09	18.46 ± 2.22	0.89 ± 0.12	–	–	[80]
Ipomoea batatas	Starch + sorbitol	50% RH, 23C, 72 h	–	7.96	77.92	–	0.212 (ng m/m^2 S Pa)	[91]
Ipomoea batatas	Starch + glycerol	50% RH, 23C, 72 h	–	3.6	85	–	0.30 (ng m/m^2 S Pa)	[91]
Dioscorea trifida	Starch + glycerol	57% RH, 25C, 7 d	13.9 ± 0.6	1.88 ± 0.08	19	–	1.8 ± 0.4 (10^{11})	[33]
Araucaria angustifolia	Starch + glycerol	75% RH, 21C, 7 d	–	–	–	35.5 ± 12.5	0.30 ± 0.02 (g µm day^{-1} m^{-2} $mmHg^{-1}$)	[32]
Pisum sativum L.	Starch + glycerol	67% HR, 25°C, 48 h	34.56 ± 4.55	3.76 ± 0.40	91.55 ± 8.22	–	–	[102]
Chenopodium quinoa Wildenow	Starch + glycerol	58% RH, 25°C, 48 h	–	7.6 ± 0.6	5.2 ± 0.6	–	0.393 ± 0.054 (g mm m^{-2} $h^{-1}Pa^{-1}$)	[113]
Musa paradisiaca var. Terra	Starch + glycerol	58% RH, 25°C, 48 h	≈850	≈9.8	≈2	–	≈7 (10^{11})	[77]
Pueraria lobata	Starch + glycerol	53% RH, 25C, 48 h	–	19.48 ± 2.80	2.85 ± 0.73	–	3.15 ± 0.25 (10^{11})	[37]

1Means ± standard deviation, where applicable. Data is indicated as published.

According to Perricone et al. and Martin-Gordo, the compounds identified as antimicrobials and antioxidants fulfill a specific biological function in the botanical species of origin [50, 51]; these compounds are usually produced by plants as secondary metabolites whose functionality includes that described by Benbettaïeb et al. [52], who refer that phytochemicals act as a means of defense or attack against competitors, predators, infections, or diseases caused by biological agents, and against cellular and nutrient oxidation, such as fatty acids and vitamins, caused by physical or chemical agents as it was also stated by Sanches-Silva et al. [53].

Many of the botanical-origin bioactive compounds used to provide antimicrobial and/or antioxidant capacity to bio-based materials have been classified as "generally recognized as safe" (GRAS) by the US Food and Drug Administration (FDA), as reported by Benbettaïeb et al. and Salgado et al. Although those not classified as such, it has been necessary to study the use of appropriate doses of these natural compounds because in many cases their toxicity level is unknown [52, 54].

The action mechanism of botanical-origin compounds is varied, and for example, it has been theorized that they can cause cellular plasmolysis, which impedes the microorganisms' vital functions, as reported by Hernández-González et al. and Mlalila et al. According to the latter authors, these mechanisms should continue to be studied, especially to optimize the migrating components of the container and the food that is intended to be preserved [55, 56].

Eça et al. indicated that natural antioxidant compounds used in the functionalization of bio-based packaging materials are employed to prevent or reduce the development of enzymatic browning in fresh foods of plant origin, with positive results [57]. On the other hand, Gómez-Estaca et al. have also reported that the use of some natural additives was effective in combating lipid oxidation in foods of animal origin, naturally rich in fat [58]. Sanches-Silva et al. on their part, indicated that these bioactive compounds are beneficial for preserving the nutrients of packaged foods, protecting them from oxidation, thus slowing down the deterioration of the contained food, while also preserving their sensory attributes [53].

As discussed by Sánchez-González et al. one of the advantages of using natural compounds with antioxidants and/or antimicrobial capacity is to avoid the use of synthetic additives that can cause certain health conditions in consumers [49, 57], side effects that have gradually disappeared over time in the population. This is the reason why markets currently strive to offer products that do not rely on their use [59], while still requiring processed foods to have an acceptable shelf life comparable to using synthetic additives or, ideally, longer [60]. Table 10.2 has compiled the various functionalizing agents of botanical origin that have been utilized to functionalize bio-based starch films in the past decade.

10.4 FILMS AND COATING BASED ON STARCHES FROM NON-CONVENTIONAL SOURCES FUNCTIONALIZED WITH NATURAL ANTIMICROBIAL AND ANTIOXIDANT AGENTS OF BOTANICAL ORIGIN

Taking advantage of the compounds with proven antimicrobial and/or antioxidant capacity of some plants, bio-based materials have been developed and further functionalized with such compounds [58, 60–62], in such a way that it has been sought to also provide them with antimicrobial and/or antioxidant capacity in accordance with what is described in Figure 10.1. This is in order to make them useful in the preparation of active and biodegradable packaging for food [30, 48] so that they can help preserve their chemical and nutritional integrity [63]. At the same time, their production is sustainable, with an emphasis on the environmental aspect [31, 64].

10.4.1 MECHANICAL, THERMAL, AND BARRIER PROPERTIES

As a result of this, other materials, equally organic, and of innocuous and inert inorganic components, have been used as reinforcement of thermoplastic starches (TPS) in order to improve their characteristics or mechanical, permeability, and thermal properties mainly, which have resulted in composites with better aptitude to be used as substitutes for conventional plastics in food packaging, as described by Umaraw & Verma [14], and bio-based starch materials from unconventional sources have also been tested in this area. Biomaterials such as cellulose derivatives, lignin, chitosan, pectins, carrageenans, gums, and even proteins and their derivatives and waxes have been used as reinforcing elements of starch-based films [54]. Table 10.3 shows several examples of these composed materials starch bio-based which have been recently published. As reported by Valdés et al. and Ávila-Martín et al. natural functionalizing agents used in many researches have also been the cause of bio-based materials' properties modification in which formulations are included, in most cases with interesting results; however, they can

TABLE 10.2 Botanical Additives Used to Functionalize Native Starch Films from Conventional and Non-Conventional Sources

Species	Common Name	Starch Origin	Film Composition1	Measurement Technique	Antioxidant Level2	Microorganism and Inhibition Level2	References
Lippia berlandieri Schauer	Mexican oregano	NS	Starch + glycerol + EO (0.50%)	Direct inoculation and plate diffusion	NM	*Aspergillus niger*, *Pennicillium* sp. (total)	[90]
Opuntia oligacantha	Xoconostle (fruits)	Chayotextle	Starch + glycerol + MEB (1.2 g)	DPPH, ABTS, plate diffusion	29.11 ± 0.48 mg (ABTS) 41.42 ± 1.81 mg (DPPH)	*Salmonella typhimurium* (2 mm halo)	[80]
Thymus capitatus	Origanum	Sweet potato	Starch + glycerol + EO (2.0%)	Plate diffusion	NM	*Salmonella enteritidis* (18.4 ± 2.8 mm^2 halo), *Escherichia coli* O157:H7 (10.1 ± 5.0 mm^2 halo), *Listeria monocytogenes* (9.0 ± 4.7 mm^2 halo)	[91]
Curcuma longa L.	Turmeric	Turmeric	Starch + glycerol	DPPH, ABTS	281.8 ± 15.1 µM TE/mg (DPPH) 612.0 ± 62.2 µM TE/mg (ABTS)	NM	[42]
Attlaea speciosa	Babassu mesocarp	Babassu mesocarp	Starch + glycerol	DPPH	63.8 ± 1.0 (%)	NM	[41]
Manguifea indica cv. "Tommy Atkins)	Mango	Mango seed	Starch + glycerol + PE (6.7%)	DPPH	97.22% RSA/mg	NM	[116]
Rubus fruticosus cv. Tupy	Blackberry	Arrowroot	Starch + glycerol + pulp (20%)	FRAP	481.16 ± 12.44 µM FS/g pulp solids	NM	[68]
Thymus serpyllum L.	Thyme	Pea	Starch + glycerol +WE (15%)	ABTS	1.7 mM TE	NM	[93]

TABLE 10.2 *(Continued)*

Species	Common Name	Starch Origin	Film Composition1	Measurement Technique	Antioxidant Level2	Microorganism and Inhibition Level2	References
Curcuma longa L.	Turmeric	Turmeric	Starch + glycerol+ gelatin + WE (35%)	PCA after 20 days storage at 10°C, without film vs. film	NM	Mesofilic bacteria (–1.57 Log CFU/g), Lactic acid bacteria (–2.14 Log CFU/g), Psychrotrophic bacteria (–1.99 Log CFU/g)	[98]
Trachyspermum ammi	Carom	Arrowroot	Starch + glycerol + pectin+ EO (1.5%)	Plate diffusion	NM	*Staphylococcus* (11.667 mm halo) *Escherichia coli* (12.667 mm halo)	[67]
Ocimum basilicum	Basil	Wheat	Water + starch + chitosan + glycerol +basil EO (10% EO/w starch)	ABTS	15.9 mg TE	NM	[73]
Thymus vulgaris	Thyme	Wheat	Water + starch + chitosan + glycerol + thyme EO (10% EO/w starch)	ABTS	9.4 mg TE	NM	[73]
Zataria multiflora Boiss	Avishan Shirazi	Corn	Water + starch + glycerol + EO (3%)	Disc-diffusion	NM	*Staphylococcus aureus* (≈ 170 mm^2 halo) *Escherichia coli* (≈ 160 mm^2 halo)	[82]
Mentha pulegium.	Pune	Corn	Water + starch + glycerol + EO (3%)	Disc-diffusion	NM	*Staphylococcus aureus* (≈ 110 mm^2 halo) *Escherichia coli* (≈ 100 mm^2 halo)	[82]
Ilex paraguariensis	Yerba mate	Cassava	Water + starch + glycerol + yerba mate infusion (20 g)	DPPH	10.4 ± 0.1 mg/g	NM	[70]
Punica granatum L.	Pomegranate	Corn	Water + starch + chitosan + glycerol + pomegranate peel extract (1%)	TBARS, direct inoculation (after 21 days of storage)	≈ –1.7 mg MDA/kg vs control	*Listeria monocytogenes* (≈ –0.98 Log10 CFU/g vs control)	[99]

TABLE 10.2 *(Continued)*

Species	Common Name	Starch Origin	Film Composition1	Measurement Technique	Antioxidant Level2	Microorganism and Inhibition Level2	References
Thymus kotschyanus	Thyme	Corn	Water + starch + chitosan + glycerol + EO (1%)	TBARS, direct inoculation (after 21 days of storage)	≈ –1.5 mg MDA/kg vs control	*Listeria monocytogenes* (≈ –1.36 Log10 CFU/g vs control)	[99]
Bixa orellana	Annatto seeds	Cassava	Water + starch + glycerol + bixin nanocapsules (1.15 g/g starch)	Peroxide value in sunflower oil (after 13 days)	6.2 ± 0.1 mEq O_2/Kg	NM	[113]
Camellia sinensis + *Elaeis guineensis*	Pune green tea + palm oil	Cassava	Water + starch + glycerol + green tea extract (4.3% w/w) + palm oil extract (0.04% w/w)	% Peroxide index (PI)	–108.1% PI vs starch film without additives, after 45 days storage.	NM	[83]
Oryza sativa	Antioxidant rice extract	NS	Water + starch + glycerol + rice extract (0.0008 g/g dried film)	EC_{50} (DPPH)	3.3717 g soluble solids/g DPPH 42.147 Kg film/g DPPH.	NM	[72]
Salvia rosmarinus	Aqueous extract	Cassava	Water + starch + glycerol + rosemary extract (5%/whater mass)	DPPH (% inhibition)	28.6 ± 0.3	NM	[85]
Ilex paraguariensis + *Manguifera indica*	Yerba mate + mango pulp	Cassava	Water + yerba mate AQ (15%/whater mass) + starch + glycerol + mango pulp (20%/w search)	Increases in peroxide value (in palm oil)	–2.79 mEq/Kg vs film without additives	NM	[63]

[1]The best concentration reported.

[2]Data is indicated as published. NS: not specified; NM: not measured; EO: essential oil; MEB: microencapsulated bioextract; PE: phenolic extract; WS: aqueous extract; PCA: plate count agar; ABTS: 2,2'-Azino-bis(3-ethylbenzothiazoline-6-sulfonic acid); DPPH: 1,1-diphenyl-2-picrylhydrazyl; TE: Trolox equivalent; %RSA: % radical scavenging activity; FRAP: ferric antioxidant power; FS: ferrous sulfate; AQ: aqueous extract; PCA: plate count agar; TBARS: Thiobarbituric acid reactive substances; MDA: Malondialdehyde; and EC_{50}: concentration required to obtain a 50% antioxidant effect.

also cause drawbacks in other formulations, so this would have to be taken into account when designing these materials [48, 65].

Taqi et al. evaluated the inclusion of laurel (*Laurus nobilis* L.) essential oil in concentrations of 0.5, 1.0, and 1.5% w/w in films based on cassava starch and apple pectin, thereby improving the WVP, increasing E% and EM value; but decreasing the FT values, with respect to the non-functionalized films [66]. Yadav et al. evaluated films of a mixture of pumpkin starches and arrowroot (*Maranta arundinacea*) functionalized with carom oil (*Trachyspermum ammi*), whose impact on the mechanical properties of the films was not significant at a concentration of 1.5% w/w, compared to the film not functionalized [67]. In another study evaluating arrowroot (*M. esculenta*) films, but functionalized with blackberry pulp (*Rubus fruticosus*), it was found that this functionalizing agent increased the film's E%; however, it caused a decrease in the FT and WVP values, as well as in their solubility [68].

On their own, Zamudio-Flores et al. obtained oxidized oatmeal (*Avena sativa* L.) and banana (*Musa paradisiaca* L.) starches added with beet betalains (*Beta vulgaris*), which effect was to increase E%, without affecting the other mechanical properties, also increasing WVP, with respect to non-functionalized films [69]. In another study, performed by Medina-Jaramillo et al. matte herb (*Ilex paraguariensis*) extracts were used as antioxidants also resulting as plasticizers in the production of cassava starch films at a concentration of 5% w/w, increasing the resulting material's hydrophobic character, which was proven by the low value of WVP, without affecting the material's biodegradability property [70, 71]. Piñeros-Guerrero et al. found in their study on cassava starch and poly-ε-caprolactone that the addition of aqueous polyphenolic extract of rice husk (*O. sativa*) did not affect the film's mechanical, thermal, and permeability properties [72].

Bonilla et al. worked with films composed of wheat starch and chitosan functionalized with basil (*Ocimum basilicum*) and thyme (*Thymus vulgaris*), using glycerol as a plasticizer. They found that a concentration of 6.7% w/w of essential oil improved the WVP values in both cases. Thyme essential oil negatively affected the mechanical properties compared to the non-functionalized control, while basil oil had a positive effect, although not significantly [73]. De Araújo et al. reported that the inclusion of propolis extracts in cassava starch films decreased the EM value by around 50% and at the same time increased the WVP value, as the extract gave the material a hydrophobic character [74]. Recently, Ortega-Toro et al. functionalized corn starch films with aloe vera (*Aloe barbadensis* M.) in a 1:1 ratio, plasticized with glycerol (1.5 g/g of starch). Their study indicated that under three formulation conditions, the WVP value was reduced compared to the non-functionalized control, and there was a decrease in the T_g values in inverse relation to the increase in Aw due to the effect of a higher plasticizer concentration [75].

Another factor to consider is the effect that functionalized materials suffer in their properties in general when including surfactants in the formulation, such as tween 20 (polyoxyethylene sorbitan monolaurate) and tween 80 (polyoxyethylene sorbitan monooleate), necessary for the integration of the hydrophobic and hydrophilic fractions of the components in filmogenic solutions [76, 77], in which non-polar components are included, such as essential oils. Table 10.3 shows functionalized bio-based films on starch from conventional and non-conventional sources developed and studied in the last decade.

10.4.2 ANTIOXIDANT AND ANTIMICROBIAL FUNCTIONAL PROPERTIES

Once the inclusion of natural functionalizing agents in bio-based polymeric matrices has been achieved, their effectiveness was evaluated in terms of their antimicrobial and antioxidant capacity, as reported by Pagno et al. [78]; Homayouni et al. [79]; or Cenobio-Galindo et al. [80]. Over the last few years, advances have been published in this field, especially using conventional starches, due to their greater availability, with encouraging results [81–87].

Various interaction mechanisms have been reported between antimicrobial and/or antioxidant compounds and the food's matrix to be preserved, but, according to Sánchez-González et al. [49], the most important ones are shown in Figure 10.2.

Several recent studies have verified this mechanism with migration tests using food simulants (95% ethanol and a mixture of ethanol and acetic acid: 96 and 3%, respectively) during their interaction with functionalized bio-based materials, where the amount of active compound transferred from the functionalized bio-based material to the simulating media has been evaluated, varying the results with respect to the different materials [55, 88, 89]. Other materials also make possible a mechanism of active compound release towards the container's internal atmosphere as mentioned by Sanches-Silva et al. [53], by easily releasing the volatile compounds contained in the material's matrix into the surrounding environment.

TABLE 10.3 Mechanical, Thermal, and Barrier Properties of Functionalized Films from Non-Conventional Native Starches

Species	Film Composition	1Additive (% w/w d.b.)	Material Conditioned Before and During Test	2Elastic Modulus (EM; MPa)	2Tensile Strength (TS; MPa)	2Elongation (%E)	2Tg (°C)	2WVP (g·m–1·s–1·Pa–1)·10n	References
Mangifera indica L. var. "Tommy Atkins"	Starch + glycerol	PE of mango (6.7%)	ASTM D882-12	430.9	6.91	42.19	–	2.882 (g mm $kPa^{-1}h^{-1}m^{-2}$)	[116]
Apple (*Malus domestica*)	Starch + glycerol	Ellagic acid (0.15%)	ASTM-882-95a	3.63 ± 0.32	8.21 ± 0.45	52.44 ± 1.21	–	6.35 ± 0.22 (10^{11})	[111]
Sweet potato (*Ipomoea batatas*)	Strach + grenetin + glycerol	Clove EO (60 mg/L)	57% RH, 25°C, 5 d	3.6 GPa	65	5	–	1×10^{-8} g/m s Pa	[92]
Chayotextle (*Sechium edule* Sw.)	Starch + glycerol	Xoconostle EMB (0.4%)	57% HR, 25°C, 14 d	1.01 ± 0.22	16.80 ± 2.80	1.02 ± 0.20	–	–	[80]
Pea (*Pisum sativum* L.)	Starch + glycerol + WPI + lecithin	Carnauba wax (0.025 g/mL)	67% HR, 25°C, 48 h	51.92 ± 6.21	0.99 ± 0.01	32.56 ± 2.23	–	–	[102]
Arrowroot (*Maranta arundinacea*)	Starch + glycerol	Blackberry pulp (20%)	75 ± 3% RH, 25°C, 7 d	–	≈7	≈4	–	≈3 (g mm/m^2 day kPa)	[68]
Turmeric (*Curcuma longa L.*)	Starch + glycerol	Turmeric residue (35%)	58% RH, 25°C, 48 h	85	5	42	T_p = 112.21°C	8.01 ± 0.03 (g $day^{-1}m^{-2}atm^{-1}$)	[98]
Avena (*Avena sativa* L. var. Bachíniva)	Oxidized starch + glycerol	Aqueous betalains solution (0.083 g/L)	50 ± 5% RH, 25°C, 48 h	0.21	1.65	135.64	–	27.90 ± 2.81 (10^{11})	[120]
Banana (*Musa paradisiaca* L.)	Oxidized starch + glycerol	Aqueous betalains solution (0.083 g/L)	50 ± 5% RH, 25°C, 48 h	5.47	5.67	28.71	–	18.50 ± 1.54 (10^{11})	[120]
Potato (*Solanum tuberosum*)	Water + starch + glycerol + additive	Limonene (2 g/100 mL)	55 ± 2%RH, 25 ± 2°C	44 ± 7	1.69 ± 0.2	69 ± 23	–	–	[17]

TABLE 10.3 *(Continued)*

Species	Film Composition	1Additive (% w/w d.b.)	Material Conditioned Before and During Test	2Elastic Modulus (EM; MPa)	2Tensile Strength (TS; MPa)	2Elongation (%E)	2Tg (°C)	2WVP (g·m–1·s–1·Pa–1)·10n	References
Potato (*Solanum tuberosum*)	Water + starch + glycerol + additive	Limonene (2 g/100 mL)	55 ± 2%RH, 25 ± 2°C	2724 ± 334	38.25 ± 4.5	3 ± 1	–	–	[17]
Wheat (*Triticum vulgare*)	Water + starch + chitosan + glycerol + EO	Tyme EO (1 starch: 0.1 EO)	53% RH, 25°C, 24 h	1,450	43	3.6	–	18.2 g mm kPa^{-1} h^{-1} m^{-2}	[73]
Corn (*Zea mays* L.)	Water + starch + glycerol + EO	Avishan Shirazi EO (3%)	30% RH, 20°C, 48 h	–	≈ 9	≈ 165	–	≈ 3.7 g mm kPa^{-1} d^{-1} m^{-2}	[82]
Corn (*Zea mays* L.)	Water + starch + glycerol + EO	Pune EO (3%)	30% RH, 20°C, 48 h	–	≈ 3	≈ 110	–	≈ 3.5 g mm kPa^{-1} d^{-1} m^{-2}	[82]
Tapioca (*Manihot esculenta*)	Water + starch + glycerol +EO	Carvacrol (100 mg/g starch)	60 ± 5% RH, 27°C, 48 h	0.054 ± 0.005	0.355 ± 0.010	745 ± 18	–	1,320 ± 30 mg mm kPa^{-1} h^{-1} m^{-2}	[79]
Cassava (*Manihot esculenta*)	Water + starch + glycerol + powder mix	Cinnamon powder (g/100 g) + clove powder (g/100 g)	75% RH, 23°C, 48 h	–	2.2 ± 0.3	101.5 ± 13.5	–	4.65 ± 0.82 g mm kPa^{-1} d^{-1} m^{-2}	[81]
Cassava (*Manihot esculenta*)	Water + starch + glycerol + infusion	Yerba mate infusion (20 g)	56.7%RH, 25°C until use	0.42 ± 0.05	0.28 ± 0.04	183 ± 7	–	4.5 ± 0.4 g/m s Pa (10^{-10})	[70]
Corn (*Zea mays* L.)	Water + starch + chitosan + glycerol + peel extract (1%)	Pomegranate peel extract (1%)	52 ± 2% RH, 25°C, until use	–	7.47 ± 1.63	10.33 ± 1.17	–	≈ 15.01 g/ m^2 h	[99]
Corn (*Zea mays* L.)	Water + starch + chitosan + glycerol + EO (1%)	Thyme EO (2%)	52 ± 2% RH, 25°C, until use	–	7.26 ± 1.16	13.79 ± 0.78	–	≈ 15.01 g/ m^2 h	[99]

TABLE 10.3 *(Continued)*

Species	Film Composition	1Additive (% w/w d.b.)	Material Conditioned Before and During Test	2Elastic Modulus (EM; MPa)	2Tensile Strength (TS; MPa)	2Elongation (%E)	2Tg (°C)	2WVP (g·m–1·s–1·Pa–1)·10n	References
Cassava (*Manihot esculenta*)	Water + starch + glycerol + bixin	Bixin nanocapsules (1.15 g/g starch)	58% RH, 25°C, 48 h	–	8.95 ± 1.32	15.55 ± 1.14	–	0.216 ± 0.007 g mm kPa^{-1} d^{-1} m^{-2}	[113]
Cassava (*Manihot esculenta*)	Water + starch + glycerol + green tea extract + palm oil extract	Palm oil extract (0.04% w/w starch) + green tea extract (4.3% w/w starch)	75% RH, 25°C, until use	–	0.96 ± 0.22	131.10 ± 0.11	Tonset = 39.26°C (1st event)	0.29 ± 0.93 g mm kPa^{-1} d^{-1} m^{-2}	[83]
Not specified	Water + starch + glycerol + PCL + rice extract	Antioxidant rice extract (0.0008, g/g dried film)	75% RH, 20°C, until use	94 ± 4	2.3 ± 0.4	15.9 ± 0.2	Tm = 60.8 ± 0.7°C (1st heating scan)	–	[72]
Cassava (*Manihot esculenta*)	Water + starch + glycerol + rosemary extract	Rosemary extract (5%/water mass)	75% RH, 20°C, until use	≈ 1.5	≈ 0.7	–	–	6.0 ± 0.2 (10^{10})	[85]
Cassava (*Manihot esculenta*)	Water + yerba mate aqueous extract + starch + glycerol + mango pulp	Yerba mate aqueous extract (15%/water mass) + mango pulp (20%/w starch)	60 ± 2% RH, 23 ± 2°C, 48 h	–	2.04	69.36	–	5.74 × 10–8 mg mm kPa^{-1} h^{-1} m^{-2}	[63]

[1]Best concentration reported based on lows WVP.
[2]Data is indicated as published.

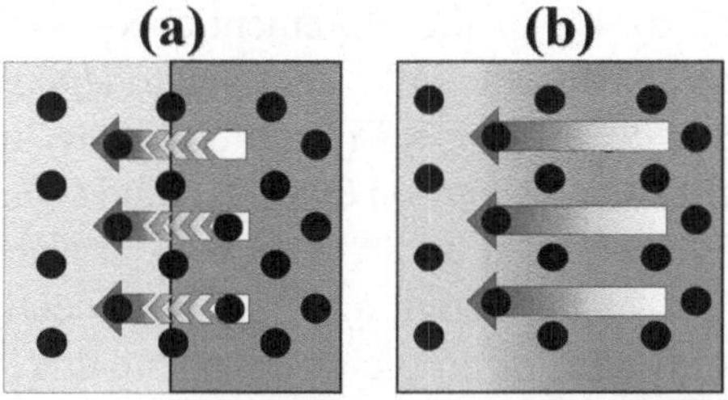

FIGURE 10.2 Interaction mechanisms between functionalized bio-based materials and the contained food´s matrix: (a) by diffusion into the food´s matrix through controlled or modulated release mechanism from the functionalized material (e.g., films); and (b) by fusion between the matrix of the functionalized material with the food´s matrix surface, which facilitates the active´s compound migration (e.g., coatings).

One of the most important limitations in the packing functionalization with bioactive antimicrobial and/or antioxidant compounds is that their structural nature prevents or favors the mass exchange in the container-food interaction since the diffusion phenomenon depends on the chemical nature of the components to be tested (container and content), the physical structure and chemical conformation of each functionalized material, as indicated by Mlalila et al. [56]. Therefore, according to these researchers, it is very important to consider the active compound concentration and the material's morphology, as well as the matrix-functionalizing agent interaction when designing any container with functionalized materials for it to be successful.

Regarding the antimicrobial capacity, interesting results have been achieved in the evaluation against food spoilage microorganisms (Table 10.2), including fungi and yeast, and both Gram-positive and Gram-negative, such as *Staphylococcus aureus, Escherichia coli, Salmonella* spp., *Listeria monocytogenes*, among others, the ones that cause foodborne illness and others. On the other side, the functionalized bio-based materials' antioxidant capacity varies according to the botanical extract used, and the evaluation method used for its analysis or determination (Table 10.2), which can be through the DPPH technique, Trolox or ABTS equivalents, commonly used for the oil and vegetable fats' antioxidant activity evaluation, or TBARS, more suitable for animal origin products.

When testing films obtained using the inverted plate technique from amaranth flour (*Amaranthus* spp.) and corn starch functionalized with oregano essential oil (*L. berlandieri* Schauer) [90] and cinnamon essential oil (*Cinnamomum verum*) [59], even total inhibition was achieved in cultures of food spoilage fungi (*Penicillium digitatum* and *Aspergillus niger*) using concentrations of 0.5 to 4% of the aforementioned essential oils. Ehivet et al. verified the antimicrobial effect of sweet potato (*Ipomoea batatas*) starch films functionalized with oregano (*Thymus capitatus*) essential oil against *Salmonella enteritidis, Escherichia coli* O157:H7, and *Listeria monocytogenes*, with growth inhibition results of up to 18.43 mm^2 in plate antibiogram using a concentration of 2% of the active agent [91]. Recently, Buso-Rios et al. evaluated in their study film of purple sweet potato starch (*I. batatas*) using gelatin as a stabilizer, glycerol as a plasticizer, and clove (*Syzygium aromaticum* L.) essential oil as a functionalizing active ingredient; their studies showed that the WVP decreased as the essential oil increased since this contributes hydrophobicity to the material, and they observed that while the TS values and E% increased (along with the essential oil concentrations), the EM decreased; so it could be said that the mechanical properties are influenced by the functional agent presence and concentration [92].

Bonilla et al. formulated films composed of wheat starch and chitosan functionalized with basil (*Ocimum basilicum*) and thyme (*Thymus vulgaris*) oils, using a concentration of 6.7% w/w. The films made with basil oil resulted in a significantly higher antioxidant content than those made with thyme oil, with values of 15.9 vs 9.4 mg TEAC (Trolox Equivalent Antioxidant Activity), respectively [73]. Talón et al. also used thyme (*Thymus vulgaris)* extracts in pea (*P. sativum*) starch films and found that the antioxidant activity depends on the film's structure and chemical composition of what it is intended to protect. In this study, the antioxidant activity showed significant differences between the treatments in which water and 20% ethanol were used as simulating liquids, with values of 1.7- and 2.26-mM TEAC, respectively [93]. Piñeros-Guerrero et al. produced films composed of cassava starch and poli-ε-caprolactone functionalized with aqueous rice husk (*O. sativa*), obtaining an EC_{50} (concentration required to obtain a 50% antioxidant effect) of 1.54 (kg film/g DPPH) [72].

De Araújo et al. evaluated the inclusion of propolis extracts (polyphenols) in cassava starch films. Their results showed that at an extract concentration of 0.5% w/w, antimicrobial activity against *S. aureus* and *E. coli* was preserved, as well as the antioxidant capacity of propolis [74]. Lozano-Navarro et al. evaluated the antimicrobial and antioxidant properties of films composed of rice starch (*O. sativa*) and chitosan, which were functionalized with different natural extracts (anthocyanins, betalains, thymol, carvacrol, and resveratrol) extracted from different fruits. They determined the antimicrobial activity of these films against aerobic mesophiles and molds; in addition, they described that the compounds used enhanced the antioxidant capacity of the resulting films [94].

Fruit pulp has also been used as a functionalizing agent in starch films. Nogira et al. elaborated cassava (*Manihot esculenta*) starch films, functionalized with blackberry pulp (*R. fruticosus*) in concentrations of 20, 30, and 40% w/w with respect to the starch weight, and found that the pulp antioxidant capacity was maintained in the films obtained. They attributed such capacity to the bioactive compounds present in it, such as anthocyanins, phenolic acids, tannins, and ascorbic acid [68]. On other occasions, the same starch source has bioactive compounds capable of granting antioxidant functionality to the films made from them. This is the case of the study reported by Hornung et al. in which films were made with starch isolated from Amazonian turmeric (*Curcuma longa* L.), resulting in higher levels of antioxidant activity (65–92%, measured by DPPH and ABTS) [42].

10.5 BIOBASED STARCH FILMS APPLICATIONS IN FOOD

Due to the disadvantages of bio-based materials coming from natural compounds, such as starches, applications that have been tested with relative success in recent studies have been limited, as mentioned by Sánchez-González et al. [49]; despite that, according to Oropeza-González et al. achievements gained so far are not demeritorious, and tests have been carried out that seek to increasingly optimize this type of material for use in third-generation packaging that can be a profitable solution compared to what the use of conventional plastics represents in handling and packaging of food products, whether fresh, minimally processed, or processed [22].

Due to their hydrophilic nature, the use of starch films functionalized and non-functionalized, has been successfully evaluated in fatty food models. Chiu et al. tested cassava starch and Hsian-Tsao gum (*Mesona procumbens* Hemsl.) films functionalized with green tea extracts (*Camellia sinensis* L.) to prolong medallions pork shelf life, achieving reductions of up to 6 logarithmic cycles during refrigerated storage in the first 48 hours [95]. Kuorwel et al. experimented with commercial maize starch films functionalized with a mixture of linalool-carvacrol-thymol (1%, 3%, and 5%, respectively), compounds present in order to evaluate their antimicrobial capacity against *S. aureus* inoculated on the surface of sterilized cheddar cheese. Their results showed that it was possible to inhibit the growth of said microorganisms, with the films functionalized with thymol obtaining the highest inhibition surfaces, achieving reductions of up to 2 logarithmic cycles in the 21 days that the test lasted at a storage temperature of 15C [96]. Medina et al. evaluated films of acetylated arracacha (*Arracacia xanthorrhiza* B.) starch in the preservation of beef to avoid contamination by fecal coliforms, obtaining results showing the absence of this type of microbiota in the packaged food [36]. Zamudio-Flores et al. used films of oxidized banana starches (*Musa paradisiaca* L.) functionalized with betalains from beets (*Beta vulgaris*) for the preservation of sausages, which resulted in no changes in terms of species reactive to TBARS for 20 days under refrigeration conditions, also preserving the characteristics of texture and color compared to control sausages [69].

In another study by u-Nisa et al., green tea extracts (*Camellia sinensis*) were used in the production of potato starch films (*Solanum tuberosum* L.) to pack cuts of beef, with positive results, since lipid oxidation was decreased (TBARS) and the production of metmyoglobin, with respect to untreated meat [97]. Tosati et al. used combined films of turmeric remains (*Curcuma longa*) starch and gelatin in frankfurters, achieving a positive effect in reducing microbial growth during 30 days of storage at 5 and 10C [98]. Mehdizadeh et al. used films composed of chitosan and cornstarch functionalized with extracts of pomegranate peel (*Punica granatum* L.) and *Thymus kotschyanus* essential oil to evaluate beef's shelf life under refrigeration conditions (4°C) for 21 days; obtaining that using 2% of *T. kotschyanus* oil inhibits the growth of *Listeria monocytogenes* and the sensory attributes of the meat are kept acceptable [99].

Regarding applications in the preservation of fruits and vegetables, Chiu et al. tested films made of cassava starch and gum from discolored leaves of Hsian-Tsao (*M. procumbens* Hemsl.) functionalized with green tea extracts (*C. sinensis* L.) in the preservation of fruit salad and Roman lettuce. They observed growth inhibition of 1 to 2 logarithmic cycles of aerobic mesophiles, as well as a reduction of up to 6 logarithmic cycles in terms of *S. aureus* and *L. monocytogenes* during lettuce storage under refrigeration conditions [95]. Pająk et al. evaluated the storage of apple slices in films made of native potato starch (*S. tuberosum*) functionalized with green and white tea extracts (*C. sinensis*) for 6 days, using polyethylene film as a control. They found that the levels of polyphenols released to food reduced oxidation, maintaining the acceptability of apple attributes, although with weight loss due to the higher starch films WVP compared to polyethylene [100]. On the other hand, Campos-Requena et al. used corn starch and clay nano-composite films functionalized with thymol and carvacrol (50:50) obtained from essential oils, by the extrusion method, as packaging for the conservation of strawberries. They observed that these films favored the inhibition of the growth of

Botrytis cinerea inoculated directly on the fruit under ambient conditions, without altering the strawberries' attributes [101]. In another study, Ortega-Toro et al. observed a delay in the proliferation of several types of fungi, especially *Fusarium oxysporum*, during the evaluation of the cherry tomato shelf life when they used a corn starch film functionalized with aloe vera (*A. barbadensis* M.) in a 1:1 ratio, using glycerol as a plasticizer compared to the control treatment (without film) [75].

Natural extracts are not limited to phytochemicals present in essential plant oils, but also other components of the same can be used as functionalizing agents. For example, in the study carried out by Mehyar et al., films formulated with pea starch (*Pisum sativum* L.), whey protein, and carnauba wax (*Copernicia prunifera*) were used to apply them as a barrier in walnuts and pine nuts. An antioxidant effect was obtained between 25 and 50°C during the storage of these products [102].

10.6 TRENDS IN FUNCTIONALIZED STARCH FILMS

According to Khan et al. [28], although in many cases it has been possible to develop functionalized bio-based materials, it should be remembered that their mechanical, thermal, and barrier properties are usually deficient compared to those of conventional petroleum-derived plastics [103]. Therefore, they do not yet represent a direct competition for those in many of their applications [104], especially speaking specifically of starch bio-based materials [76]; however, in the case of food with low moisture content, it has been possible to verify their usefulness, at least in part, as mentioned by Rossi et al. [64]. Thus, the nature of these products achieved so far could lead to other developments in the future, as their technology is in constant research and development [104].

Even so, and with all the major advantages of using entirely bio-based materials, it is important to mention that care must be taken when choosing which food matrix to pack in them, depending on their composition and characteristics, to avoid unwanted results. For example, Suoza et al. used functionalized cassava films with mango pulp (*Mangifera indica* L.) and acerola *(Malpighia emarginata* D.C.) with negative results in the packaging of palm oil, since the ascorbic acid contained in said functionalizing agents acted rather as a pro-oxidant under the conditions in which this test was carried out (63% RH at 30°C for 45 days) [105].

On the other side, these materials' functionalization is not limited to the use of phytochemicals but has also resorted to other natural elements, such as some antibiotics [106–108], organic acids (44, 65, 66, 109–111], protein derivatives [112], polysaccharides, such as lignin [39] and chitosan [40], and recently, minerals in the form of metal nanoparticles such as gold [113], silver [47, 114], and zinc and copper oxides [115], with interesting results, but that should continue to be investigated in order to define whether they are equally safe or not, with respect to natural compounds of botanical origin, as well as the effects that they could exert on the host matrix, which will indicate, at least in theory, the practical application that they may offer.

Other functionalization options for bio-based starch films have been developed so that, instead of working as a material for active packaging, they function as smart packaging. An example of this is part of the work carried out by Luchese et al. functionalizing starch films from pinhão (*A. angustifolia*) with 1% w/w of jaboticaba (*Myrciaria cauliflora*) residue flour, rich in monomeric anthocyanins that, when in contact with food matrices of acid pH, changes color, noting an important change in the possible food's deterioration, thereby helping the consumer to make an informed consumption [32].

In several first-world countries, especially in Europe, technologies have been developed that make it possible to obtain commercial packaging from starches extracted from conventional sources that have already been used in the handling of a variety of foods, including packaging films, sheets, plates, containers, trays, salad bowls, disposable cups, beverage bottles, cutlery, among other items of common use [29], so that starches from non-conventional sources could be used in the future with the same success in less developed or developing countries.

10.7 CONCLUSION

The trend in the development of functionalized starch-based materials for possible use in food packaging has advanced in recent years towards materials that actively interact with the contained food matrix, delaying or eliminating oxidative and/or microbial activity. However, the materials developed so far are still at a disadvantage against conventional materials used in food packaging in terms of their mechanical, thermal, and permeability properties. Even so, the continuous improvement of bio-based materials is encouraging, the main development of which is the reduction of contamination caused by materials with low or no biodegradability, also providing the possibility of extending the food's shelf life, with the advantages that would derive from this, and with the use of natural raw materials not used up to now or that have been explored.

KEYWORDS

- **biodegradability**
- **films**
- **food matrices**
- **food packaging**
- **food safety**
- **microbial activity**
- **permeability**
- **shelf-life**

REFERENCES

1. Smith, A. M., & Zeeman, S. C. (2020). Starch: A flexible, adaptable carbon store coupled to plant growth. *Annual Review of Plant Biology, 71*, 217–245. https://doi.org/10.1146/annurev-arplant-050718-100241.
2. Lovegrove, A., Edwards, C. H., De Noni, I., Patel, H., El, S. N., Grassby, T., & Shewry, P. R. (2017). Role of polysaccharides in food, digestion, and health. *Critical Reviews in Food Science and Nutrition, 57*(2), 237–253. DOI: 10.1080/10408398.2014.939263.
3. Parmar, A., Sturm, B., & Hensel, O. (2017). Crops that feed the world: Production and improvement of cassava for food, feed, and industrial uses. *Food Security, 9*(5), 907–927. doi:10.1007/s12571-017-0717-8.
4. Tesfaye, T., Gibril, M., Sithole, B., Ramjugernath, D., Chavan, R., Chunilall, V., & Gounden, N. (2018). valorisation of avocado seeds: extraction and characterization of starch for textile applications. *Clean Technologies and Environmental Policy, 20*(9), 2135–2154. doi:10.1007/s10098-018-1597-0.
5. Orsini, A., & Godet, C. (2018). Food security and biofuels regulations: the emulsifying effect of international regime complexes. *Journal of Contemporary European Research, 14*(1), 4–22. DOI: https://doi.org/10.30950/jcer.v14i1.
6. You, C., & Zhang, Y. H. P. (2016). Ex Vivo enzymatic conversion of non-food cellulose biomass to starch. In *Quality Living Through Chemurgy and Green Chemistry* (pp. 129–142). Springer. doi:10.1007/978-3-662-53704-6_6.
7. Alcázar-Alay, S. C., & Meireles, M. A. A. (2015). Physicochemical properties, modifications and applications of starches from different botanical sources. *Food Science and Technology (Campinas), 35*(2), 215–236. doi:10.1590/1678-457x.6749.

8. Singh, J., Colussi, R., McCarthy, O. J., & Kaur, L. (2016). Potato starch and its modification. In *Advances in Potato Chemistry and Technology* (2nd ed., pp. 195–247). Elsevier – Academic Press. doi:10.1016/b978-0-12-800002-1.00008-x.
9. Ai, Y., & Jane, J. (2016). Understanding starch structure and functionality. In *Starch in Food, Structure, Function and Applications* (2nd ed., pp. 151–178). Elsevier – Woodhead Publishing. doi:10.1016/b978-0-08-100868-3.00003-2.
10. Spada, J. C., Marczak, L. D. F., Tessaro, I. C., & Noreña, C. P. Z. (2011). Microencapsulation of β-carotene using native pinhão starch, modified pinhão starch and gelatin by freeze-drying. *International Journal of Food Science & Technology, 47*(1), 186–194. doi: 10.1111/j.1365-2621.2011.02825.x.
11. Li, G., & Zhu, F. (2018). Quinoa starch: Structure, properties, and applications. *Carbohydrate Polymers, 181*, 851–861. doi: 10.1016/j.carbpol.2017.11.067.
12. Marques, S., Moreno, A. D., Ballesteros, M., & Gírio, F. (2017). Starch biomass for biofuels, biomaterials, and chemicals. In *Biomass and Green Chemistry* (pp. 69–94). Springer. doi:10.1007/978-3-319-66736-2_4.
13. Oluwasina, O. O., Olaleye, F. K., Olusegun, S. J., Oluwasina, O. O., & Mohallem, N. D. S. (2019). Influence of oxidized starch on physic mechanical, thermal properties, and atomic force micrographs of cassava starch bioplastic film. *International Journal of Biological Macromolecules, 135*, 282–293. doi: 10.1016/j.ijbiomac.2019.05.150.
14. Umaraw, P., & Verma, A. K. (2015). Comprehensive review on application of edible film on meat and meat products: An eco-friendly approach. *Critical Reviews in Food Science and Nutrition, 57*(6), 1270–1279. doi: 10.1080/10408398.2014.986563.
15. Ogunsona, E., Ojogbo, E., & Mekonnen, T. (2018). Advanced material applications of starch and its derivatives. *European Polymer Journal, 108*, 570–581. doi: 10.1016/j.eurpolymj.
16. Santana, A. L., & Meireles, M. A. A. (2014). New starches are the trend for industry applications: A review. *Food and Public Health, 4*, 229–241. DOI: 10.5923/j.fph.20140405.04.
17. Antosik, A., Wilpiszewska, K., Wróblewska, A., Markowska-Szczupak, A., & Malko, M. (2017). Fragrant starch-based films with limonene. *Current Chemistry Letters, 6*, 41–48. DOI: 10.5267/j.ccl.2017.2.002.
18. Gadhave, R. V., Das, A., Mahanwar, P. A., & Gadekar, P. T. (2018). Starch based bio plastics: The future of sustainable packaging. *Open Journal of Polymer Chemistry, 8*, 21–33. https://doi.org/10.4236/ojpchem.2018.82003.
19. Luchese, C. L., Spada, J. C., & Tessaro, I. C. (2017). Starch content affects physicochemical properties of corn and cassava starch-based films. *Industrial Crops and Products, 109*, 619–626. doi: 10.1016/j.indcrop.2017.09.020.
20. Machado, C. M., Benelli, P., & Tessaro, I. C. (2017). Sesame cake incorporation on cassava starch foams for packaging use. *Industrial Crops and Products, 102*, 115–121. doi: 10.1016/j.indcrop.2017.03.007.
21. Sanyang, M. L., Ilyas, R. A., Sapuan, S. M., & Jumaidin, R. (2017). Sugar palm starch-based composites for packaging applications. In *Bionanocomposites for Packaging Applications* (pp. 125–147). Springer. doi: 10.1007/978-3-319-67319-6_7.
22. Oropeza-González, R. A., Montes-Hernández, A. I., & Padrón-Pereira, C. A. (2016). Biodegradable films based on starch: mechanical/functional properties and biodegradation. *Revista Venezolana de Ciencia y Tecnología de Alimentos, 7*, 065–093. https://sites.google.com/site/2rvcta/v7-n1-2016/r5 (accessed on 27 May 2024).

23. Broeren, M. L. M., Kuling, L., Worrell, E., & Shen, L. (2017). Environmental impact assessment of six starch plastics focusing on wastewater-derived starch and additives. *Resources, Conservation and Recycling, 127*, 246–255. doi: 10.1016/j.resconrec.2017.09.001.
24. Roohi, Srivastava, P., Bano, K., Zaheer, M. R., & Kuddus, M. (2018). Biodegradable smart biopolymers for food packaging: Sustainable approach toward green environment. In *Bio-Based Materials for Food Packaging* (pp. 197–216). Springer. doi:10.1007/978-981-13-1909-9_9.
25. Hernández-Centeno, F., López-De la Peña, H. Y., Hernández-González, M., Rodríguez-González, C. A., Tirado-Gallegos, J. M., Rios-Velasco, C., & Zamudio-Flores, P. B. (2020). Physicochemical, thermal, rheological and morphological characteristics of flour and starch from a non-conventional source: *Cucurbita foetidissima* Kunth roots. *Journal of Food Measurement and Characterization, 14*, 1976–1985. https://doi.org/10.1007/s11694-020-00444-x.
26. Pfister, B., & Zeeman, S. C. (2016). Formation of starch in plant cells. *Cellular and Molecular Life Sciences, 73*, 2781–2807. https://doi.org/10.1007/s00018-016-2250-x.
27. Jiménez, A., Fabra, M. J., Talens, P., & Chiralt, A. (2012). Edible and biodegradable starch films: A review. *Food and Bioprocess Technology, 5*, 2058–2076. doi:10.1007/s11947-012-0835-4.
28. Khan, B., Bilal Khan Niazi, M., Samin, G., & Jahan, Z. (2012). Thermoplastic starch: a possible biodegradable food packaging material-A Review. *Journal of Food Process Engineering, 40*, e12447. doi:10.1111/jfpe.12447.
29. Tadini, C. C. (2017). Bio-based materials from traditional and nonconventional native and modified starches. In *Starch-Based Materials in Food Packaging, Processing, Characterization and Applications* (pp. 19–36). Elsevier – Academic Press. doi:10.1016/b978-0-12-809439-6.00002-9.
30. Abdul-Khalil, H. P. S., Banerjee, A., Saurabh, C. K., Tye, Y. Y., Suriani, A. B., Mohamed, A., Karim A. A., Rizal, S., Paridah, M. T. (2018). Biodegradable films for fruits and vegetables packaging application: Preparation and properties. *Food Engineering Reviews, 10*, 139-153. doi:10.1007/s12393-018-9180-3.
31. Hammam, A. R. A. (2019). Technological applications and characteristics of edible films and coatings: a review. *SN Applied Sciences, 1*, 632. doi:10.1007/s42452-019-0660-8.
32. Luchese, C. L., Frick, J. M., Patzer, V. L., Spada, J. C., & Tessaro, I. C. (2015). Synthesis and characterization of biofilms using native and modified pinhão starch. *Food Hydrocolloids, 45*, 203–210. doi: 10.1016/j.foodhyd.2014.11.015.
33. Gutiérrez, T. J., Tapia, M. S., Pérez, E., & Famá, L. (2015). Structural and mechanical properties of edible films made from native and modified cush-cush yam and cassava starch. *Food Hydrocolloids, 45*, 211–217. doi: 10.1016/j.foodhyd.2014.11.017.
34. Ginting, M. H. S., Kristiani, M., Amelia, Y., & Hasibuan, R. (2016). The effect of chitosan, sorbitol and heating temperature bioplastic solution on mechanical properties of bioplastic from durian seed starch. *International Journal of Engineering Research and Applications, 6*, 33–38. http://ijera.com/papers/Vol6_issue1/Part%20-%205/D61053338.pdf (accessed on 27 May 2024).
35. Atarés, L., & Chiralt, A. (2016). Essential oils as additives in biodegradable films and coatings for active food packaging. *Trends in Food Science & Technology, 48*, 51–62. doi: 10.1016/j.tifs.2015.12.001.

36. Medina V., Oscar J., Pardo C., Oscar H., Ortiz M., & Cesar A. (2012). Modified arracacha starch films characterization and its potential utilization as food packaging. *Vitae, 19*, 186–196. http://www.scielo.org.co/scielo.php?script=sci_arttext&pid=S0121-40042012000200005&lng=en&tlng=pt (accessed on 27 May 2024).
37. Zhong, Y., & Li, Y. (2014). Effects of glycerol and storage relative humidity on the properties of kudzu starch-based edible films. *Starch – Stärke, 66*, 524–532. doi:10.1002/star.201300202.
38. Zhang, Y., Rempel, C., & Liu, Q. (2014). Thermoplastic starch processing and characteristics—A Review. *Critical Reviews in Food Science and Nutrition, 54*, 1353–1370. doi: 10.1080/10408398.2011.636156.
39. Espinoza-Acosta, J. L., Torres-Chávez, P. I., Ramírez-Wong, B., Bello-Pérez, L. A., Vega-Ríos, A., Carvajal-Millán, E., Plascencia-Jatomea, M., & Ledesma-Osuna, A. I. (2015). Mechanical, thermal, and antioxidant properties of composite films prepared from durum wheat starch and lignin. *Starch – Stärke, 67*(6), 502–511. doi: 10.1002/star.201500009.
40. Zhong, Y., Song, X., & Li, Y. (2011). Antimicrobial, physical and mechanical properties of kudzu starch–chitosan composite films as a function of acid solvent types. *Carbohydrate Polymers, 84*, 335–342. doi: 10.1016/j.carbpol.2010.11.041.
41. Maniglia, B. C., Tessaro, L., Lucas, A. A., & Tapia-Blácido, D. R. (2017). Bioactive films based on babassu mesocarp flour and starch. *Food Hydrocolloids, 70*, 383–391. doi: 10.1016/j.foodhyd.2017.04.022.
42. Hornung, P. S., Masisi, K., Malunga, L. N., Beta, T., & Ribani, R. H. (2018). Natural bioactive starch film from Amazon turmeric (Curcuma longa L.). *Polymer Bulletin, 75*, 4735–4752. doi: 10.1007/s00289-018-2285-2.
43. Hornung, P. S., Ávila, S., Masisi, K., Malunga, L. N., Lazzarotto, M., Schnitzler, E., Ribani, R. H., & Beta, T. (2018). Green development of biodegradable films based on native yam (Dioscoreaceae) starch mixtures. *Starch – Stärke, 70*(7–8), 1700234. doi: 10.1002/star.201700234.
44. Menzel, C., Olsson, E., Plivelic, T. S., Andersson, R., Johansson, C., Kuktaite, R., Järnström; L., & Koch, K. (2013). Molecular structure of citric acid cross-linked starch films. *Carbohydrate Polymers, 96*, 270–276. doi: 10.1016/j.carbpol.2013.03.044.
45. Gutiérrez, T. J. (2018). Biological macromolecule composite films made from sagu starch and flour/poly(ε-caprolactone) blends processed by blending/thermo molding. *Journal of Polymers and the Environment, 26*(10), 3902–3912. doi: 10.1007/s10924-018-1268-6.
46. Avérous, L., & Fringant, C. (2001). Association between plasticized starch and polyesters: Processing and performances of injected biodegradable systems. *Polymer Engineering and Science, 41*(4), 727–734. doi:10.1002/pen.10768.
47. Ortega, F., Giannuzzi, L., Arce, V. B., & García, M. A. (2017). Active composite starch films containing green synthetized silver nanoparticles. *Food Hydrocolloids, 70*, 152–162. doi: 10.1016/j.foodhyd.2017.03.036.
48. Valdés, A., Mellinas, A. C., Ramos, M., Burgos, N., Jiménez, A., & Garrigós, M. C. (2015). Use of herbs, spices and their bioactive compounds in active food packaging. *RSC Advances, 5*, 40324–40335. doi:10.1039/c4ra17286h.
49. Sánchez-González, L., Vargas, M., González-Martínez, C., Chiralt, A., & Cháfer, M. (2011). Use of essential oils in bioactive edible coatings: A review. *Food Engineering Reviews, 3*(1), 1–16. doi:10.1007/s12393-010-9031-3.

50. Perricone M., Arace E., Corbo M. R., Sinigaglia M., & Bevilacqua A. (2015). Bioactivity of essential oils: A review on their interaction with food components. *Frontiers in Microbiology, 6*, 76. doi:10.3389/fmicb.2015.00076.
51. Martin-Gordo, D. A. (2018). The phenolic compounds: An approach to their biosynthesis, synthesis and biological activities. *Revista de Investigacion Agraria y Ambiental, 9*, 81–104. DOI: 10.22490/21456453.1968.
52. Benbettaïeb, N., Debeaufort, F., & Karbowiak, T. (2018). Bioactive edible films for food applications: Mechanisms of antimicrobial and antioxidant activities. *Critical Reviews in Food Science and Nutrition, 59*, 3431–3455. doi:10.1080/10408398.2018.1494132.
53. Sanches-Silva, A., Costa, D., Albuquerque, T. G., Buonocore, G. G., Ramos, F., Castilho, M. C., Machado, A. V., & Costa, H. S. (2014). Trends in the use of natural antioxidants in active food packaging: A review. *Food Additives & Contaminants: Part A, 31*, 374–395. doi: 10.1080/19440049.2013.879215.
54. Salgado, P. R., Ortiz, C. M., Musso, Y. S., Di Giorgio, L., & Mauri, A. N. (2015). Edible films and coatings containing bioactives. *Current Opinion in Food Science, 5*, 86–92. doi: 10.1016/j.cofs.2015.09.004.
55. Hernández-González, M., Pérez-Berumen, C. M., Sánchez-Ruíz, H., Ruíz-Salazar, C. V., Hernández- Paz, J. F., Olivas-Armendáriz, I., Martel-Estrada, S. A., & Rodríguez-González, C. A. (2017). Polysuccinimide functionalized with oregano's essential oil extracts, an antimicrobial extended release bio-material. *Materials Letters, 191*, 73–76. doi: 10.1016/j.matlet.2017.01.035.
56. Mlalila, N., Hilonga, A., Swai, H., Devliéghere, F., & Ragaert, P. (2018). Antimicrobial packaging based on starch, poly(3-hydroxybutyrate) and poly(lactic-co-glycolide) materials and application challenges. *Trends in Food Science & Technology, 74*, 1–11. doi: 10.1016/j.tifs.2018.01.015.
57. Eça, K. S., Sartori, T., & Menegalli, F. C. (2014). Films and edible coatings containing antioxidants – A review. *Brazilian Journal of Food Technology, 17*, 98–112. doi:10.1590/bjft.2014.017.
58. Gómez-Estaca, J., López-de-Dicastillo, C., Hernández-Muñoz, P., Catalá, R., Gavara, R. (2014). Advances in antioxidant active food packaging. *Trends in Food Science & Technology, 35*, 42–51. doi: 10.1016/j.tifs.2013.10.008.
59. Avila-Sosa, R., Palou, E., Jiménez Munguía, M. T., Nevárez-Moorillón, G. V., Navarro Cruz, A. R., López-Malo, A. (2012). Antifungal activity by vapor contact of essential oils added to amaranth, chitosan, or starch edible films. *International Journal of Food Microbiology, 153*, 66–72. doi: 10.1016/j.ijfoodmicro.2011.10.017.
60. Carocho, M., Morales, P., & Ferreira, I. C. F. R. (2015). Natural food additives: Quo vadis? *Trends in Food Science & Technology, 45*, 284–295. doi: 10.1016/j.tifs.2015.06.007.
61. Maisanaba, S., Llana-Ruiz-Cabello, M., Gutiérrez-Praena, D., Pichardo, S., Puerto, M., Prieto, A. I., Jos, A., & Cameán, A. M. (2016). New advances in active packaging incorporated with essential oils or their main components for food preservation. *Food Reviews International, 33*, 447–515. doi:10.1080/87559129.2016.1175010.
62. Ribeiro-Santos, R., Andrade, M., Melo, N. R. de, & Sanches-Silva, A. (2017). Use of essential oils in active food packaging: Recent advances and future trends. *Trends in Food Science & Technology, 61*, 132–140. doi: 10.1016/j.tifs.2016.11.021.
63. Reis, L. C. B., de Souza, C. O., da Silva, J. B. A., Martins, A. C., Nunes, I. L., & Druzian, J. I. (2015). Active biocomposites of cassava starch: The effect of yerba mate extract and mango pulp as antioxidant additives on the properties and the stability of

a packaged product. *Food and Bioproducts Processing, 94,* 382–391. doi: 10.1016/j.fbp.2014.05.004.
64. Rossi, V., Cleeve-Edwards, N., Lundquist, L., Schenker, U., Dubois, C., Humbert, S., & Jolliet, O. (2015). Life cycle assessment of end-of-life options for two biodegradable packaging materials: Sound application of the European waste hierarchy. *Journal of Cleaner Production, 86,* 132–145. doi: 10.1016/j.jclepro.2014.08.049.
65. Ávila-Martín, L., & Beltrán-Osuna, Á. A., Perilla, J. E. (2020). Effect of the addition of citric acid and whey protein isolate in *Canna indica* L. starch films obtained by solvent casting. *Journal of Polymers and the Environment, 28,* 871–883. doi:10.1007/s10924-019-01648-z.
66. Taqi, A., Mutihac, L., & Stamatin, I. (2013). Physical and barrier properties of apple pectin/cassava starch composite films incorporating *Laurus nobilis* L. oil and oleic acid. *Journal of Food Processing and Preservation, 38,* 1982–1993. doi:10.1111/jfpp.12174.
67. Yadav, U., Aarati P., & Anchal, Y. V. (2018). Effect of carom seed oil on the antimicrobial, physicochemical and mechanical properties of starch-based edible film. *International Journal of Environmental Agriculture and Biotechnology, 3,* 1258–1263. doi:10.22161/ijeab/3.4.16.
68. Nogueira, G. F., Soares, C. T., Cavasini, R., Fakhouri, F. M., & de Oliveira, R. A. (2019). Bioactive films of arrowroot starch and blackberry pulp: Physical, mechanical and barrier properties and stability to pH and sterilization. *Food Chemistry, 275,* 417–425. doi: 10.1016/j.foodchem.2018.09.054.
69. Zamudio-Flores, P. B., Ochoa-Reyes, E., Ornelas-Paz, J. D. J., Aparicio-Saguilán, A., Vargas-Torres, A., Bello-Pérez, L. A., rubio Rios, A., & Cárdenas-Félix, R. G. (2015). Effect of storage time on physicochemical and textural properties of sausages covered with oxidized banana starch film with and without betalains. *CyTA – Journal of Food, 13,* 456–463. doi:10.1080/19476337.2014.998713.
70. Medina Jaramillo, C., González Seligra, P., Goyanes, S., Bernal, C., & Famá, L. (2015). Biofilms based on cassava starch containing extract of yerba mate as antioxidant and plasticizer. *Starch – Stärke, 67,* 780–789. doi:10.1002/star.201500033.
71. Medina Jaramillo, C., Gutiérrez, T. J., Goyanes, S., Bernal, C., & Famá, L. (2016). Biodegradability and plasticizing effect of yerba mate extract on cassava starch edible films. *Carbohydrate Polymers, 151,* 150–159. doi: 10.1016/j.carbpol.2016.05.025.
72. Piñeros-Guerrero, N., Piñeros-Castro, Y., & Ortega-Toro, R. (2019). Active biodegradable films based on thermoplastic starch and poly (ε-caprolactone): Technological application of antioxidant extracts from rice husk. *Revista Mexicana de Ingeniería Química, 19,* 1095–1101. https://doi.org/10.24275/rmiq/poli898.
73. Bonilla, J., Talón, E., Atarés, L., Vargas, M., & Chiralt, A. (2013). Effect of the incorporation of antioxidants on physicochemical and antioxidant properties of wheat starch–chitosan films. *Journal of Food Engineering, 118,* 271–278. doi: 10.1016/j.jfoodeng.2013.04.008.
74. De Araújo, G. K. P., de Souza, S. J., da Silva, M. V., Yamashita, F., Gonçalves, O. H., Leimann, F. V., & Shirai, M. A. (2015). Physical, antimicrobial and antioxidant properties of starch-based film containing ethanolic propolis extract. *International Journal of Food Science & Technology, 50,* 2080–2087. doi:10.1111/ijfs.12869.
75. Ortega-Toro, R., Collazo-Bigliardi, S., Roselló, J., Santamarina, P., & Chiralt, A. (2017). Antifungal starch-based edible films containing *Aloe vera. Food Hydrocolloids, 72,* 1–10. doi: 10.1016/j.foodhyd.2017.05.023.

76. Herera-Brandelero, R. P., Yamashita, F., & Grossmann, M. V. E. (2010). The effect of surfactant Tween 80 on the hydrophilicity, water vapor permeation, and the mechanical properties of cassava starch and poly(butylene adipate-co-terephthalate) (PBAT) blend films. *Carbohydrate Polymers, 82,* 1102–1109. doi: 10.1016/j.carbpol.2010.06.034.
77. Sartori, T., & Menegalli, F. C. (2016). Development and characterization of unripe banana starch films incorporated with solid lipid microparticles containing ascorbic acid. *Food Hydrocolloids, 55,* 210–219. doi: 10.1016/j.foodhyd.2015.11.018.
78. Pagno, C. H., de Farias, Y. B., Costa, T. M. H., Rios, A. de O., & Flôres, S. H. (2016). Synthesis of biodegradable films with antioxidant properties based on cassava starch containing bixin nanocapsules. *Journal of Food Science and Technology, 53,* 3197–3205. doi: 10.1007/s13197-016-2294-9.
79. Homayouni, H., Kavoosi, G., & Nassiri, S. M. (2017). Physicochemical, antioxidant and antibacterial properties of dispersion made from tapioca and gelatinized tapioca starch incorporated with carvacrol. *LWT – Food Science and Technology, 77,* 503–509. doi: 10.1016/j.lwt.2016.12.007.
80. Cenobio-Galindo, A., Pimentel-González, D. J., Del Razo-Rodríguez, O. E., Medina-Pérez, G., Carrillo-Inungaray, M. L., Reyes-Munguía, A., & Campos-Montiel, G. (2019). Antioxidant and antibacterial activities of a starch film with bioextracts microencapsulated from cactus fruits (*Opuntia oligacantha*). *Food Science and Biotechnology, 28,* 1553–1561. https://doi.org/10.1007/s10068-019-00586-9.
81. Kechichian, V., Ditchfield, C., Veiga-Santos, P., & Tadini, C. C. (2010). Natural antimicrobial ingredients incorporated in biodegradable films based on cassava starch. *LWT – Food Science and Technology, 43,* 1088–1094. doi: 10.1016/j.lwt.2010.02.014.
82. Ghasemlou, M., Aliheidari, N., Fahmi, R., Shojaee-Aliabadi, S., Keshavarz, B., Cran, M. J., & Khaksar, R. (2013). Physical, mechanical and barrier properties of corn starch films incorporated with plant essential oils. *Carbohydrate Polymers, 98,* 1117–1126. doi: 10.1016/j.carbpol.2013.07.026.
83. Perazzo, K. K. N. C. L., Conceição, A. C. d. V., Santos, J. C. P. d., Assis, D. d. J., Souza, C. O., & Druzian, J. I. (2014). Properties and antioxidant action of actives cassava starch films incorporated with green tea and palm oil extracts. *Plos One, 9*, e105199. https://doi.org/10.1371/journal.pone.0105199.
84. Medina-Jaramillo, C., Ochoa-Yepes, O., Bernal, C., & Famá, L. (2017). Active and smart biodegradable packaging based on starch and natural extracts. *Carbohydrate Polymers, 176,* 187–194. doi: 10.1016/j.carbpol.2017.08.079.
85. Piñeros-Hernandez, D., Medina-Jaramillo, C., López-Córdoba, A., & Goyanes, S. (2017). Edible cassava starch films carrying rosemary antioxidant extracts for potential use as active food packaging. *Food Hydrocolloids, 63,* 488–495. doi: 10.1016/j.foodhyd.2016.09.034.
86. Valencia-Sullca, C., Vargas, M., Atarés, L., & Chiralt, A. (2018). Thermoplastic cassava starch-chitosan bilayer films containing essential oils. *Food Hydrocolloids, 75,* 107–115. doi: 10.1016/j.foodhyd.2017.09.008.
87. Talón, E., Vargas, M., Chiralt, A., & González-Martínez, C. (2019). Antioxidant starch-based films with encapsulated eugenol. Application to sunflower oil preservation. *LWT – Food Science and Technology, 113,* 108290. doi: 10.1016/j.lwt.2019.108290.
88. Kuorwel, K. K., Cran, M. J., Sonneveld, K., Miltz, J., & Bigger, S. W. (2013). Migration of antimicrobial agents from starch-based films into a food simulant. *LWT – Food Science and Technology, 50,* 432–438. doi: 10.1016/j.lwt.2012.08.023.

89. Luchese, C. L., Garrido, T., Spada, J. C., Tessaro, I. C., & de la Caba, K. (2018). Development and characterization of cassava starch films incorporated with blueberry pomace. *International Journal of Biological Macromolecules, 106,* 834–839. doi: 10.1016/j.ijbiomac.2017.08.083.
90. Avila-Sosa, R., Hernández-Zamoran, E., López-Mendoza, I., Palou, E., Munguía, M. T. J., Nevárez-Moorillón, G. V., & López-Malo, A. Fungal Inactivation by Mexican Oregano (*Lippia berlandieri* Schauer) Essential Oil Added to Amaranth, Chitosan, or Starch Edible Films. *Journal of Food Science, 75,* M127–M133. doi: 10.1111/j.1750-3841.2010.01524.x.
91. Ehivet, F. E., Min, B., Park, M. -K., & Oh, J. -H. (2011). Characterization and antimicrobial activity of sweetpotato starch-based edible film containing origanum (*Thymus capitatus*) Oil. *Journal of Food Science, 76,* C178–C184. doi: 10.1111/j.1750-3841.2010.01961.x.
92. Buso-Rios, O. I., Velazquez, G., Jarquín-Enríquez, L., & Flores-Martínez, N. L. (2020). Effect of the concentration of starch and clove essential oil on the physicochemical properties of biodegradable films. *Revista Mexicana de Ingeniería Química, 19,* 1315–1326. https://doi.org/10.24275/rmiq/Alim965.
93. Talón, E., Trifkovic, K. T., Vargas, M., Chiralt, A., & González-Martínez, C. (2017). Release of polyphenols from starch-chitosan based films containing thyme extract. *Carbohydrate Polymers, 175,* 122–130. doi: 10.1016/j.carbpol.2017.07.067.
94. Lozano-Navarro, J., Díaz-Zavala, N., Velasco-Santos, C., Martínez-Hernández, A., Tijerina-Ramos, B., García-Hernández, M., & Reyes-de la Torre, A. (2017). Antimicrobial, optical and mechanical properties of chitosan–starch films with natural extracts. *International Journal of Molecular Sciences, 18,* 997. doi:10.3390/ijms18050997.
95. Chiu, P. -E., & Lai, L. -S. (2010). Antimicrobial activities of tapioca starch/decolorized hsian-tsao leaf gum coatings containing green tea extracts in fruit-based salads, romaine hearts and pork slices. *International Journal of Food Microbiology, 139,* 23–30. doi: 10.1016/j.ijfoodmicro.2010.01.010.
96. Kuorwel, K. K., Cran, M. J., Sonneveld, K., Miltz, J., & Bigger, S. W. (2011). Antimicrobial Activity of Natural Agents Coated on Starch-Based Films against *Staphylococcus aureus*. *Journal of Food Science, 76,* M531–M537. doi: 10.1111/j.1750-3841.2011.02344.x.
97. u-Nisa, I., Ashwar, B. A., Shah, A., Gani, A., Gani, A., & Masoodi, F. A. (2015). Development of potato starch based active packaging films loaded with antioxidants and its effect on shelf life of beef. *Journal of Food Science and Technology, 52,* 7245–7253. doi: 10.1007/s13197-015-1859-3.
98. Tosati, J. V., Messias, V. C., Carvalho, P. I. N., Rodrigues Pollonio, M. A., Meireles, M. A. A., & Monteiro, A. R. (2017). Antimicrobial effect of edible coating blend based on turmeric starch residue and gelatin applied onto fresh frankfurter sausage. *Food Bioprocess Technology, 10,* 2165–2175. doi:10.1007/s11947-017-1985-1.
99. Mehdizadeh, T., Tajik, H., Langroodi, A. M., Molaei, R., & Mahmoudian, A. (2020). Chitosan-starch film containing pomegranate peel extract and *Thymus kotschyanus* essential oil can prolong the shelf life of beef. *Meat Science, 163,* 108073. doi: 10.1016/j.meatsci.2020.108073.
100. Pająk, P., Socha, R., Łakoma, P., & Fortuna, T. (2016). Antioxidant properties of apple slices stored in starch-based films. *International Journal of Food Properties, 20,* 1117–1128. doi:10.1080/10942912.2016.1203931.

101. Campos-Requena, V. H., Rivas, B. L., Pérez, M. A., Figueroa, C. R., Figueroa, N. E., & Sanfuentes, E. A. (2017). Thermoplastic starch/clay nanocomposites loaded with essential oil constituents as packaging for strawberries – In vivo antimicrobial synergy over *Botrytis cinerea*. *Postharvest Biology and Technology, 129,* 29–36. doi: 10.1016/j.postharvbio.2017.03.005.
102. Mehyar, G. F., Al-Ismail, K., Han, J. H., & Chee, G. W. (2012). Characterization of edible coatings consisting of pea starch, whey protein isolate, and carnauba wax and their effects on oil rancidity and sensory properties of walnuts and pine nuts. *Journal of Food Science, 77,* E52–E59. doi: 10.1111/j.1750-3841.2011.02559.x.
103. Enríquez C., M., Velasco M., R., & Ortiz G., V. (2012). Películas biodegradables basadas en almidón. *Biotecnología en el Sector Agropecuario y Agroindustrial, 10,* 182–192. http://www.scielo.org.co/pdf/bsaa/v10n1/v10n1a21.pdf (accessed on 27 May 2024).
104. González-Seligra, P. (2016). Nanocompuestos de base almidón con aplicaciones en envases biodegradables. PhD dissertation. Universidad de Buenos Aires, Ciudad de Buenos Aires, Argentina.
105. Souza, C. O., Silva, L. T., Silva, J. R., López, J. A., Veiga-Santos, P., & Druzian, J. I. (2011). Mango and acerola pulps as antioxidant additives in cassava starch bio-based film. *Journal of Agricultural and Food Chemistry, 59,* 2248–2254. doi:10.1021/jf1040405.
106. Ollé-Resa, C. P., Gerschenson, L. N., & Jagus, R. J. (2012). Effect of natamycin on physical properties of starch edible films and their effect on *Saccharomyces cerevisiae* Activity. *Food Bioprocess Technology, 6,* 3124–3133. doi:10.1007/s11947-012-0960-0.
107. Ollé Resa, C. P., Jagus, R. J., & Gerschenson, L. N. (2014). Effect of natamycin, nisin and glycerol on the physicochemical properties, roughness and hydrophobicity of tapioca starch edible films. *Materials Science and Engineering: C., 40*, 281–287. https://doi.org/10.1016/j.msec.2014.04.005.
108. Ollé Resa, C. P., Gerschenson, L. N., & Jagus, R. J. (2016). Starch edible film supporting natamycin and nisin for improving microbiological stability of refrigerated Argentinian Port Salut cheese. *Food Control, 59*, 737–742. https://doi.org/10.1016/j.foodcont.2015.06.056.
109. Martínez-Ortiz, M. A., Vargas-Torres, A., Román-Gutiérrez, A. D., Chavarría-Hernández, N., Zamudio-Flores, P. B., Meza-Nieto, M., & Palma-Rodríguez, H. M. (2017). Partial characterization of chayotextle starch-based films added with ascorbic acid encapsulated in resistant starch. *International Journal of Biological Macromolecules, 98*, 341–347. https://doi.org/10.1016/j.ijbiomac.2017.02.016.
110. Pyla, R., Kim, T. -J., Silva, J. L., & Jung, Y. -S. (2010). Enhanced antimicrobial activity of starch-based film impregnated with thermally processed tannic acid, a strong antioxidant. *International Journal of Food Microbiology, 137*, 154–160. https://doi.org/10.1016/j.ijfoodmicro.2009.12.011.
111. Tirado-Gallegos, J. M., Zamudio-Flores, P. B., Ornelas-Paz, J. J., Rios-Velasco, C., Olivas Orozco, G. I., Espino-Díaz, M., Baeza-Jiménez, R., Buenrostro-Figueroa, J. J., Aguilar-González, M. A., Lardizábal-Gutiérrez, D., Hernández-González, M., Hernández-Centeno, F., & López-De la Peña, H. Y. (2018). Elaboration and characterization of active apple starch films incorporated with ellagic acid. *Coatings, 8*, 384. https://doi.org/10.3390/coatings8110384.

112. Jiménez, A., Fabra, M. J., Talens, P., & Chiralt, A. (2013). Physical properties and antioxidant capacity of starch–sodium caseinate films containing lipids. *Journal of Food Engineering, 116*(3), 695–702. https://doi.org/10.1016/j.jfoodeng.2013.01.010.
113. Pagno, C. H., Costa, T. M. H., de Menezes, E. W., Benvenutti, E. V., Hertz, P. F., Matte, C. R., Tosati, J. V., Monteiro, A. R., Rios, A. O., & Flôres, S. H. (2015). Development of active biofilms of quinoa (*Chenopodium quinoa* W.) starch containing gold nanoparticles and evaluation of antimicrobial activity. *Food Chemistry, 173*, 755–762. https://doi.org/10.1016/j.foodchem.2014.10.068.
114. Marie-Arockianathan, P., Sekar, S., Sankar, S., Kumaran, B., & Sastry, T. P. (2012). Evaluation of biocomposite films containing alginate and sago starch impregnated with silver nanoparticles. *Carbohydrate Polymers, 90*, 717–724. https://doi.org/10.1016/j.carbpol.2012.06.003.
115. Peighambardoust, S. J., Peighambardoust, S. H., Pournasir, N., & Mohammadzadeh Pakdel, P. (2019). Properties of active starch-based films incorporating a combination of Ag, ZnO and CuO nanoparticles for potential use in food packaging applications. *Food Packaging and Shelf Life, 22*, 100420. https://doi.org/10.1016/j.fpsl.2019.100420.
116. Melo, P. E. F., Silva, A. P. M., Marques, F. P., Ribeiro, P. R. V., Filho, M. de Sá M. S., Brito, E. S., Lima, J. R., & Azeredo, H. M. C. (2019). Antioxidant films from mango kernel components. *Food Hydrocolloids, 95*, 487–495. https://doi.org/10.1016/j.foodhyd.2019.04.061.
117. Daudt, R. M., Avena-Bustillos, R. J., Williams, T., Wood, D. F., Külkamp-Guerreiro, I. C., Marczak, L. D. F., & McHugh, T. H. (2016). Comparative study on properties of edible films based on pinhão (*Araucaria angustifolia*) starch and flour. *Food Hydrocolloids, 60*, 279–287. https://doi.org/10.1016/j.foodhyd.2016.03.040.
118. Terrazas-Hernandez, J. A., Berrios, J. D. J., Glenn, G. M., Imam, S. H., Wood, D., Bello-Pérez, L. A., & Vargas-Torres, A. (2014). Properties of cast films made of chayote (*Sechium edule* Sw.) tuber starch reinforced with cellulose nanocrystals. *Journal of Polymers and the Environment, 23*(1), 30–37. https://doi.org/10.1007/s10924-014-0652-0.
119. Martínez-Ortiz, M. A., Palma-Rodríguez, H. M., Montalvo-González, E., Sáyago-Ayerdi, S. G., Utrilla-Coello, R., & Vargas-Torres, A. (2019). Effect of using microencapsulated ascorbic acid in coatings based on resistant starch chayotextle on the quality of guava fruit. *Scientia Horticulturae, 256*, 108604. https://doi.org/10.1016/j.scienta.2019.108604.
120. Zamudio-Flores, P. B., Ochoa-Reyes, E., Ornelas-Paz, J. de J., Tirado-Gallegos, J. M., Bello-Pérez, L. A., Rubio-Ríos, A., & Cárdenas-Felix, R. G. (2015). Caracterización fisicoquímica, mecánica y estructural de películas de almidones oxidados de avena y plátano adicionadas con betalaínas. *Agrociencia, 49*, 483–498. Retrieved from: http://www.scielo.org.mx/scielo.php?script=sci_arttext&pid=S1405-31952015000500002 (accessed on 27 May 2024).

CHAPTER 11

Agrochemicals as a Risk to Food Security in Mexico: A Case Study

CRESCENCIO URÍAS-GARCÍA,[1*] LAURA MARÍA NAVARRO FLORES,[1] SANDRA PÉREZ ÁLVAREZ,[2] and EDUARDO FIDEL HÉCTOR ARDISANA[3]

[1]*Universidad Vizcaya de las Américas, Campus Delicias, Cd. Delicias Chihuahua, Mexico*

[2]*Facultad de Ciencias Agrícolas y Forestales, Universidad Autónoma de Chihuahua, Cd. Delicias Chihuahua, Mexico*

[3]*Instituto de Posgrado, Universidad Técnica de Manabí, Portoviejo, Ecuador*

*Corresponding author.

ABSTRACT

Agrochemicals, fertilizers, and fuels are the main expense items for farmers in different regions, representing 19% of the production costs of annual crops according to the 2017 Agricultural Census. These inputs are becoming increasingly expensive, highlighting the relationship of each crop with the use or application of agrochemicals and fertilizers. This is why close to 670 million people will continue to suffer from hunger in 2030, which is 8% of the world's population. Approximately 3.7% of the Mexican population lives in severe food insecurity (almost a quarter of the population – 33 million people) due to the increase in food prices, contributing to the levels of extreme poverty. The pandemic crisis caused by the SARS-CoV-2 virus has also led to an increase in severe or extreme malnutrition by 10.3%. Similarly, the use of beneficial organisms such as antagonistic and entomopathogenic fungi in agriculture is relevant to reducing the applications of agricultural pesticides. Additionally, the use of phytoremediation has been implemented using technologies based on the use of plants to restore contaminated environments

Food Security, Safety, and Sustainability: Validated Innovative Approaches and Technological Advancements. José Juan Buenrostro-Figueroa and Julio César Tafolla-Arellano (Eds.)

such as water, soil, and air. These sustainable technologies aim to reduce the dangerous concentration of organic and inorganic contaminants in the soil through biochemical processes carried out by plants and microorganisms, helping to produce healthy crops.

11.1 INTRODUCTION

La demand for agricultural industries has increased and with it the increase in the use of agricultural inputs such as pesticides and fertilizers that are currently generating an impact on the environment and consumer health [1]. The application of pesticides continues to be the most effective and accepted practice for crop protection, contributing significantly to the increase in agricultural productivity. On average, each producer uses 3.7 ± 3 active ingredients (AI) of pesticides [2]. In tomato cultivation and to a lesser extent in flowers, corn, beans, and alfalfa, a diversity of pesticides from different groups and chemical categories are used. In Mexico, the main agricultural states that register the highest use of pesticides are Sinaloa, Chiapas, Colima, Jalisco, Nayarit, Sonora, Tamaulipas, and Veracruz [3, 4]. Fertilizers are the main agricultural input used to increase crop productivity, whose chemical components are heavy metals [5]. According to Saynes-Santillan et al. [6], in Mexico, more than half of the soils have low fertility, which makes it necessary to apply fertilizers in 68% of the cultivated soils [6].

It has been investigated that the degree of pesticide uptake within crops, vegetables, and fruits depends on the nature of the pesticide functional group and the surrounding environmental conditions. Aguilar et al. [1] describe that after application, some fruits and vegetables retain pesticides on the surface of the skin. In some cases, washing with water and disinfectant could help reduce the amount of agrochemicals. However, there are agrochemicals that generate a very strong bond between the functional group and the fruits, which makes washing less effective. In cereal grains and legumes, the presence of agrochemicals in the bran has been found, and in some cases where the pesticide has lipophilic properties, it could move inside the grain [1].

The excessive application of fertilizers has generated problems of water, soil, and air contamination. For example, the contamination of water and soil allows the bioaccumulation of metals in plants. These accumulate in the tissues and organs of the animals that ingest them. These animals later go on to produce meat, milk, eggs, and fish, becoming consumed by man. On the other hand, industrial emissions and waste are the main causes of air contamination of metals in food [3].

The health of the Mexican population is exposed through water, air, and food intake contaminated by AIs from agrochemicals and fertilizers. The toxic effects on humans depend on the chemical group used, the concentration, and in some cases the age or physiological condition of the exposed population [7]. Sánchez et al. [8] observed the bioaccumulation of pesticides in breast milk and adipose tissue of Mexican women exposed to pesticides, with higher concentrations found in blood, plasma, and urine in exposed occupational workers than in unexposed controls [8].

Therefore, technical factors, such as the use of agrochemicals in food production, could be an obstacle that prevents sustainable food security. The foregoing demonstrates the importance of knowing which AI of the agro-inputs are used for the care and production of crops and the impact it causes on the food security of Mexicans, as well as bioremediation and phytoremediation strategies for soils to reduce contamination from agro-inputs.

11.2 FOOD SECURITY IN MEXICO

11.2.1 CURRENT STATUS OF MEXICO'S FOOD SECURITY INDICATORS

According to the Food and Agriculture Organization of the United Nations (FAO), food security can be defined in terms of the availability, accessibility, and biological use of food plus the stability of these three factors [9]. The following is an in-depth explanation of the dimensions taken into account when analyzing the food security of a country.

11.2.2 FOOD AVAILABILITY IN MEXICO

Availability means that there is food in sufficient quantity and quality (whether through local production and/or imports) to meet the country's demand [10]. According to the information reported in 2020 in the FAO's balance sheet, Mexico's food production, exports, and imports add up to a total of 204,202,740 metric tons of food, of which 90,720,540 metric tons are used for human consumption [11]. These are mainly composed of corn-derived products, dairy, fruits, and vegetables, and represent an average daily food intake of 103.7 g of lipids (29.4% of total calories), 91.16 g of protein (11.5%), and 468.3 g of carbohydrates (59.1%) [3] for a total of 3,171 kilocalories, showing that the available food allows for a caloric intake 40% higher than the minimum requirements and with a balanced macronutrient intake [12].

11.2.3 FOOD ACCESSIBILITY IN MEXICO

Accessibility means that people live under conditions that allow them to get enough food to meet their needs, be it by its purchase or production. People's ability to access food is analyzed from both a physical and economic standpoint. Economic accessibility means people can acquire enough food to maintain adequate intake while meeting their other needs. In most households, the highest monthly expense was food, drink, and tobacco, adding up to 3,793 Mexican pesos per family, an average of 1,581 Mexican pesos per family member [13]. In rural communities, the average monthly expense was lower by 329 Mexican pesos [14]. In order to meet a baseline of 2,191 kcal/day for rural areas and 2073.8 kcal/day for urban areas, an average expense of 1,194 and 1,666 Mexican pesos was needed [15].

Comparing the estimated monthly expense per person and the needed amount to meet basic food intake, it can be seen that in the rural areas the expenditure per person was enough to meet the minimum requirements, while in the urban areas, the expenditure per person was below what was needed to meet the minimum requirements. Supporting this, the results of the 2020 National Household Income and Expense Poll (ENIGH) show that 47% of all households had some difficulty meeting their food requirements due to a lack of money or resources [15]. This could be due to an increase in immigration from indigenous communities to cities in the last 20 years. People from indigenous communities tend to have irregular jobs with low salaries and no benefits, which keeps them poor and makes them more vulnerable to any rise in food prices [16].

Physical accessibility, on the other hand, is about people having the physical ability to access the food they need [17]. In a rural context, it is difficult to access food, especially if the community is small and far away from a city [18]. In 2021, around 14,043 communities had some form of disadvantage in this regard, meaning 52.84% of the total 26,573 communities surveyed [19]. Given the fragility of the market and logistics in rural areas, accessibility difficulties can cause nutritional effects.

11.2.4 BIOLOGICAL USE OF FOOD IN MEXICO

The third parameter, biological use, means the appropriate use of food and nutrients, and at the same time, it is about the biological ability to make use of food for physical well-being [20]. Effective food utilization depends largely on in-home knowledge about food storage and processing techniques and the basic principles of nutrition [21].

The amount of energy that a person requires per day depends on different factors, such as age, weight, height, physical activity, and health status. Therefore, there must be a balance between food intake and energy expenditure; an imbalance in this equation is reflected in an increase or decrease in the person's weight. The average calorie needs for an average healthy adult in Mexico are 2,500 kcal. In addition, the diet requires daily consumption of fruits and vegetables, seeds, and the choice of whole grains, as well as legumes, to cover the requirement of micronutrients, essential fatty acids, and fiber, and to obtain the correct functioning of the body. In 2018, the National Institute of Public Health reported that around 50% of adults aged 20 years and over consumed legumes, fruits, and vegetables. In addition, 85% reported the consumption of sugary drinks and 64.6% reported the consumption of meat [11].

The inappropriate use of food can cause a biological risk leading to the appearance of metabolic diseases. In 2018, data showed that more than 70% of the population aged 20 years and over were overweight and obese, with higher rates in women. Around 18.4% and 10.3% had a medical diagnosis of arterial hypertension and diabetes, respectively. Additionally, 19.5% had high concentrations of cholesterol and triglycerides. These data indicate that the majority of Mexicans have a positive energy balance, inadequate food consumption, and a distribution of monetary spending on food and drinks within the home according to certain rules [11]. The National Institute of Statistics and Geography reports that around 50% of food spending is allocated to the purchase of meat, cereals, and other diverse foods such as rice, oatmeal cereal, bananas, and apples. The lack of consumption of fruits and vegetables in the diet, as well as other protein sources like eggs, could contribute to the onset of these diseases [15].

11.2.5 STABILITY OF FOOD SECURITY FACTORS IN MEXICO: A LANDSCAPE OF VULNERABILITY

Stability means constant and uninterrupted access to food; there shouldn't be a possibility or risk of running out of food as a consequence of a crisis. Because of this, stability means that the other parameters we talked about should be sustainable [10]. In recent years, Mexico has increased external dependence to meet the demands of the most consumed products. Official data show that the annual tons of imported staple foods, such as corn, wheat, eggs, beans, rice, barley, and poultry meat, grew, respectively by 104%, 12.4%, 260%, 24%, 31%, 54%, and 45% from 2010 to 2020 [21]. In 2021, the variation in the importing of agricultural goods was 17.7% higher than last year [22]. Food security in Mexico is vulnerable to the growth of food imports in the country since an increase in world prices could compromise the availability of food in the country, depending on national food production [23]. In 2020, the National Index of Consumer Prices registered a yearly inflation rate of 6.8% on food, drink, and tobacco. Agricultural goods prices increased by 3.6% in the same period [24]. During 2021, the yearly inflation rate and the increase in agricultural goods were 7.40% and 15.78%, respectively [21]. In 2021, 20.6% of Mexican households reduced the amount of food due to the decrease in economic income because of the global contingency [21]. Compared to 2018, in 2020, the average income fell by 8.0% in urban communities; a significant portion of this fall was due to a decrease in wages. In rural communities, income rose by 3.6%, mainly due to donations and government spending [15]. This circumstance discourages the possibility of developing a dynamic in which decreases in monetary resources are compensated by means other than adjusting public spending.

11.2.6 FOOD SECURITY IN MEXICO

Food Insecurity (FI) means that the availability of nutritionally adequate food or access to it is limited [20]. If we analyze the information we have available, we can see that Mexico has enough supply to cover the energy needs of its population; however, food distribution and access are not equal due to economic and logistic factors. Price volatility can also produce disturbances in supply chains, making access to food more difficult. If these disturbances last for a long time, dietary quality could be sacrificed. According to the 2020 National Health and Nutrition Survey, 59.1% of Mexican households showed some sign of Food Insecurity, of which 38.5% were mild, 11.4% moderate, and 9.4% severe [14]. Compared to 2018, FI increased by 8% in urban households in 2020. Rural FI showed no significant changes [25].

The goal of food security is to produce enough food at a low cost, guaranteeing food supply to reduce dependence on imported food and generate self-sufficiency. Generating agricultural programs that support small and medium-sized producers with access to low-cost inputs and the strengthening of infrastructure, as well as improving the economy of the country, could be strategies to increase the national production of quality basic foods at low cost for the consumer, generate self-sufficiency, and reduce food insecurity in the country [26]. It should also place emphasis on easing access to more diverse and healthy diets with the objective of improving the health of the Mexican population [27].

11.3 AGROCHEMICALS USED IN MEXICO AND THEIR RISKS

The use of agrochemicals in the agricultural sector is one of the most common practices in the primary activity of the region, being the cause of effects on environmental health and society, in this case, that of rural day laborers, mainly [28]. Pesticides (insecticides, fungicides, and herbicides) and synthetic fertilizers have increased in a worrying way. Likewise, the promotion of this model has fostered corporate control of the entire food system, leaving behind the production of crops with zero tillage for the production of healthy foods. Making an indiscriminate use of synthetic fertilizers caused great damage to soils and ecosystems [29]. Therefore, the consumption of food and the intake of water with the presence of toxic residues have caused a series of accidental poisonings, and even the routine use of pesticides that can generate risks to human health either in the short or long term [30, 31].

11.3.1 PESTICIDES AND FUNGICIDES

According to Räsänen [173]; and López Dávila et al. [32], pesticides have been a factor in the fight against many human diseases and in the fight against hunger. Given that the positive effects of pesticides have allowed access to more quality foods to a broader sector of the population [32, 33]. There is a wide diversity of pesticides and each of them has a different mechanism of action; likewise, its negative impact on health is related to its toxicological potential, which generates effects associated with signs and symptoms of acute pesticide poisoning (IAP) [2, 34].

In Mexico, the chemical groups and active ingredients used in various crops that stand out for their high toxicity are organophosphates (methyl parathion, methamidophos, chlorpyrifos, dimethoate, monocrotophos), carbamates (carbofuran, methomyl), dithiocarbamates (mancozeb), organochlorines (endosulfan), pyrethroids (cypermethrin, lambda cyhalothrin), bipyridyls (paraquat), phosphomethyl-glycine (glyphosate), and chlorophenoxy (2,4-D), among others [3].

11.3.2 TYPES AND CHARACTERISTICS OF PESTICIDES

According to Rodríguez et al. [2], Table 11.1 identifies 37 Active Ingredients (AI) of pesticides. On average, each producer uses 3.7 ± 3.0 different AI. Around 5.4% correspond to Toxicological Category (TC) I, 16.2% to TC II and III, respectively, 54% to CT IV, and 8.1% to CT V. Regarding chemical classification, 18% correspond to organophosphates; 10% to pyrethroids and benzimidazoles, respectively; 8% to carbamates, and 5% to organochlorines. The rest belong to a diversity of chemical groups (36%), while 13% use commercial products that contain two active ingredients from different chemical groups. Regarding the application of crops in the municipality of San Baltazar Chihicápam, Oaxaca, the crops with the most biocidal applications are the tomato crop, where the greatest diversity of pesticides is applied (23 AI), followed by corn (17 AI), beans, and alfalfa (15 AI for each), flowers (12 AI), chickpea (3 AI), and avocado (2 AI). The use of the organophosphate insecticide methamidophos (CT II) predominates in corn, beans, alfalfa, and flowers. In corn, the use of herbicides paraquat, glyphosate, and 2,4-D also stands out. Mainly in tomato cultivation and to a lesser extent in flowers, corn, beans, and alfalfa, a diversity of pesticides from different chemical groups and categories are used. Toxicologically, insecticides stand out: carbofuran (carbamate, CT II), abamectin (pentacycline, CT II), bifenthrin (pyrethroid, CT IV), methomyl (carbamate CT II), Lambda-cyhalothrin (pyrethroid, CT IV), and fungicides: mancozeb (dithiocarbamate, CT IV), and captan (carboxamide, CT IV).

Table 11.1 shows the classification of pesticides according to compound type (structure) [35].

11.3.3 DAMAGE TO HEALTH

According to Ordoñes-Beltrán et al. [36], they carried out an investigation to determine the use of pesticides and their damage to human health, revealing that some phosphate compounds have been associated with different health issues due to pesticide exposure. Thus, different groups such as children, pregnant women, or aging populations may be more susceptible to the effects of various pesticides.

Studies conducted by Mill [175] cited by Ordoñes-Beltrán et al. [36] indicated that frequent use of the fungicide Captan leads to an increased incidence of leukemia and prostate cancer, also showing associations between this fungicide and the occurrence of multiple myeloma (plasma cell cancer).

TABLE 11.1 Characteristics of pesticides and type of toxicity. M (Corn) n = 39; F (Bean) n = 39; A (alfalfa) n = 34; G (Chickpea) n = 14; T (Tomato) n = 10; F (Flowers) n = 10; A (Avocado) n = 1; Total n = 50; TC: Toxicological Category; Type of Action: H = Herbicide, I = Insecticide, F = fungicide; (1) High Acute Toxicity; (2) Chronic Effects on Human Health; (3) Environmental Toxicity; and (4) Restricted or Prohibited by Environmental Agreements; Pesticides Authorized in Mexico That are Prohibited in Other Countries [2]

					Cultivo					1				Número de
		H/I/			M	F	A	G	T	F	A	Total	Lista	Países
CT2	IA	Nombre Comercial	Grupo Químico	3	(%)	(%)	(%)	(%)	(%)	(%)	(%)	(%)	PAP4	prohi-bidos5
I	Paration metilico	Foley, paratión metílico,	Organofosforado	I	5.1	33.3	29.4	7.1	20.0	10.0	100	34.0	1,4	59
I	Azinfos metilo	Gusatión	Organofosforado	I				7.1				2.0	1,3,4	39
II	Metamidofos	Monitor 600, tamaron,	Organofosforado	I	38.4	58.9	67.6		30.0	50.0		74.0	1,3,4	49
II	Oxamyl	Vydate	Carbamato	I					20.0			4.0	1,3	3
II	Endosulfán	Tridane	Organoclorado	I						10.0		2.0	1,4	75
II	Carbofuran	Furadan	Carbamato	I	15.3	7.6	5.8		20.0	30.0	100	26.0	1,3,4	49
II	Metomilo	Lannate	Carbamato	I	5.1	5.1	5.8					4.0	1,3	13
II	Abamectina	Agrimec, biomec,	Pentaciclina	I					40.0			8.0	1,3	
III	Paraquat	Diabloquat, gramoxone	Bipiridilo	H	23.0	5.1						20.0	1,4	38
III	2-4 D	Hierbamina, herbipol	Clorofenoxi	H	15.3	7.6						14.0		3
III	Clorpirifos etil + permetrina	Disparo	Organofosforado + Piretroide	I	2.5							2.0	3/2,3	2/29
III	Cipermetrina	Arrivo	Piretroide	I		2.5	5.8			10.0		4.0	3	
III	Deltametrina	Decis	Piretroide	I					10.0	10.0		2.0	2,3	
III	Dimetoato	Dimetoato, rogueor	Organofosforado	I	2.5		5.8		10.0			6.0	3	4
IV	Tiocianometiltio benzotiazol	Busan 30 WB	Benzimidazol	F					20.0			4.0		
IV	Thiametoxam + lambda	Engeo	Neonicotinoide + piretroide	I					20.0	20.0		4.0	3/1,2,	
IV	Imidacloprid + betacyflutrin	Muralla max	Coronicotinilo + piretroide	I		2.5			20.0			8.0	3/1,3	–/29

TABLE 11.1 *(Continued)*

					Cultivo					1			Número de	
				H/I/	M	F	A	G	T	F	A	Total	Lista	Países
CT2	IA	Nombre Comercial	Grupo Químico	3	(%)	(%)	(%)	(%)	(%)	(%)	(%)	(%)	PAP4	prohi-bidos5
IV	Lambda cyhalotrina	Karate	Piretroide	I	2.5		2.9		20.0	20.0		8.0	1,2,3	
IV	Clorpirifos etil	Lorsban	Organofosforado	I					10.0	10.0		2.0	3	2
IV	Bifentrina	Talstar	Piretroide	I	2.5	7.6	2.9		60.0	60.0		18.0	2,3	2
IV	Diazinon	Diazinon	Organofosforado	I			2.9					2.0	2,3	30
IV	Imidacloprid	Confidor	Imida	I						10.0		4.0	3	
IV	Tiabendazol	Tecto	Benzimidazol	F	2.5				10.0	10.0		6.0	-	1
IV	Glifosato	Coloso, faena	Fosfonometil-glicina	H	23	15.3	2.9					22.0	2	1
IV	Malathión	Malathión	Organofosforado	I	2.5	2.5	5.8	7.1				6.0	2,3	2
IV	Azoxistrobin	Amistar	Pirimidina	F	5.1	2.5	2.9			10.0		6.0	-	-
IV	Captan	Captan	Carboxamida	F	2.5	10.2	5.8		20.0			14.0	-	6
IV	Carbendazim	Prozycar	Benzimidazol	F		2.5				10.0		4.0	2	29
IV	Cymoxanil	Curzate	Sal inorgánica de cobre	F					30.0			6.0	-	-
IV	Mancozeb	Manzate, ridomil	Ditiocarbamato	F	10.2	10.2	8.8		50.0	10.0		24.0	2	1
IV	Clorotalonil	Bravo 720, Balear	Cloronitrilos	F					30.0			2.0	1,2	3
IV	Benomilo	Promyl	Benzimidazol	F			2.9		10.0			2.0	1	33
IV	Tebuconazole + trifloxystrobin	Consist max	Triazol + estrobilurinas	F					10.0	10.0		4.0	-	-
IV	Myclobutanil	Rally	Triazol	F	2.5							2.0	-	-
V	Imidacloprid + deltametrina	New leverage	Cloronicotinilo + piretroide	I					20.0			2.0	3/2,3	-/-
V	Spinosad	Spintor	Spinosines	I					20.0	20.0		4.0	3	-
V	Spirotetramat	Movento	Ácido tetrónico	F					10.0			2.0	-	-

Among the main agricultural states that report higher pesticide use are Sinaloa, Chiapas, Colima, Jalisco, Nayarit, Sonora, Tamaulipas, and Veracruz [3, 4].

According to Díaz-Vallejo et al. [4], they identified various agrochemicals that have been banned in different countries and that, following the Stockholm and Rotterdam conventions, should be prohibited in Mexico due to their effects on human health and the environment. However, these chemicals still have current health registrations, such as Carbofuran (30 registrations with indeterminate validity), Endosulfan (importation prohibited as of 2019), Methamidophos (79 registrations with indeterminate validity), Monocrotophos (36 with indeterminate validity), Methyl Parathion (importation prohibited as of 2019, but with 175 records with indeterminate validity), among others.

11.3.4 AGROCHEMICALS USED IN MEXICO AND THEIR RISKS

The use of agrochemicals in agriculture is a common practice in the region, leading to environmental and societal effects, particularly on rural day laborers [28]. The increase in pesticides (insecticides, fungicides, and herbicides) and synthetic fertilizers has raised concerns. This model promotes corporate control of the food system, shifting away from zero tillage crop production for healthier foods. Indiscriminate use of synthetic fertilizers damages soils and ecosystems [29]. The presence of toxic residues in food and water has led to accidental poisoning. Routine pesticide use poses short and long-term health risks [30, 31].

11.4 SOIL CONTAMINATION BY AGROCHEMICALS

Agrochemicals can be defined as products such as fertilizers, hormones, fungicides, nematicides, insecticides, etc., that are used to develop plant growth and yield [37]. These products accumulate in soil, crops, underground water, and can be sprayed into the air, causing toxic effects on humans, animals, plants, and soil microorganisms (Figure 11.1) [38].

Agriculture should focus on meeting the needs of a world population that continues to increase, thus reducing the number of malnourished people and promoting health and well-being. Therefore, more agrochemicals, intensive irrigation, genetically modified crops, among other advances, will be used. Organic agriculture could be a solution to the problem of contamination

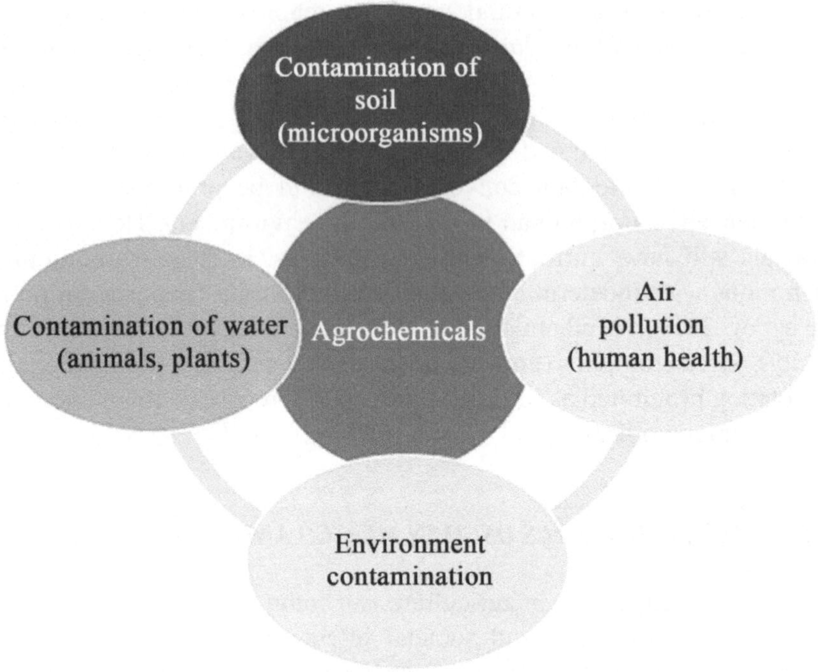

FIGURE 11.1 Agrochemicals in the ecosystem.

caused by agrochemicals because it avoids their use and is based on obtaining food "free" of chemicals and, therefore, healthier. However, this agriculture does not conceive the production of crops in mass or the quantity of calories necessary to feed humankind. Therefore, organic agriculture contributes to improving food security, but it does not help to feed the entire world population [39].

11.4.1 FERTILIZERS

Chemical fertilizers contribute to the increase in crop yields, which leads to an increase in food production in the world [40].

Nitrogen is one of the major soil contaminants applied through chemical fertilization of crops, since it is a necessary component of all living organisms. The accessible forms of nitrogen change according to the particular organism. Non-reactive compounds (nitrogen gas N_2) can be assimilated by microorganisms. Crops require compounds with greater chemical availability like NO_3^- (nitrate) and NH_4^+ (ammonium), whereas animals need more complex forms (nucleic acids and amino acids) [41].

Phosphorus (P) is another contaminant, and it is a macronutrient very important for all living entities. It is part of RNA and DNA and is used to carry cellular energy via ATP (adenosine triphosphate) [42].

Excessive applications of fertilizers lead to increased soil salinity, eutrophication of water, increased nitrates, accumulation of heavy metals, and can become a cause of environmental contamination and risk to human health. Fertilizers are also considered a source of heavy metals like Pb, Hg, Ni, Cu, As, and Cd, and radionuclides such as 238U (uranium), 232Th (Thorium), and 210Po (Polonium). Appropriate management and administration of fertilizers is essential to avoid soil contamination [43].

According to Fan et al. [44], the application of organic fertilizers compared with the application of 100% chemical fertilizers can increase the yield of tomato crop (*Solanum lycopersicum* L.) by 3.48%, soil organic matter by 24.43%, soil bacteria by 5.94%, acidic soil (pH < 6) by 7.98%, total P by 23.97%, total N_2 by 32.79%, total potassium (K) by 44.91%, available K by 16.21%, available P by 14.46%, catalase by 17.68%, and urease by 22.32%; all this translates into higher crop yields and improved soil quality.

Phaseolus vulgaris L. is a legume of great significance for human ingestion around the world and it is also a source of vegetal protein, bioactive compounds, antioxidants, and minerals. This crop develops symbiosis with bacteria like *Rhizobium,* which allows the N_2-fixation and decreases the request for synthetic N fertilizers to enhance quality and yield. However, fertilization is not the only way; yield and quality of this crop can be improved by several agronomic practices like planting density, use of rhizobia, irrigation, etc. [45].

11.4.2 PESTICIDES

Pesticides are substances or mixtures of substances that prevent or control any pest that damages crops, their products, their processing, or storage [46]. The application of pesticides helps the growth of food production to supply the increasing population; however, their excessive use triggers negative effects not only on the environment but also on human health [47, 48].

It is also important to highlight the negative effects of pesticides on soil organisms [49, 50], among which glyphosate stands out, an herbicide that alters the ability of mycorrhization of arbuscular fungi and affects the regulation of gene expression related to the metabolism of amino acids and other processes. Similarly, it harms various species of earthworms that live in the soil, affecting their ability to recycle organic matter [51] (Table 11.2).

TABLE 11.2 Classification of the Main Pesticides According to the Type of Pest or Target Group

Pesticides	Classification
Insecticides	• Organophosphate
	• Organochlorines
	• Carbamates
	• Pyrethroids
Fungicides	• Organochlorines
	• Mercurial organ
Herbicides	• Bipyridyls
	• Organochlorines
	• Organophosphates

According to the chemical nature of the pesticides they can be classified in a more specific way [52], as shown in Table 11.3 [53, 54].

TABLE 11.3 Classification of Pesticides According to the Chemical Nature

Chemical Structure	Examples
Organophosphates	Diazinon, omethoate, glyphosate, malathion, or parathion
Organochlorines	Endosulfan, hexachlorobenzene
Thiocarbamide derivatives and carbamic	Carbaryl, aldicarb, carbofuran
Urea derivatives	Fenuron, metoxuron, diuron, linuron, monuron
Carboxylic acids and their derivatives	Pentanal, butanamide, butanamide
Phenol and nitrophenol derivatives	Dinocap, dinoseb
Heterocyclic compounds	Triazole derivatives, etc.
Copper-containing compounds	Macc 80, caocobre
Fluorine-containing compounds	Dichlofluanid, cryolite
Synthetic pyrethroids	Cypermethrin, fluvalinate
Hydrocarbons, ketones, and some others	Benzene, cerenox, toluene
Metal-organic compounds	Maneb, Nabam, mancozeb

11.4.3 ORGANOCHLORINE PESTICIDES (OCPS)

Organochlorines are the most used pesticides. They are persistent in the environment and have slow biodegradability. Their half-life is up to five years, depending on the type of product [35]. These compounds cause damage to the soil due to direct application to crops, causing the loss of soil productivity by altering its chemical properties (pH, acidity, fertility, and organic matter) [46].

Peanuts (*Arachis hypogaea* L.) can be contaminated with organochlorine pesticides (OCPs) in China. Some of the most used pesticides during the cultivation phase of the crops are metolachlor, quizalofop-p-ethyl, and pirimicarb, resulting in higher concentrations of those products in the vegetable. However, when evaluating the risk for the health of age groups that ingested peanuts and peanut oil, it turned out that the risk was very low [55].

The proportions of 11 OCPs were studied in several vegetables, and authors found that some of the pesticides analyzed exceeded maximum residue limits in Welsh onion (*Allium cepa* L.) (22.50%), cucumber (*Cucumis sativus* L.) (3.23%), Chinese cabbage (*Brassica rapa* L. subsp. Pekinensis) (14.29%), tomato (*Solanum lycopersicum* L.) (2.78%), eggplant (*Beta vulgaris* L.) (2.94%), radish (*Raphanus sativus* L.) (18.75%), and pepper (*Capsicum annuum* L.) (6.90%). The sources of OCPs were identified, where DDTs originated mostly from recently imported dicofol, whereas hexachlorocyclohexanes (HCHs) came mostly from lindane [56].

Shaibu et al. [57] studied the OCP levels in tomato fruits (fresh and dried) in Nigeria, revealing the presence of OCP residues including isodrin 0.129 mg kg^{-1}, α HCH 3.669 mg kg^{-1}, β HCH 0.0100 mg kg^{-1}, DDM 0.0067 mg kg^{-1}, Endosulfan I and II 0.1714 mg kg^{-1}, mitotane 0.2734 mg kg^{-1}, and dieldrin 0.0054 mg kg^{-1}. Three studied products (isodrin, mitotane, and endosulfan) exceeded the limits established for OCPs in Nigeria in fresh samples but not in dry samples, whereas dieldrin, mitotane, and endosulfan II did not exceed the limits in dried tomato extracts.

11.4.4 ORGANOPHOSPHORUS COMPOUNDS

Organophosphorus compounds are esters of phosphoric acid and are derived from the chemical structure of phosphorus; they are biodegradable and easily hydrolyzed in a humid environment and alkaline pH [58].

According to Rivera et al. [59], corn is one of the most cultivated crops in Mexico, so the health of farmers may be affected by the wide use of pesticides. Regarding this situation, the effects of pesticides on several farmers were studied in Tlaxcala, Mexico, through the analysis of the buccal micronucleus cytome (an indicator of cytotoxicity and genotoxicity) and nuclear abnormalities. Around 21 commercial products were identified, mostly organophosphate compounds, triazine, and chlorophenoxy. It was also detected that some farmers use carbofuran, as well as fungicides and insecticides. Similarly, farmers who used more pesticides had more cytotoxic and genotoxic damage compared to farmers who used agroecological practices.

A study on the absorption, translocation, and accumulation of organophosphate esters (OPE) in corn (*Zea mays* L.) and rice (*Oryza sativa* L.) showed that the concentration of the product was higher in rice (221 ± 62.0 ng g^{-1}) than in maize (149 ± 31.6 ng g^{-1}). The OPE in rice increased linearly according to the growth period, and the translocation was from roots to leaves and finally to grains [60].

11.4.5 CARBAMIC AND THIOCARBAMIDE DERIVATIVES

Carbamic acids and thiocarbamides are compounds that contain an amide group and a carbamate or thiocarbamate group, respectively. Carbamic acids are derived from the reaction of an amine and an acid, while thiocarbamides are derived from the reaction of a thiol and an acid. These compounds are used in many different industrial applications, such as catalysts, preservatives, and agricultural chemicals [61–63].

The health risks of carbofuran pesticide exposure were investigated in 485 farmers in Thailand by Boonkhao et al. [64]. The quantity of carbofuran in several samples of soil was below 0.01 mg kg^{-1}. The daily dose of carbofuran ingested by farmers in the studied villages oscillated between 3.9×10^{-9} mg kg^{-1}-day to 5 × 10–9 mg kg^{-1}-day, being an acceptable level of risk for health according to the hazard quotient (HQ). However, it was established that 98.97% of the farmers consumed food and 97.53% drank water while cultivating.

Cabbage (*Brassica oleracea* L.) is a vegetable that, when harvested, is subjected to cold storage (refrigerated rooms) to maintain shelf life, but pests frequently occur during storage, so they are treated with insecticides and pesticides that contain carbofuran. This insecticide is also used to improve the shine of the fruit and to make it better marketed. For these reasons, the effect of storage with low temperature (cold) and the degradation of carbofuran on cabbage was studied. In the investigation, it was found that the level of carbofuran residues declined with an intensification in the period of storage and that 30 days are needed for the product degradation [65].

11.4.6 FUNGICIDES

The use of systemic fungicides has increased greatly in recent years, despite the fact that pathogens have developed resistance and negative effects on the environment and human health [66].

This type of agrochemicals includes a large number of products with various modes of action that cause toxicity in living organisms (aquatic or terrestrial). Exposure experiments have been carried out in rats, finding biochemical, endocrine, hematological, and histopathological disruptive effects [33]. Several fungicides are forbidden in the European Union, and they are considered dangerous chemicals by the World Health Organization (WHO) [66].

Copper (Cu) and sulfur (S) are the most used and studied inorganic fungicides. Both can be found in different forms, and they are used in crops to control not only foliage but also fruit diseases. These fungicides can be absorbed by sediments and soils and run off into groundwater and surface water. Both Cu and S are important nutrients for various functions in crops, but high concentrations can be detrimental.

Copper can harm beneficial fungi and bacteria found in the environment [67]. For example, the use of high doses of Cu-based pesticides can harm arbuscular mycorrhizal fungi (AMF) [68].

Seed germination and seedling development can be reduced by Cu, for example in sunflower (*Helianthus annuus* L.) Cu is toxic to seedlings because it can induce oxidative stress [69]. The germination of bean seeds (*Phaseolus vulgaris* L.) can be reduced by Cu because it affects the cotyledonary carbohydrate status [70].

Sulfur, like Cu, can harm fungi and bacteria that are beneficial to the environment, meaning they are not crop pests, and it can also be phytotoxic to several plants, such as apricots (*Prunus armeniaca* L.), cucurbits, and raspberries (*Rubus idaeus* L.) [71, 72]. *Vibrio fischeri* (gram-negative bacteria) is sensitive to sulfur exposure [73].

The environmental fate of organic fungicides, such as phenylamides, succinate-dehydrogenase inhibitors, dicarboximides, phosphonates, depends on several physicochemical parameters, including water solubility, ionization (pKa), volatility, half-life in soil and water (DT_{50}), and K_{ow} [74].

Two of the fungi that mainly affect tomato crop development and growth are *Alternaria alternata* (Fr.) Keissl. (1912) and *Stemphylium lycopersici* (*Enjoji*) W. Yamam. (1960). Mancozeb is an organic contact fungicide with a broad spectrum that displays great fungicidal activity in several horticultural crops, including tomatoes. This product interrupts respiration and lipid metabolism by affecting the sulfhydryl groups of enzymes and amino acids in fungal cells. Due to its solubility in water, it can contaminate water bodies [75]. Mancozeb (dithiocarbamates) in living matter cannot exceed the limit of 0.01–25 ppm [76]; however, residue levels higher than the permissible limit have been found in tomatoes [77].

11.4.7 INSECTICIDES

Insecticides are defined as substances that kill insects [78]. They include larvicides and ovicides used against larvae and insect eggs, respectively. Insecticides have many uses, such as agriculture, industry, consumers, and medicine. Also, they have contributed to the increase in crop production in the 20th century [79].

Several authors have reported the resistance of some pathogens to chemical insecticides. *Spodoptera frugiperda* Walker is an insect pest of economically important crops in many countries, but in some of them, like Brazil, this species in *Z. mays* is considered the most destructive pest, so spinosyn is an insecticide frequently used, causing insect resistance [80]. Also, resistance of this insect to organophosphate and pyrethroid [81], chlorantraniliprole [82], spinetoram [83], and emamectin benzoate [84] has been reported.

Pyrethroids are very common insecticides used in urban sites to control ants, cockroaches, mosquitoes, and spiders [85]. Contamination with this product in urban sites has been reported for several countries such as the USA, China, Australia, and Vietnam [86–90]. Also, pollution with pyrethroids and atrazine has been reported in China, mostly in greenhouses (80% and 60% respectively) compared with open-field soils [91].

In the Portuguese cabbage crop, after applying deltamethrin, the effect of this product on the soil was determined, and the study showed that deltamethrin application appears to have caused a change in the soil microbiota community because the bacteria *Sphingomonas* sp. and *Nocardioides* sp. increased exponentially; both bacteria degrade the pyrethroid and its metabolites [92]. Similarly, the negative impact of pyrethroids (cypermethrin, cyhalothrin, and deltamethrin) and its metabolite phenoxybenzoic acid (3-PBA) in a concentration range between 50 and 500 μg kg^{-1} was evaluated on *C. sativus,* specifically in seed germination, shoot length, root elongation, and leaf length; 3-PBA affected seed germination, deltamethrin only affected root length, cypermethrin impacted seed germination, root elongation, and leaf length, and cypermethrin and deltamethrin increased chlorophylls and total carotenoids median values [93].

11.4.8 NEMATICIDES

Nematodes that parasitize plants cause enormous economic losses in important crops around the world and also have a negative impact on modern agricultural development. Chemical nematicides are the most used for the effective control of nematodes, but their continuous use, especially carbamates and organophosphates, has caused resistance in these pathogens [94].

Nematodes belonging to *Meloidogyne* induce root-knot and are responsible for the maximum pathogenicity to most hosts. They have been mostly controlled by the use of synthetic nematicides such as carbamates (oxamyl), methyl bromide, 1,3-dichloropropene, and organophosphates, as well as by agronomic practices like resistant varieties and crop rotation [95]. On the other hand, the use of synthetic nematicides is being monitored due to their high toxicity [96, 97]. The current requirements for the low presence or absence of these nematicides in the food chain have led to a greater demand for the use of alternatives with low or no toxicity [95, 98].

Aldicarb is a very toxic nematicide, specifically its residual metabolite (aldicarb sulfoxide). Between 1985 and 1988, three outbreaks of food poisoning occurred in California, USA, induced by the ingestion of watermelons (*Citrullus lanatus* (Thunb.) Matsum. and Nakai, 1916) or cucumbers (*C. sativus*) treated with aldicarb, with estimated doses between 0.0011 and 0.06 mg kg^{-1} of body weight, most of which were well below the lowest observed effect level (LOEL) of 0.025 mg kg^{-1} [99].

Several nematicides such as fenamiphos and cadusafos have higher acute LD_{50} values (lower toxicity) of 8.0 mg kg^{-1} and 37.1 mg kg^{-1}, respectively, for rats by oral administration [100], but are still categorized as highly toxic to humans.

Fluopyram is a non-fumigant nematicide applied to crops to reduce yield losses produced by parasitic nematodes [101]. Unfortunately, free-living nematodes can be affected too [102, 103] and one example is the application of this nematicide in turfgrass (*Cynodon dactylon* L.) giving as a result negative effects on free-living nematode population densities [104].

11.5 SOIL BIOREMEDIATION AND PHYTOREMEDIATION TO REDUCE CONTAMINATION BY AGROCHEMICALS

According to the most recent data from FAO [105], the world consumption of pesticides in that year was 1.81 kg/ha and 0.37 kg per capita (Figure 11.2). However, this is a world average, closely related to the magnitude of the national economies and the level of wealth; thus, very poor countries such as Tanzania, Haiti, and Lesotho (not shown) barely use these products in their agriculture, while others such as Japan and Germany use high volumes per cultivated hectare. The figures for per capita consumption obviously depend on the population of each country, and on this basis, data from very poor countries such as Vanuatu are roughly similar to those of Japan.

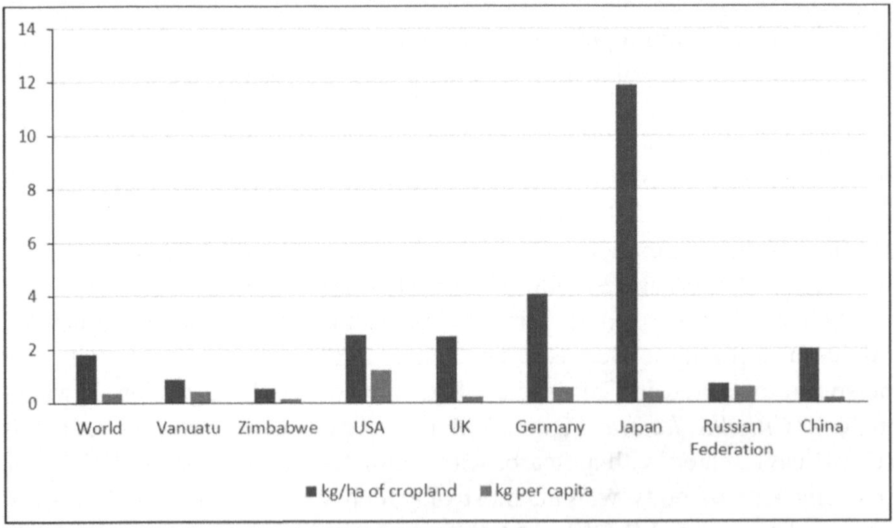

FIGURE 11.2 Consumption of pesticides (global and some countries).

Source: Adapted from Ref. [105].

In any case, the use of high volumes of pesticides is worrying because a significant part of these products never reaches the target organism and instead remains in the soil [106] and can be integrated into the food chain. An example of this is organochlorine pesticides (OCPs), among which are some already banned such as dichlorodiphenyltrichloroethane (DDT) and hexachlorocyclohexane (HCH), and others whose use is authorized, such as aldrin, dieldrin, and endosulfan, among others; their permanence in the soil in some cases can exceed 30 years, mainly due to their liposolubility [107]. Herbicides such as atrazine and trifluralin can remain in the soil from a few days to a year, depending on humidity and temperature conditions, but instead they dissipate easily in the soil, contaminating large areas and penetrating underground water [108].

The high costs of physicochemical processes to mitigate soil contamination and their low relative efficiency have led to the use of plants, microorganisms (bacteria, fungi, algae), and combinations of them for this purpose [109].

11.5.1 BIOREMEDIATION

Bioremediation is a process in which soil microorganisms totally or partially break down contaminants. The pathways by which this decomposition occurs [110] are:

1. **Bioattenuation:** The degradation of pollutants occurs naturally, with the intervention of endogenous soil microbes.
2. **Biostimulation:** It is stimulated by creating favorable conditions for the development of degrading microorganisms, such as the supply of carbon, nitrogen, phosphorus, oxygen, and the control of pH and temperature.
3. **Bioaugmentation:** The degrading action is increased by adding an inoculum of microorganisms to the soil, which generally comes from contaminated soil.

In general, bioremediation is achieved by the degradation of the contaminant to less toxic forms until it becomes carbon dioxide and water. This process occurs by oxidation, in which electrons are transferred from the pesticide to various acceptors, with the intervention of microbial enzymes [111].

Plant growth-promoting rhizobacteria (PGPR) can be useful in the bioremediation of contaminated soils in two ways. PGPRs directly degrade or transform contaminants through hydrolytic enzymes, chelation, or biotransformation; indirectly, they make contaminants more accessible for absorption by plants, or produce substances that stimulate growth, which increases vigor and biomass production by plants and, as a consequence, phytoremediation potential [112]. Additionally, PGPR synthesize 1-amino-cyclopropane-1-carboxylate (ACC) deaminase, an enzyme that degrades ACC (ethylene precursor); by reducing ethylene levels in the rhizosphere, the plant suffers less from the stress that results from exposure to the pollutant [113]. Several bacterial genera have the ability to degrade pesticides, Pseudomonas being the most active (Table 11.4).

TABLE 11.4 Some Bacterial Genera That Degrade Pesticides

Pesticide	Bacterial Genera	References
Chlorpyrifos	*Aeromonas*, *Microbacterium*, *Paenibacillus*, *Bacillus*, *Paracoccus*, *Pseudomonas*	[107, 113–116]
Atrazine	*Bacillus*, *Nocardioides*, *Arthrobacter*, *Bradyrhizobium*, *Methylobacterium*	[117, 118]
Glyphosate	*Burkholderia*, *Providencia*, *Chryzeobacterium*	[119–121]
Metolachlor	*Bacillus*	[122]
Aldrin	*Pseudomonas*	[123, 124]
Dieldrin	*Pseudomonas*	[124]
Endosulfan	*Pseudomonas*, *Pseudonocardia*, *Alcaligenes*	[124–126]

Source: Various.

In bioremediation, the potential uses of fungi are less recognized [127]. However, some genera and species with this property have been identified. Under in vitro conditions, *Phlebia acanthocystis, Phlebia brevispora,* and *Phlebia aurea* removed 50% of the dieldrin and 90% of the supplied aldrin [128]. *Cordyceps militaris* and *Cordyceps brongniartii* hydrolyze dieldrin to less toxic products [129]. *Lentinula edodes* is capable of degrading the agro-chemicals pendimethalin, terbutylazine, difenoconazole, and diflufenican [130]. *Pleurotus ostreatus* completely removes aldrin and between 3 and 18% of dieldrin in the culture medium [131]. *Mucor racemosus* degrades *in vitro* more than 90% of the dieldrin existing in the culture medium, and *Mortiriella* sp. reduces endosulfan by 50–70% [132]. *Trichoderma harzianum* degrades endosulfan [133]. *Pleurotus eryngii* and *Coprinus comatus* extracted 87% of endosulfan from soil contaminated with both this pesticide and cadmium [134].

Laccase, manganese peroxidases, and lignin peroxidases act in the fungal enzymatic degradation of pesticides and other contaminants [135]. The synthesis of ACC deaminase is also carried out by plant growth-promoting fungi (PGPF) [136]. Also, indirectly, fungi of various genera such as *Trichoderma* can contribute to the reduction of pesticides in the soil since they can be used as biological pesticides [137].

11.5.2 PHYTOREMEDIATION

For almost 300 years, the ability of plants to renew polluted soil and water has been perceived [138]. The term phytoremediation is used to define the ability of plants to remove contaminants from the soil and is appreciated as an ecological alternative to physical-chemical methods [139]. Plants have various mechanisms for phytoremediation (Table 11.5).

TABLE 11.5 Phytoremediation Mechanisms

Mechanism	Description
Phytoextraction	The contaminants are absorbed by the roots and mobilized to the aerial parts.
Phytodegradation	Pollutants are broken down by plant enzymes.
Phytostabilization	The pollutants are absorbed by the roots (mainly grasses) avoiding their dispersion in the ecosystem.
Phytovolatilization	Pollutants are eliminated in gaseous form through transpiration.
Rhizodegradation	Pollutants are broken down in the rhizosphere of plants, generally with the contribution of microorganisms.

Source: Adapted from: Srivastav et al. [139].

Various genera and species of plants have demonstrated their efficacy in the remediation of soils contaminated with insecticides such as DDT [140], cypermethrin [141], lindane [142], HCH [143], and endosulfan [144].

Despite the fact that the terms "phytoremediation" and "bioremediation" appear in the literature on these topics, in practice there is no clear dividing line between the two. This is because one of the most prominent mechanisms in the removal of contaminants is rhizodegradation, in which microorganisms living in the rhizosphere play an important role. Plants improve the living conditions of these microbes through several processes: they oxygenate the rhizosphere, which favors aerobic organisms; they provide organic carbon, they contribute enzymes, carbohydrates, amino acids, and other molecules in radical exudates [145]. Mycorrhizal fungi and bacteria grow in the rhizosphere and use these substances in their metabolism, degrading various types of pollutants, including pesticides, so they can be used in soil remediation [146].

On the other hand, the potential of endophytic microorganisms is also recognized as a source for phytoremediation. The bacteria and fungi that live inside the plant collaborate with it in numerous detoxification mechanisms, which, once the corresponding associations have been established, can be used in soil phytoremediation [147].

In such a way, phytoremediation is actually a combined effect resulting from the interaction of plants and microbes in the rhizosphere and endosphere, and which is based on beneficial effects for both types of organisms [148].

11.5.3 NOVEL TECHNOLOGIES USING LIVING ORGANISMS FOR SOIL REMEDIATION GENETIC ENGINEERING

The potential of genetic engineering has been tested for more than 20 years to increase the phytoremediation capacity of plants with the introduction of foreign genes. In this direction, two main pathways have been worked on: the synthesis of glutathione, which participates in the detoxification of pesticides and organic pollutants [149], and the monooxygenases of mammalian cytochrome P450.

Resistance to the herbicides atrazine and pyriminobac-methyl has been induced in transgenic potato plants expressing CYP1A1/yeast NADPH-cytochrome P450 reductase (YR)-fused enzyme [150]; in tobacco, alfalfa, and *Arabidopsis thaliana*, a bacterial gene (atzA) converts them into atrazine degraders [151]; the introduction of human cytochrome P450 genes into rice plants increases their tolerance to various herbicides [152].

More recently, other advances have been made in this field. Resistance to the herbicide simazine was induced in *Arabidopsis thaliana* by the introduction of the human P450 gene CYP1A2 [153]. Also, a transgenic *A. thaliana* was created from genetic information of the bacterium *Sphingobium* sp. This transgenic plant is capable of expressing the N-demethylase PdmAB of the bacteria, and with it demethylate the herbicide isoproturon to 3-(4-isopropylphenyl)-1-methyl-urea; this product, once released by the roots, continues to be degraded by *Sphingobium* sp. in the rhizosphere [154]. The introduction of the ginseng PgCYP76B93 gene into *Arabidopsis thaliana* resulted in increased tolerance of transgenic plants to the herbicide chlorotoluron under *in vitro* conditions [155].

However, the ethical barrier represented by the introduction of animal and human genes in plants has put a brake on the practical application of these results [156], as well as the difficulties encountered to maintain the expression of the genes in the hosts [157], and the fact that a mixture of contaminants often occurs in soils, so the use of a transgenic organism specialized in the degradation of one of them is usually not enough [158].

In exchange, there is the possibility of obtaining transgenic microorganisms and using them for remediation instead of traditional ones, which would overcome ethical and practical limitations, and would expand the possibilities of collaboration between microbes and plants [159]. For example, a transgenic *Sphingobium japonicum* is capable of degrading cadmium and the herbicides chlorpyrifos and hexachlorocyclohexane simultaneously [160].

11.5.4 NANOTECHNOLOGY

The nanoproducts exhibit a high reaction efficiency due to their small size and their high surface/volume ratio. The interest in obtaining bio-nanoparticles from microorganisms and applying them in soil remediation has increased in recent years.

In soil remediation, the use of nanomaterials can contribute in three ways [161]:

- Direct removal of contaminants, through adsorption of contaminants by nanoparticles or their destruction by redox reactions.
- Promotion of plant growth by nanomaterials, which increases plant biomass and with it its phytoremediation capacity.
- Increased accessibility of plants to contaminants, by conversion of contaminants to assimilable forms.

The photocatalysis of pesticides such as hexachlorobenzene, dicofol, and cypermethrin is increased with the use of nano-TiO_2 particles [162]. Similar results have been reported with multiwalled carbon nanotubes for metolachlor degradation [163] and with zerovalent iron nanoparticles for DDT degradation [164]. The application of multiwalled carbon nanotubes to the *Lactuca sativa* planting substrate significantly reduced the uptake of chlordane, transchlordane, and trans-nonachlor [165]. The addition of zerovalent iron nanoparticles (1,000 mg Kg^{-1} of soil) in soil contaminated with endosulfan facilitated the removal of 82% of the pesticide by *Alpinia calcarata* plants. The nanoparticles helped most of the degradation to occur in the soil itself, without significant accumulation of endosulfan in the plant [166].

Although it is a technology in full development and is appreciated as useful in the remediation process, further studies will be required for its reasonable use, preventing nanoparticles from becoming contaminants or entering the food chain [167].

KEYWORDS

- **bio-nanoparticles**
- **cypermethrin**
- **endosulfan**
- **hexachlorobenzene**
- **multiwalled carbon nanotubes**
- **photocatalysis**
- **phytoremediation capacity**
- **transgenic plants**

REFERENCES

1. Aguilar, M., & Salgado, E. (2020). Obstaculos a la Seguridad Alimentaria Sostenible en América Latina. *Revista Centroamericana de Administración Pública, 78*, 134–148. https://doi.org/10.35485/rcap78_5.
2. Rodríguez, M. K., Zavaleta, D., Torres, H., Reyes, L., & Bernardino, H. U. (2020). Use of pesticides and acute poisoning in the rural population of San Baltazar Chichicápam, Oaxaca, Mexico. *AIDIS Magazine of Engineering and Environmental Sciences. Research, Development & Practice, 13*(2), 616–629. https://doi.org/10.22201/iingen.0718378xe.2020.13.2.68117.
3. García-Hernández, J., Leyva, M. J. B., Martínez, R. I. E., Hernández, O. M. I., Aldana, M. M. L., & Rojas, G. A. E. (2018). Current state of research on pesticides in Mexico. *Revista Internacional de Contaminación Ambiental, 34*, 29–60. Retrieved from: https://www.revistascca.unam.mx/rica/index.php/rica/article/view/RICA.2018.34.esp01.03/46727 (accessed on 27 May 2024).
4. Díaz-Vallejo, J., Barraza-Villarreal, A., Yáñez-Estrada, L., & Hernández-Cadena, L. (2021). Pesticides in food: Health risk and regulatory framework in Veracruz, Mexico. *Public Health Mex, 63*, 86–497. https://doi.org/10.21149/12297.

5. García-Salazar, J. A., Borja-Bravo, M., & Rodríguez-Licea, G. (2018). Consumo de fertilizantes en el sector agrícola de México: Un estudio sobre los factores que afectan la tasa de adopción. *Interciencia, 43*, 505–510.
6. Saynes-Santillan, V., Turrent-Fernández, A., & Etchevers-Barraet, J. (2019). Fertilización sostenible con nitrógeno y fósforo en México. Elementos para apolíticas públicas. *Revista Mexicana de Ciencias Agrícolas, 3*, 1–15.
7. Reyes, Y. C., Vergara, I., Torres, O. E., Díaz, M., & Gonzalez, E. E. (2016). Contaminación por metales pesados: implicaciones en salud, ambiente y seguridad alimentaria. *Revista Ingeniería, Investigación y Desarrollo, 16*, 66–77.
8. Sánchez-Alarcón, J., Milic, M., Kasuba, V., Tenorio-Arvide, M. G., Montiel-González, J. M. R., Bonassi, S., & Valencia-Quintana, R. (2021). A systemic review of studies on genotoxicity and related biomarkers in populations exposed to pesticides in Mexico. *Toxics, 9*(11), 272. https://doi.org/10.3390/toxics9110272.
9. Novotny, I. P., Tittonell, P., Fuentes-Ponce, M. H., López-Ridaura, S., & Rossing, W. A. H. (2021). The importance of the traditional milpa in food security and nutritional self-sufficiency in the highlands of Oaxaca, México. *PLoS ONE, 16*(1), e0246281. https://doi.org/10.1371/journal.pone.0246281.
10. Pérez, O. A. (2020). La competitividad y la seguridad alimentaria en México. *Economía coyuntural, 5*, 119–142. https://doi.org/10.5281/zenodo.3746608.
11. Food and Agriculture Organization of the United Nations. (2020). *Food Security 2020*. Retrieved from https://www.fao.org/3/y5061e/y5061e08.htm (accessed on 24 May 2024).
12. Urquia-Fernández, N. (2014). La seguridad alimentaria en México. *Salud Pública de México, 56*, 92–98.
13. Instituto Nacional de Estadística y Geografía. (2020). *Estadística a Propósito del día Mundial de la Alimentación*. Retrieved from: https://www.inegi.org.mx/contenidos/saladeprensa/aproposito/2020/EAPAlimentacion.pdf (accessed on 27 May 2024).
14. Encuesta Nacional de Salud y Nutrición. (2021). *Encuesta Nacional de Salud y Nutrición 2020 sobre Covid-19*. Retrieved from: https://ensanut.insp.mx/encuestas/ensanutcontinua2020/doctos/informes/ensanutCovid19ResultadosNacionales.pdf (accessed on 27 May 2024).
15. Instituto Nacional de Estadística y Geografía. (2020). *Índice de Precios al Consumidor de la Canasta de Consumo Mínimo*. Retrieved from: https://www.inegi.org.mx/contenidos/productos/prod_serv/contenidos/espanol/bvinegi/productos/nueva_estruc/702825196929.pdf (accessed on 27 May 2024).
16. Mundo-Rosas, V., Unar-Munguía, M., Hernández-F, M., Pérez-Escamilla, R., & Shamah-Levy, T. (2019). La seguridad alimentaria en los hogares en pobreza de México: Una mirada desde el acceso, la disponibilidad y el consumo. *Salud pública de México, 61*, 866–875. https://doi.org/10.21149/10579.
17. Consejo Nacional de Evaluación de la Política de Desarrollo Social. (2012). *Informe de Evaluación de la Política de Desarrollo Social en México*. Retrieved from: https://www.senado.gob.mx/comisiones/desarrollo_social/docs/CONEVAL1.pdf (accessed on 27 May 2024).
18. Santana-Cardenas, S., & López-Uriarte, P. (2021). Inseguridad alimentaria y calidad de vida en México: una revisión de estudios con enfoque cualitativo. *Journal de Ciencias Sociales, 16*, 4–20. https://doi.org/10.18682/jcs.vi16.4373.

19. Secretaria de Agricultura y Desarrollo Rural. (2021). Comercio exterior agropecuario y oportunidades de exportación. https://www.gob.mx/cms/uploads/attachment/file/640419/Balanza_por_paises_feb_2021.pdf (accessed on 27 May 2024).
20. Romero, N. M., Sangerman-Jarquín, D. M., Hernández, M., León, A., & Escalon, M. J. (2019). Vulnerabilidad alimentaria en hogares rurales y su relación con la política alimentaria en México. *Revista Mexicana de Ciencias Agrícolas, 10*, 935–945. https://doi.org/10.29312/remexca.v10i4.1746.
21. Instituto Nacional de Estadística y Geografía. (2019). *Encuesta Nacional de Salud y Nutrición 2018 Presentación de Resultados 2019*. Retrieved from: https://ensanut.insp.mx/encuestas/ensanut2018/doctos/informes/ensanut_2018_presentacion_resultados.pdf (accessed on 27 May 2024).
21. Organización de las Naciones Unidas para la Alimentación y la Agricultura. (2020). Balances de alimentos. https://www.fao.org/faostat/es/#data/FBS (accessed on 27 May 2024).
22. Secretaria de Agricultura y Desarrollo Rural. (2021). Programa Institucional 2020-2024 de DICONSA S.A. de C.V. Avance y resultados. https://www.gob.mx/cms/uploads/attachment/file/640419/Balanza_por_paises_feb_2021.pdf (accessed on 27 May 2024).
23. Torres, F., & Rojas, A. (2020). Seguridad alimentaria y sus desequilibrios regionales en México. *Revista Latinoamericana de Economía, 51*, 57–83. https://doi.org/10.22201/iiec.20078951e.2020.201.69521.
24. Clavenilla, J. L. (2022). Choques de oferta, estanflación y credibilidad en la post-pandemia. *Economía UNAM, 19*, 41–52. https://doi.org/10.22201/fe.24488143e.2022.57.742.
25. Instituto Nacional de Salud Pública. (2018). *Seguridad Alimentaria en Hogares Mexicanos*. Retrieved from: https://insp.mx/assets/documents/webinars/2021/CIEE_Seguridad_alimentaria.pdf (accessed on 27 May 2024).
26. Shamah Levy, T., Méndez-Gómez, I., Mundo-Rosas, V., Rodríguez-Ramírez, S., & Gaona-Pineda, E. B. (2021). Factores asociados con el cambio en la inseguridad alimentaria en México: Ensanut 2012 y 2018-19. *Salud Publica Mex, 63*, 350–358. https://doi.org/10.21149/12145.
27. Sosa, A., & Ruíz, G. (2017). La disponibilidad de alimentos en México: un análisis de la producción agrícola de 35 años y su proyección para 2050. *Papeles de Población, 93*, 207–230. https://doi.org/10.22185/24487147.2017.93.027.
28. Martínez, C., Romano, G., Cuadras, A. A., & Ortega, L. D. (2019). Pesticides, health and environmental impact in Sinaloa (Mexico): Implications and challenges in environmental governance. *Transcontinental Human Trajectories, 4*. https://doi.org/10.25965/trahs.1615.
29. Arellana-Aguilar, O., Ponce de León, H. C., Rendon Von Osten, J., Gonzáles, M., Dzul, R., & Mercado, R. (2016). The footprint of pesticides in Mexico. Greenpeace. Retrieved from: https://www.greenpeace.org/static/planet4-mexicostateless/2018/11/30b49459-30b49459-plaguicidas_en_agua_ok_em.pdf (accessed on 27 May 2024).
30. Damalas, C. A., & Eleftherohorinos, I. G. (2011). Pesticide exposure, safety issues, and risk assessment indicators. *International Journal of Environmental Research and Public Health, 8*, 1402–1419. https://doi.org/10.3390/ijerph8051402.
31. González, P. (2019). Effect of pesticides on human health, exposure and impacts. *Library of the National Congress of Chile/BCN. Parliamentary Technical Advisory*, 1–8. Retrieved from: https://obtienearchivo.bcn.cl/obtienearchivo?id=reposit

orio/10221/26823/2/Efecto_de_los_plaguicidas_en_la_Salud.pdf (accessed on 27 May 2024).

32. López, E., Martínez, Y., & Romero, O. (2022). Characteristics and adverse consequences to human health of agrochemicals used in Cuban agriculture. *Cuban Journal of Public Health, 48*, 1–24. https://revsaludpublica.sld.cu/index.php/spu/article/view/2810/1791 (accessed on 27 May 2024).
33. Nicolopoulou-Stamati, P., Maipas, S., Kotampasi, C., Stamatis, P., & Hens, L. (2016). Chemical pesticides and human health: The urgent need for a new concept in agriculture. *Frontiers in Public Health, 4*, 148. https://doi.org/10.3389/fpubh.2016.00148.
34. Plenge, T., & Vargas, M. (2003). Toxic effect of agricultural pesticides on muscle relaxation: Sarcoplasmic reticulum Ca2+-ATPase studies (SERCA). *Ciencia en Frontera: science and technology magazine of the UACJ, 2*, 75–79.
35. Ramírez, J. Y. (2001). Plaguicidas: clasificación, uso, toxicología y medición. *Archivos de Prevención de Riesgos Laborales, 4*, 67–75.
36. Ordoñes-Beltrán, V., Frías-Moreno, M., Parra, H., & Martínez-Tapia, M. (2019). Study on the use of pesticides and their possible relationship with damage to health. *Journal of Toxicology, 36*(3), 148–153. https://www.redalyc.org/journal/919/91967023011/html/ (accessed on 27 May 2024).
37. Biswas, S. K., Rahman, S., Kobir, S. A., Ferdous, T., & Banu, N. A. (2014). A review on impact of agrochemicals on human health and environment: Bangladesh perspective. *Plant Environment Development, 3*, 31–35.
38. Gnanaprakasam, P. D., & Vanisree, A. J. (2022). Recurring detrimental impact of agrochemicals on the ecosystem, and a glimpse of organic farming as a possible rescue. *Environmental Science and Pollution Research, 29*, 75103–75112. https://doi.org/10.1007/s11356-022-22750-1.
39. Carvalho, F. P. (2006). Agriculture, pesticides, food security and food safety. *Environmental Science and Policy, 9*, 685–692.
40. Wang, Y., Zhang, B. W., Chen, N. J., Wang, C., Feng, S., & Xu, H. (2018). Combined bioremediation of soil co-contaminated with cadmium and endosulfan by *Pleurotus eryngii* and *Coprinus comatus*. *Journal of Soils and Sediments, 18*(6), 2136–2147. https://doi.org/10.1007/s11368-017-1762-9.
41. Yaron, B., Dror, I., & Berkowitz, B. (2012). *Soil-Subsurface Change*. Kindle Edition. *Cambios en el suelo y el subsuelo: impactos de los contaminantes químicos* . Springer Verlag. https://doi.org/10.1007/978-3-642-24387-5.
42. Johan, P. D., Ahmed, O. H., Omar, L., & Hasbullah, N. A. (2021). Phosphorus transformation in soils following co-application of charcoal and wood ash. *Agronomy, 11*, 2010. https://doi.org/10.3390/agronomy11102010.
43. Stewart, W. M., Dibb, D. W., Johnston, A. E., & Smyth, T. J. (2005). The contribution of commercial fertilizer nutrients to food production. *Agronomy Journal, 97*(1), 1–6. https://doi.org/10.2134/agronj2005.0001.
44. Fan, H., Zhang, Y., Li, J., Jiang, J., Waheed, A., Wang, S., Rasheed, S. M., Zhang, L., & Zhang, R. (2023). Effects of organic fertilizer supply on soil properties, tomato yield, and fruit quality: A global meta-analysis. *Sustainability, 15*, 2556. https://doi.org/10.3390/su15032556.
45. Karavidas, I., Ntatsi, G., Vougeleka, V., Karkanis, A., Ntanasi, T., Saitanis, C., Agathokleous, E., Ropokis, A., Sabatino, L., Tran, F., Iannetta, P. P. M., & Savvas, D. (2022). Agronomic practices to increase the yield and quality of common bean

(*Phaseolus vulgaris* L.): A systematic review. *Agronomy, 12*, 271. https://doi.org/10.3390/agronomy12020271.
46. FAO. (2006). *International Code of Conduct on the Distribution and Use of Pesticides: Guidelines on Efficacy Evaluation for the Registration of Plant Protection Products*. Rome, Italy: Food and Agriculture Organization of the United Nations. Retrieved from: http://www.fao.org/fileadmin/templates/agphome/documents/Pests_Pesticides/Code/Efficacy.pdf (accessed on 27 May 2024).
47. Popp, J., Pető, K., & Nagy, J. (2013). Pesticide productivity and food security. A review. *Agronomy for Sustainable Development, 33*, 243–255. https://doi.org/10.1007/s13593-012-0105-x.
48. FAO. (2017). *Voluntary Guidelines for Sustainable Soil Management*. Rome, Italy: Food and Agriculture Organization of the United Nations. Retrieved from: http://www.fao.org/3/a-bl813e.pdf (accessed on 24 May 2024).
49. Jacobsen, C. S., & Hjelmsø, M. H. (2014). Agricultural soils, pesticides, and microbial diversity. *Current Opinion in Biotechnology, 27*, 15–20. https://doi.org/10.1016/j.copbio.2013.09.003.
50. Ockleford, C., Adriaanse, P., Berny, P., Brock, T., Duquesne, S., Grilli, S., & Rob, S. (2017). Scientific opinion addressing the state of the science on risk assessment of plant protection products for in-soil organisms. *EFSA Journal, 15*(2), e04690. https://doi.org/10.2903/j.efsa.2017.4690.
51. González-Ortega, E., & Fuentes-Ponce, M. H. (2022). Dinámica del glifosato en el suelo y sus efectos en la microbiota. *Revista Internacional de Contaminación Ambiental, 38*, 127–144. https://doi.org/10.20937/RICA.54197.
52. Franco, R., Li, S., Rodriguez-Rocha, H., Burns, M., & Panayiotidis, M. I. (2010). Molecular mechanisms of pesticide-induced neurotoxicity: Relevance to Parkinson's disease. *Chemico-Biological Interactions, 188*, 289–300. https://doi.org/10.1016/j.cbi.2010.06.003.
53. Bolognesi, C., & Merlo, F. D. (2011). Pesticides: human health effects. *Journal of Nriagu ed.*, 438–453. Retrieved from: https://www.scirp.org/(S(vtj3fa45qm1ean45vvffcz55))/reference/ReferencesPapers.aspx?ReferenceID=2007778 (accessed on 27 May 2024).
54. Katagi, T. (2010). Bioconcentration, bioaccumulation, and metabolism of pesticides in aquatic organisms. *Reviews of Environmental Contamination and Toxicology, 204*, 1–132. https://doi.org/10.1007/978-1-4419-1440-8_1.
55. Yang, C., Zhenlan, X., Shanshan, T., Yawei, W., & Guibin, J. (2022). Organochlorine pesticides and other pesticides in peanut oil: Residue level, source, household processing factor, and risk assessment. *Journal of Hazardous Materials, 429*, 128272. https://doi.org/10.1016/j.jhazmat.2022.128272.
56. Wang, N., Cui, Z., Wang, Y., & Zhang, J. (2022). Characteristics and residual health risk of organochlorine pesticides in fresh vegetables in the suburb of Changchun, Northeast China. *International Journal of Environmental Research and Public Health, 19*(19), 12547. https://doi.org/10.3390/ijerph191912547.
57. Shaibu, M., Suleiman, M. A. T., Lafia-Araga, R. A., & Salihu, S. O. (2020). Study of organochlorine pesticide residues level in fresh and dried tomatoes from selected farmlands in Zamfara state, Nigeria. *International Journal of Engineering Applied Sciences and Technology, 5*, 91–96. https://doi.org/10.33564/IJEAST.2020.v05i08.013.

58. Hurtado, M., & Gutierrez, M. (2005). Intoxicación aguda por plaguicidas organofosforados. *Revista de la Facultad de Medicina, Universidad Nacional de Colombia, 53*.
59. Rivera, A., Cedillo Ramírez, L., Parraguirre Lezama, C., Baez Simon, A., Laug Garcia, B., & Romero-Arenas, O. (2022). Evaluation of cytotoxic and genotoxic risk derived from exposure to pesticides in corn producers in Tlaxcala, Mexico. *Applied Sciences, 12*(18), 9050. https://doi.org/10.3390/app12189050.
60. Wang, R., Zhang, S., Xiao, K., Cai, M., & Liu, H. (2023). Occurrence, sources, and risk assessment of pyrethroid insecticides in surface water and tap water from Taihu Lake, China. *Journal of Environmental Management, 325*, 116565. https://doi.org/10.1016/j.jenvman.2022.116565.
61. Gupta, D., Soman, G., & Dev, S. (1982). Thiourea, a convenient reagent for the reductive cleavage of olefin ozonolysis products. *Tetrahedron, 38*, 3013–3018. https://doi.org/10.1016/0040-4020(82)80187-7.
62. Khanna, R. K., & Moore, M. H. (1999). Carbamic acid: Molecular structure and IR spectra. *Spectrochimica Acta Part A: Molecular and Biomolecular Spectroscopy, 55*, 961–967. https://doi.org/10.1016/S1386-1425(98)00228-5.
63. Bossa, J. B., Theulé, P., Duvernay, F., Borget, F., & Chiavassa, T. (2008). Carbamic acid and carbamate formation in NH_3: CO2 ices – UV irradiation versus thermal processes. *Astronomy and Astrophysics, 492*, 719–724. https://doi.org/10.1051/0004-6361:200810536.
64. Boonkhao, L., Phonkaew, S., Kwonpongsagoon, S., & Rattanachaikunsopon, P. (2022). Carbofuran residues in soil and consumption risks among farmers growing vegetables in Ubon Ratchathani Province, Thailand. *AIMS Environmental Science, 9*, 593–602. https://doi.org/10.3934/environsci.2022035.
65. Shafiee, M. N., Masoodi, F. A., Wani, S. M., & Ahad, T. (2022). Effect of cold storage on carbofuran dissipation in cabbages (*Brassica oleracea*). *BMC Chemistry, 16*, 78. https://doi.org/10.1186/s13065-022-00867-1.
66. Gikas, G. D., Parlakidis, P., Mavropoulos, T., & Vryzas, Z. (2022). Particularities of fungicides and factors affecting their fate and removal efficacy: A review. *Sustainability, 14*, 4056. https://doi.org/10.3390/su14074056.
67. Lamichhane, J. R., Osdaghi, E., Behlau, F., Kohl, J., Jones, J. B., & Noel, J. (2018). Thirteen decades of antimicrobial copper compounds applied in agriculture: A review. *Agronomy for Sustainable Development, 38*, 28. https://doi.org/10.1007/s13593-018-0503-9.
68. Ambrosini, V. G., Voges, J. G., Canton, L., da Rosa Couto, R., Ferreira, P. A. A., Comin, J. J., & Soares, C. R. F. S. (2015). Effect of arbuscular mycorrhizal fungi on young vines in copper-contaminated soil. *Brazilian Journal of Microbiology, 46*, 1045–1052. https://doi.org/10.1590/S1517-838246420140622.
69. Pena, L. B., Azpilicueta, C. E., & Gallego, S. M. (2011). Sunflower cotyledons cope with copper stress by inducing catalase subunits less sensitive to oxidation. *Journal of Trace Elements in Medicine and Biology, 25*, 125–129. https://doi.org/10.1016/j.jtemb.2011.05.001.
70. Sfaxi-Bousbih, A., Chaoui, A., & El Ferjani, E. (2010). Copper affects the cotyledonary carbohydrate status during the germination of bean seeds. *Biological Trace Element Research, 137*, 110–116. https://doi.org/10.1007/s12011-009-8556-x.

71. Cetkauskaite, A., Pessala, P., & Södergren, A. (2004). Elemental sulphur: Toxicity in vivo and in vitro to bacterial luciferase, in vitro yeast alcohol dehydrogenase, and bovine liver catalase. *Environmental Toxicology, 19*, 372–386. https://doi.org/10.1002/tox.20025.
72. Huber, B., Herzog, B., Drewes, J. E., Koch, K., & Muller, E. (2016). Characterization of sulfur oxidizing bacteria related to biogenic sulfuric acid corrosion in sludge digesters. *BMC Microbiology, 16*, 153. https://doi.org/10.1186/s12866-016-0767-7.
73. Kuklińska, K., Wolska, L., Namieśnik, J., & Monika Cieszynska, M. (2013). Analytical and bioanalytical problems associated with the toxicity of elemental sulphur in the environment. *Trends in Analytical Chemistry, 48*, 14–21. https://doi.org/10.1016/j.trac.2013.03.006.
74. Marsala, R. Z., Capri, E., Russo, E., Bisagni, M., Colla, R., Lucini, L., Gallo, A., & Suciu, N. A. (2020). First evaluation of pesticide occurrence in groundwater of Tidone Valley, an area with intensive viticulture. *Science of the Total Environment, 736*, 139730. https://doi.org/10.1016/j.scitotenv.2020.139730.
75. Tudi, M., Daniel, R. H., Wang, L., Lyu, J., Sadler, R., Connell, D., Chu, C., & Phung, D. T. (2021). Agriculture development, pesticide application, and its impact on the environment. *International Journal of Environmental Research and Public Health, 18*(3), 1112. https://doi.org/10.3390/ijerph18031112.
76. Sharma, D., & Choudhury, P. P. (2018). Pesticide use and residue management in vegetables. *Journal of Vegetable Science, 40*, 1–10.
77. Atuhaire, A., Kaye, E., Mutambuze, I. L., Matthews, G., Friedrich, T., & Jørs, E. (2017). Assessment of dithiocarbamate residues on tomatoes conventionally grown in Uganda and the effect of simple washing to reduce exposure risk to consumers. *Environmental Health Insights, 11*, 1–8. https://doi.org/10.1177/117863021771.
78. IUPAC. (2006). Glossary of terms relating to pesticides. *Pure and Applied Chemistry, 78*, 2075–2154.
79. Van Emden, H. F., & Peakall, D. B. (1996). *Beyond Silent Spring: Integrated Pest Management and Chemical Safety*. Springer, 320. ISBN 978-0-412-72800-6.
80. Okuma, D. M., Bernardi, D., Horikoshi, R. J., Bernardi, O., Silva, A. P., & Omoto, C. (2018). Inheritance and fitness costs of *Spodoptera frugiperda* (*Lepidoptera*: Noctuidae) resistance to spinosad in Brazil. *Pest Management Science, 74*(7), 1441–1448. https://doi.org/10.1002/ps.4829.
81. Carvalho, R. A., Omoto, C., Field, L. M., Williamson, M. S., & Bass, C. (2013). Investigating the molecular mechanisms of organophosphate and pyrethroid resistance in the fall armyworm *Spodoptera frugiperda*. *PLoS ONE, 8*, e62268. https://doi.org/10.1371/journal.pone.0062268.
82. Bolzan, A., Padovez, F. E. O., Nascimento, A. R. B., Kaiser, I. S., Lira, E. C., Amaral, F. S. A., Kanno, R. H., Malaquias, J. B., & Omoto, C. (2019). Selection and characterization of the inheritance of resistance of *Spodoptera frugiperda* (Lepidoptera: Noctuidae) to chlorantraniliprole and cross-resistance to other diamide insecticides. *Pest Management Science, 75*, 2682–2689. https://doi.org/10.1002/ps.5376.
83. Lira, E. C., Bolzan, A., Nascimento, A. R. B., Amaral, F. S. A., Kanno, R. H., Kaiser, I. S., & Omoto, C. (2020). Resistance of *Spodoptera frugiperda* (*Lepidoptera*: Noctuidae) to spinetoram: Inheritance and cross-resistance to spinosad. *Pest Management Science, 76*(8), 2674–2680. https://doi.org/10.1002/ps.5812.

84. Muraro, D. S., Abbade Neto, D. O., Kanno, R. H., Kaiser, I. S., Bernardi, O., & Omoto, C. (2021). Inheritance patterns, cross-resistance, and synergism in *Spodoptera frugiperda* (*Lepidoptera*: Noctuidae) resistant to emamectin benzoate. *Pest Management Science, 77*(12), 5049–5057. https://doi.org/10.1002/ps.6545.
85. Budd, R., Wang, D., Ensminger, M., & Phillips, B. (2020). An evaluation of temporal and spatial trends of pyrethroid concentrations in California surface waters. *Science of The Total Environment, 718*, 137402. https://doi.org/10.1016/j.scitotenv.2020.137402.
86. Wang, Y., Li, J., Xu, Y., Rodgers, F. M. T., Bao, M., & Tan, F. (2023). Uptake, translocation, bioaccumulation, and bioavailability of organophosphate esters in rice paddy and maize fields. *Journal of Hazardous Materials, 446*, 130640. https://doi.org/10.1016/j.jhazmat.2022.130640.
87. Tang, W., Wang, D., Wang, J., Wu, Z., Li, L., Huang, M., Xu, S., & Yan, D. (2018). Pyrethroid pesticide residues in the global environment: An overview. *Chemosphere, 191*, 990–1007. https://doi.org/10.1016/j.chemosphere.2017.10.115.
88. Allinson, G., Zhang, P., Bui, A., Allinson, M., Rose, G., Marshall, M., & Pettigrove, V. (2015). Pesticide and trace metal occurrence and aquatic benchmark exceedances in surface waters and sediments of urban wetlands and retention ponds in Melbourne, Australia. *Environmental Science and Pollution Research, 22*, 10214–10226. https://doi.org/10.1007/s11356-015-4206-3.
89. Duong, H. T., Kadokami, K., Pan, S., Matsuura, N., & Nguyen, T. Q. (2014). Screening and analysis of 940 organic micro-pollutants in river sediments in Vietnam using an automated identification and quantification database system for GC–MS. *Chemosphere, 107*, 462–472. https://doi.org/10.1016/j.chemosphere.2014.01.064.
90. Hintzen, E. P., Lydy, M. J., & Belden, J. B. (2009). Occurrence and potential toxicity of pyrethroids and other insecticides in bed sediments of urban streams in central Texas. *Environmental Pollution, 157*, 110–116. https://doi.org/10.1016/j.envpol.2008.07.023.
91. Dou, R., Sun, J., Deng, F., Wang, P., Zhou, H., Wei, Z., Chen, M., He, Z., Lai, M., Ye, T., & Zhu, L. (2020). Contamination of pyrethroids and atrazine in greenhouse and open-field agricultural soils in China. *Science of The Total Environment, 701*, 134916. https://doi.org/10.1016/j.scitotenv.2019.134916.
92. Bragança, I., Mucha, A. P., Tomasino, M. P., Santos, F., Lemos, P. C., Delerue-Matos, C., & Domingues, V. F. (2019). Deltamethrin impact in a cabbage planted soil: Degradation and effect on microbial community structure. *Chemosphere, 220*, 1179–1186. https://doi.org/10.1016/j.chemosphere.2019.01.004.
93. Bragança, I., Lemos, P. C., Barros, P., Delerue-Matos, C., & Domingues, V. F. (2018). Phytotoxicity of pyrethroid pesticides and its metabolite towards *Cucumis sativus*. *Science of the Total Environment, 619–620*, 685–691. https://doi.org/10.1016/j.scitotenv.2017.11.164.
94. Chen, J., Li, Q. X., & Song, B. (2020). Chemical nematicides: Recent research progress and outlook. *Journal of Agricultural and Food Chemistry, 68*(44), 12175–12188. https://doi.org/10.1021/acs.jafc.0c02871.
95. El-Saadony, M. T., Abuljadayel, D. A., Shafi, M. E., Albaqami, N. M., Desoky, E. S. M., El Tahan, A. M., Mesiha, P. K., Elnahal, A. S. M., Almakas, A., & Taha, A. E., et al. (2021). Control of foliar phytoparasitic nematodes through sustainable natural materials: Current progress and challenges. *Saudi Journal of Biological Science, 28*, 7314–7326. https://doi.org/10.1016/j.sjbs.2021.08.035.

96. Walia, S., Saha, S., Tripathi, V., & Sharma, K. K. (2018). Phytochemical biopesticides: Some recent developments. *Phytochemistry Review, 16*, 989–1007. https://doi.org/10.1007/s11101-017-9512-6.
97. Nyczepir, A. P., & Thomas, S. H. (2009). Current and future management strategies in intensive crop production systems. In Perry, R. N., Moens, M., & Starr, J. (Eds.), *Root-Knot Nematodes* (pp. 412–433). https://www.cabdirect.org/cabdirect/abstract/20093330198 (accessed on 27 May 2024).
98. Gómez-Ramos, M. D. M., Nannou, C., Martínez Bueno, M. J., Goday, A., Murcia-Morales, M., Ferrer, C., & Fernández-Alba, A. R. (2020). Pesticide residues evaluation of organic crops. A critical appraisal. *Food Chemistry, 5*, 100079. https://doi.org/10.1016/j.fochx.2020.100079.
99. Goldman, L. R., Beller, M., & Jackson, R. J. (1990). Aldicarb food poisonings in California, 1985-1988: Toxicity estimates for humans. *Archives of Environmental Health, 45*, 141–147. https://doi.org/10.1080/00039896.1990.9936707.
100. PubChem. (2019). National Library of Medicine, National Center for Biotechnology Information, Bethesda, MD, USA. https://pubchem.ncbi.nlm.nih.gov/ (accessed on 27 May 2024).
101. Faske, T. R., & Hurd, K. (2015). Sensitivity of Meloidogyne incognita and Rotylenchulus reinforms to fluopyram. *Journal of Nematology, 47*, 316–321.
102. Hodson, A. K., Milkereit, J., John, G. C., Doll, D. A., & Duncan, R. A. (2019). The effect of fumigation on nematode communities in California almond orchards. *Nematology, 27*, 899–912. https://doi.org/10.1163/15685411-00003262.
103. Grabau, Z. J., Mauldin, M. D., Habteweld, A., & Carter, E. T. (2020). Nematicide efficacy at managing Meloidogyne arenaria and non-target effects on free-living nematodes in peanut production. *Journal of Nematology, 52*, 1–10. https://doi.org/10.21307/jofnem-2020-028.
104. Waldo, B. D., Grabau, Z. J., Mengistu, T. M., & Crow, W. T. (2019). Nematicide effects on non-target nematodes in bermudagrass. *Journal of Nematology, 51*, 1–12.
105. FAOSTAT. (2020). *Statistics Division. Pesticides Indicators*. Retrieved from: https://www.fao.org/faostat/en/#data/EP (accessed on 24 May 2024).
106. Niti, C., Sunita, S., Kamlesh, K., & Rakesh, K. (2013). Bioremediation: an emerging technology for remediation of pesticides. *Research Journal of Chemistry and Environment, 17*(1), 88–105.
107. Yadav, I. C., Devi, N. L., Syed, J. H., Cheng, Z., Li, J., Zhang, G., & Jones, K. C. (2015). Current status of persistent organic pesticides residues in air, water, and soil, and their possible effect on neighboring countries: A comprehensive review of India. *Science of the Total Environment, 511*, 123–137. https://doi.org/10.1016/j.scitotenv.2014.12.041.
108. Chowdhury, I. F., Rohan, M., Stodart, B. J., Chen, C., Wu, H., & Doran, G. S. (2021). Persistence of atrazine and trifluralin in a clay loam soil undergoing different temperature and moisture conditions. *Environmental Pollution, 276*, 116687. https://doi.org/10.1016/j.envpol.2021.116687.
109. Pérez, S., Héctor, E. F., Guerrero, S., & Palacios, A. (2021). Mitigation of hazardous contaminants: A phyto-microbiome approach. In Kumar, V., Prasad, R., & Kumar, M. (Eds.), *Rhizobiont in Phytoremediation of Hazardous Waste* (pp. 363–390). Springer, Singapore.

110. Megharaj, M., Ramakrishnan, B., Venkateswarlu, K., Sethunathan, N., & Naidu, R. (2011). Bioremediation approaches for organic pollutants: A critical perspective. *Environmental International, 37*, 1362–1375. https://doi.org/10.1016/j.envint.2011.06.003.
111. Varshney, K. (2019). Bioremediation of pesticide waste at contaminated sites. *Journal of Emerging Technologies and Innovative Research, 6*, 128–134.
112. Kaur, J. (2021). PGPR in management of soil toxicity. In V. Kumar, R. Prasad, & M. Kumar (Eds.), *Rhizobiont in Bioremediation of Hazardous Wastes* (pp. 317–344). Springer. https://doi.org/10.1080/21655979.2021.2004645.
113. Eevers, N., Whitex, J. C., Vangronsveld, J., & Weyens, N. (2017). Bio- and phytoremediation of pesticide-contaminated environments: A review. *Advances in Botanical Research, 83*, 277–318.
114. Chen, M., Xu, P., Zeng, G., Yang, C., Huang, D., & Zhang, J. (2015). Bioremediation of soils contaminated with polycyclic aromatic hydrocarbons, petroleum, pesticides, chlorophenols and heavy metals by composting: applications, microbes and future research needs. *Biotechnology Advances, 33*, 745–755. https://doi.org/10.1016/j.biotechadv.2015.05.003.
115. Gilani, R. A., Rafique, M., Rehman, A., Munis, M. F., Rehman, S. U., & Chaudhary, H. J. (2016). Biodegradation of chlorpyrifos by bacterial genus Pseudomonas. *Journal of Basic Microbiology, 56*, 105–119. https://doi.org/10.1002/jobm.201500336.
116. Kumar, P. S., Carolin, F. C., & Varjani, S. J. (2018). Pesticides bioremediation. In S. J. Varjani, A. K. Agarwal, E. Gnansounou, & B. Gurunathan (Eds.), *Bioremediation: Applications for Environmental Protection and Management* (pp. 197–222). Springer Nature.
117. Fang, H., Lian, J., Wang, H., Cai, L., & Yu, Y. (2015). Exploring bacterial community structure and function associated with atrazine biodegradation in repeatedly treated soils. *Journal of Hazardous Materials, 286*, 457–465. https://doi.org/10.1016/j.jhazmat.2015.01.006.
118. Zhu, J., Fu, L., Meng, Z., & Jin, C. (2019). Characteristics of an atrazine degrading bacterium and the construction of a microbial agent for effective atrazine degradation. *Water and Environment Journal, 0*, 1–11. https://doi.org/10.1111/wej.12491.
119. Manogaran, M., Shukor, M. Y., Yunus, M., Yasid, N. A., Johari, W. L. W., & Ahmad, S. A. (2017). Isolation and characterization of glyphosate-degrading bacteria isolated from local soils in Malaysia. *Rendiconti Lincei, 28*, 471–479. https://doi.org/10.1007/s12210-017-0620-4.
120. Xu, B., Sun, Q. J., Lan, J. C. W., Chen, W. M., Hsueh, C. C., & Chen, B. Y. (2019). Exploring the glyphosate-degrading characteristics of a newly isolated, highly adapted indigenous bacterial strain, *Providencia rettgeri* GDB 1. *Journal of Bioscience and Bioengineering, 128*(1), 80–87. https://doi.org/10.1016/j.jbiosc.2019.01.012.
121. Zhang, W., Li, J., Zhang, Y., Wu, X., Zhou, Z., Huang, Y., Zhao, Y., Mishra, S., Bhatt, P., & Chen, S. (2022). Characterization of a novel glyphosate-degrading bacterial species, *Chryseobacterium* sp. Y16C, and evaluation of its effects on microbial communities in glyphosate-contaminated soil. *Journal of Hazardous Materials, 432*, 128689. https://doi.org/10.1016/j.jhazmat.2022.128689.
122. Kanissery, R. G., Welsh, A., Gomez, A., Connor, L., & Sims, G. K. (2017). Identification of metolachlor mineralizing bacteria in aerobic and anaerobic soils using DNA-stable isotope probing. *Biodegradation, 29*, 117–128. https://doi.org/10.1007/s10532-017-9817-6.

123. Birolli, W. G., Lima, R. N., & Porto, A. L. M. (2019). Applications of marine-derived microorganisms and their enzymes in biocatalysis and biotransformation, the underexplored potentials. *Frontiers in Microbiology, 10*, 1453. https://doi.org/10.3389/fmicb.2019.01453.
124. Doolotkeldieva, T., Bobusheva, S., & Konurbaeva, M. (2021). The improving conditions for the aerobic bacteria performing the degradation of obsolete pesticides in polluted soils. *Air, Soil and Water Research, 14*, 1–14. https://doi.org/10.1177/1178622120982590.
125. Kong, L., Zhu, S., Zhu, L., Xie, H., Su, K., Yan, T., Wang, J., Wang, J., Wang, F., & Sun, F. (2013). Biodegradation of organochlorine pesticide endosulfan by bacterial strain Alcaligenes faecalis JBW4. *Journal of Environmental Sciences, 25*, 2257–2264. https://doi.org/10.1016/S1001-0742(12)60288-5.
126. Takagi, K. (2020). Study on the biodegradation of persistent organic pollutants (POPs). *Journal of Pesticide Science, 45*, 119–123. https://doi.org/10.1584/jpestics.J19-06.
127. Wang, X., Sial, M. U., Bashir, M. A., Bilal, M., Raza, Q. U. A., Ali Raza, H. M., Rehim, A., & Geng, Y. (2022). Pesticides xenobiotics in soil ecosystem and their remediation approaches. *Sustainability, 14*(6), 3353. https://doi.org/10.3390/su14063353.
128. Xiao, P., Mori, T., Kamei, I., Kiyota, H., Takagi, K., & Kondo, R. (2011). Novel metabolic pathways of organochlorine pesticides dieldrin and aldrin by the white rot fungi of the genus Phlebia. *Chemosphere, 85*(2), 218–224. https://doi.org/10.1016/j.chemosphere.2011.06.028.
129. Xiao, P., & Kondo, R. (2013). Biodegradation of dieldrin by Cordyceps fungi and detection of metabolites. *Applied Mechanics and Materials, 295–298*, 30–34. https://doi.org/10.4028/www.scientific.net/AMM.295-298.30.
130. Pinto, A. P., Rodrigues, S. C., Caldeira, A. T., & Teixeira, D. M. (2016). Exploring the potential of novel biomixtures and *Lentinula edodes* fungus for the degradation of selected pesticides: Evaluation for use in biobed systems. *Science of the Total Environment, 541*, 1372–1381. https://doi.org/10.1016/j.scitotenv.2015.10.046.
131. Purnomo, A., Nawfa, R., Martak, F., Shimizu, K., & Kamei, I. (2017). Biodegradation of aldrin and dieldrin by the white-rot fungus *Pleurotus ostreatus*. *Current Microbiology, 74*, 320–324. https://doi.org/10.1007/s00284-016-1184-8.
132. Kataoka, R. (2018). Biodegradability and biodegradation pathways of chlorinated cyclodiene insecticides by soil fungi. *Journal of Pesticide Science, 43*, 314–320. https://doi.org/10.1584/jpestics.J18-03.
133. Gaur, N., Narasimhulu, K., & Pydisetty, Y. (2018). Recent advances in the bio-remediation of persistent organic pollutants and its effect on environment. *Journal of Cleaner Production, 198*, 1602–1631. https://doi.org/10.1016/j.jclepro.2018.07.076.
134. Wang, Y., Zhu, Y., Zhang, S., & Wang, Y. (2018). What could promote farmers to replace chemical fertilizers with organic fertilizers? *Journal of Cleaner Production, 199*, 882–890. https://doi.org/10.1016/j.jclepro.2018.07.222.
135. Bhattacharya, S., Angayarkanni, J., Das, A., & Palaniswamy, M. (2012). Mycoremediation of benzo [a] pyrene by Pleurotus ostreatus isolated from Wayanad district in Kerala, India. *International Journal of Pharmacy and Biological Sciences, 2*, 84–93. https://doi.org/10.1007/s13205-013-0148-y.
136. Naziya, B., Murali, M., & Amruthesh, K. N. (2020). Plant growth-promoting fungi (PGPF) instigate plant growth and induce disease resistance in *Capsicum annuum* L. upon infection with *Colletotrichum capsici* (Syd.). Butler & Bisby. *Biomolecules, 10*(1), 41. https://doi.org/10.3390/biom10010041.

137. Murali, M., Naziya, B., & Amruthesh, K. N. (2021). Bioprospecting of rhizosphere-resident fungi: Their role and importance in sustainable agriculture. *Journal of Fungi, 7*, 314. https://doi.org/10.3390/jof7040314.
138. Tripathy, S., Paul, S., & Khalua, R. K. (2014). Phytoremediation: proficient to prevent pesticide pollution. *International Journal of Innovative Science, Engineering & Technology, 1*, 282–287.
139. Srivastav, A., Yadav, K. K., Yadav, S., Gupta, N., Singh, J. K., Katiyar, R., & Kumar, V. (2018). Nano-phytoremediation of pollutants from contaminated soil environment: current scenario and prospects. In Ansari, A. A., Gill, S. S., Gill, R., Lanza, G. R., & Newman, L. (Eds.), *Phytoremediation* (pp. 383–401). Springer, Cham.
140. Nurzhanova, A., Kalugin, S., & Zhambakin, K. (2013). Obsolete pesticides and application of colonizing plant species for remediation of contaminated soil in Kazakhstan. *Environmental Science and Pollution Research International, 20*(4), 2054–2063. https://doi.org/10.1007/s11356-012-1111-x.
141. Dubey, K. K., & Fulekar, M. H. (2013). Investigation of potential rhizospheric isolate for cypermethrin degradation. *Biotech, 3*, 33–43. https://doi.org/10.1007/s13205-012-0067-3.
142. Abhilash, P. C., Singh, B., Srivastava, P., Schaeffer, A., & Singh, N. (2013). Remediation of lindane by *Jatropha curcas* L: utilization of multipurpose species for rhizoremediation. *Biomass and Bioenergy, 51*, 189–193. https://doi.org/10.1016/j.biombioe.2013.01.028.
143. Rissato, S. R., Galhiane, M. S., Fernandes, J. R., Gerenutti, M., Gomes, H. M., Ribeiro, R., & Almeida, M. V. (2015). Evaluation of *Ricinus communis* L. for the phytoremediation of polluted soil with organochlorine pesticides. *BioMed Research International, 2015*, 549863. https://doi.org/10.1155/2015/549863.
144. Mitton, F. M., Gonzalez, M., Monserrat, J. M., & Miglioranza, K. S. (2016). Potential use of edible crops in the phytoremediation of endosulfan residues in soil. *Chemosphere, 148*, 300–306. https://doi.org/10.1016/j.chemosphere.2016.01.028.
145. Dowling, D. K., Maklakov, A. A., Friberg, U., & Hailer, F. (2009). Applying the genetic theories of ageing to the cytoplasm: Cytoplasmic genetic covariation for fitness and lifespan. *Journal of Evolutionary Biology, 22*, 818–827. https://doi.org/10.1111/j.1420-9101.2009.01692.x.
146. Abdel-Shafy, H. I., & Mansour, M. S. M. (2018). Phytoremediation for the elimination of metals, pesticides, PAHs, and other pollutants from wastewater and soil. In V. Kumar, M. Kumar, & R. Prasad (Eds.), *Phytobiont and Ecosystem Restitution* (pp. 91–110). Springer. https://doi.org/10.1007/978-981-13-1187-1_5.
147. Afzal, M., Khan, Q. M., & Sessitsch, A. (2014). Endophytic bacteria: prospects and applications for the phytoremediation of organic pollutants. *Chemosphere, 117*, 232–242. https://doi.org/10.1016/j.chemosphere.2014.06.078.
148. Yang, Y., Liu, Y., Li, Z., Wang, Z., Li, C., & Wei, H. (2020). Significance of soil microbe in microbial-assisted phytoremediation: an effective way to enhance phytoremediation of contaminated soil. *International Journal of Environmental Science and Technology, 17*(5), 2477–2484. https://doi.org/10.1007/s13762-020-02668-2.
149. Hussain, I., Aleti, G., Naidu, R., Puschenreiter, M., Mahmood, Q., Rahman, M. M., Wang, F., Shaheen, S., Syed, J. H., & Reichenauer, T. G. (2018). Microbe and plant assisted -remediation of organic xenobiotics and its enhancement by genetically modified organisms and recombinant technology: A review. *Science of the Total Environment, 628–629*, 1582–1599.

150. Inui, H., Ueyama, Y., Shiota, N., Ohkawa, Y., & Ohkawa, H. (1999). Herbicide metabolism and cross-tolerance in transgenic potato plants expressing human CYP1A1. *Pesticide Biochemistry and Physiology, 64*, 33–46. https://doi.org/10.1006/pest.1999.2407.
151. Wang, L., Samac, D. A., Shapir, A., Wackett, L. P., Vance, C. P., Olszewski, N. E., & Sadowsky, M. J. (2005). Biodegradation of atrazine in transgenic plants expressing a modified bacterial atrazine chlorohydrolase (atzA) gene. *Plant Biotechnology Journal, 3*(4), 475–486. https://doi.org/10.1111/j.1467-7652.2005.00138.x.
152. Kawahigashi, H., Hirose, S., Ohkawa, H., & Ohkawa, Y. (2006). Phytoremediation of herbicide atrazine and metolachlor by transgenic rice plants expressing human CYP1A1, CYP2B6, and CYP2C19. *Journal of Agriculture and Food Chemistry, 54*, 2985–2991. https://doi.org/10.1021/jf052610u.
153. Azab, E., Hegazy, A. K., El-Sharnouby, M. E., & Abd Elsalam, H. E. (2016). Phytoremediation of the organic xenobiotic simazine by P450-1A2 transgenic Arabidopsis thaliana plants. *International Journal of Phytoremediation, 18*, 738–746. https://doi.org/10.1080/15226514.2015.1133559.
154. Yan, X., Huang, J. W., Xu, X. H., Chen, D., Xie, X. T., Tao, Q., He, J., & Jiang, J. D. (2018). Enhanced and complete removal of phenyl urea herbicides by combinational transgenic plant-microbe remediation. *Applied and Environmental Microbiology, 84*(9), e00273–e00318. https://doi.org/10.1128/AEM.00273-18.
155. Jang, J., Khanom, S., Moon, Y., Shin, S., & Lee, O. R. (2020). PgCYP76B93 docks on phenylurea herbicides and its expression enhances chlorotoluron tolerance in Arabidopsis. *Applied Biological Chemistry, 63*, 14. https://doi.org/10.1186/s13765-020-00498-x.
156. Eapen, S., Singh, S., & D´Souza, S. F. (2007). Advances in development of transgenic plants for remediation of xenobiotic pollutants. *Biotechnology Advances, 25*, 442–451. https://doi.org/10.1016/j.biotechadv.2007.05.001.
157. Tripathi, S., Singh, V. K., Srivastava, P., Singh, R., Devi, R. S., Kumar, A., & Bhadouria, R. (2019). Phytoremediation of organic pollutants: current status and future directions. In P. Singh, A. Kumar, & A. Borthakur (Eds.), *Abatement of Environmental Pollutants* (pp. 81–105). Elsevier, Amsterdam.
158. Asante-Badu, B., Kgorutla, L. E., Li, S. S., Danso, P. O., Xue, Z., & Qiang, Z. (2020). Phytoremediation of organic and inorganic compounds in a natural and an agricultural environment: a review. *Applied Ecology and Environmental Research, 18*, 6875–6904.
159. Ijaz, A., Imran, A., Ul Haq, M. A., Khan, Q. M., & Afzal, M. (2015). Phytoremediation: Recent advances in plant-endophytic synergistic interactions. *Plant and Soil, 405*, 179–195. Retrieved from: https://www.jstor.org/stable/43872707.
160. Yang, C., Yu, H., Jiang, H., Qiao, C., & Liu, R. (2015). An engineered microorganism can simultaneously detoxify cadmium, chlorpyrifos, and γ-hexachlorocyclohexane. *Journal of Basic Microbiology, 55*(1), 1–7. https://doi.org/10.1002/jobm.201500559.
161. Song, B., Xu, P., Chen, M., Tang, W., Zeng, G., Gong, J., Zhang, P., & Ye, S. (2019). Using nanomaterials to facilitate the phytoremediation of contaminated soil. *Critical Reviews in Environmental Science and Technology, 49*, 791–824. https://doi.org/10.1080/10643389.2018.1558891.
162. Yu, B., Zeng, J., Gong, L., Zhang, M., Zhang, L., & Chen, X. (2007). Investigation of the photocatalytic degradation of organochlorine pesticides on a nano-TiO2 coated film. *Talanta, 72*(4), 1667–1674. https://doi.org/10.1016/j.talanta.2007.03.013.
163. Restivo, J., Órfão, J. J. M., Armenise, S., Garcia-Bordejé, E., & Pereira, M. F. R. (2012). Catalytic ozonation of metolachlor under continuous operation using nanocarbon

materials grown on a ceramic monolith. *Journal of Hazardous Materials, 239–240*, 249–256. https://doi.org/10.1016/j.jhazmat.2012.08.073.
164. El-Temsah, Y. S., Sevcu, A., Bobcikova, K., Cernik, M., & Joner, E. J. (2016). DDT degradation efficiency and ecotoxicological effects of two types of nano-sized zero-valent iron (nZVI) in water and soil. *Chemosphere, 144*, 2221–2228. https://doi.org/10.1016/j.chemosphere.2015.10.122.
165. Hamdi, H., De La Torre-Roche, R., Hawthorne, J., & White, J. C. (2014). Impact of non-functionalized and amino-functionalized multiwall carbon nanotubes on pesticide uptake by lettuce (*Lactuca sativa* L.). *Nanotoxicology, 9*, 172–180. https://doi.org/10.3109/17435390.2014.907456.
166. Pillai, H. P. S., & Kottekottil, J. (2016). Nano-phytotechnological remediation of endosulfan using zero-valent iron nanoparticles. *Journal of Environmental Protection, 7*, 734–744. https://doi.org/10.4236/jep.2016.75066.
167. Bakshi, M., & Abhilash, P. C. (2020). Nanotechnology for soil remediation: revitalizing the tarnished resource. In P. Singh, A. Borthakur, P. K. Mishra, & D. Tiwary (Eds.), *Nano-Materials as Photocatalysts for Degradation of Environmental Pollutants* (pp. 343–365). Elsevier. https://doi.org/10.1016/B978-0-12-818598-8.00017-1.
168. Dalvie, M., Cairncross, E., Solomon, A., & London, L. (2003). Contamination of rural surface and ground water by endosulfan in farming areas of the Western Cape, South Africa. *Environmental Health: A Global Access Science Source, 2*, 1–15. https://doi.org/10.1186/1476-069X-2-1.
169. Instituto Nacional de Estadística y Geografía. (2022). *Indice Nacional de Precios al Consumidor Deciembre de 2021*. Retrieved from: https://www.inegi.org.mx/contenidos/saladeprensa/boletines/2022/inpc_2q/inpc_2q2022_01.pdf (accessed on 27 May 2024).
170. Instituto Nacional de Estadística y Geografía. (2020). *Ingresos y Gastos de los Hogares*. Retrieved from: https://www.inegi.org.mx/temas/ingresoshog/ (accessed on 27 May 2024).
171. Papadakis, E. N., Vryzas, Z., Kotopoulou, A., Kintzikoglou, K., Makris, K. C., & Papadopoulou-Mourkidou, E. (2015). A pesticide monitoring survey in rivers and lakes of northern Greece and its human and ecotoxicological risk assessment. *Ecotoxicology and Environmental Safety, 116*, 1–9. https://doi.org/10.1016/j.ecoenv.2015.02.033.
172. Perez, M. A., Navarro, H., & Miranda, E. (2013). Pesticide residues in vegetables: Problems and risk in Mexico. *International Journal of Environmental Pollution, 29*, 45–64. https://www.revistascca.unam.mx/rica/index.php/rica/article/view/414236 (accessed on 27 May 2024).
173. Räsänen, K., Mattila, T., Porvari, P., Kurppa, S., & Tiilikkala, K. (2015). Estimating the development of ecotoxicological pressure on water systems from pesticides in Finland 2000-2011. *Journal of Cleaner Production, 89*, 65–77. https://doi.org/10.1016/j.jclepro.2014.11.008.
174. Rostami, S., & Azhdarpoor, A. (2019). The application of plant growth regulators to improve phytoremediation of contaminated soils: a review. *Chemosphere, 220*, 818–827. https://doi.org/10.1016/j.chemosphere.2018.12.203.
175. Mills, PK (1998). Análisis de correlación de los datos sobre el uso de pesticidas y las tasas de incidencia de cáncer en los condados de California. *Archives of Environmental Health: An International Journal, 53* (6), 410- 413. https://doi.org/10.1080/00039899809605729.

Index

C

F

G

J

K

L

Q

R

S

T

U

V